Green Energy and Technology

More information about this series at http://www.springer.com/series/8059

Jesús Polo · Luis Martín-Pomares ·
Antonio Sanfilippo

Editors

Solar Resources Mapping

Fundamentals and Applications

 Springer

Editors
Jesús Polo
Photovoltaic Solar Energy Unit
Renewable Energy Division of CIEMAT
Madrid, Spain

Luis Martín-Pomares
Qatar Environment and Energy Research
Institute
Hamad Bin Khalifa University
Doha, Qatar

Antonio Sanfilippo
Qatar Environment and Energy Research
Institute
Hamad Bin Khalifa University
Doha, Qatar

ISSN 1865-3529 ISSN 1865-3537 (electronic)
Green Energy and Technology
ISBN 978-3-319-97483-5 ISBN 978-3-319-97484-2 (eBook)
https://doi.org/10.1007/978-3-319-97484-2

Library of Congress Control Number: 2018964230

This Springer imprint is published by the registered company Springer Nature Switzerland AG
The registered company address is: Gewerbestrasse 11, 6330 Cham, Switzerland

Preface

Solar photovoltaic (PV) represented 55% of the new renewable energy installations in 2017, with a total global capacity of 402 GW, exceeding combined fossil fuels and nuclear power additions, according to the Renewables 2018 Global Status Report (REN2018). At the same time, the global operating commercial capacity of Concentrating Solar Power (CSP) reached 4.9 GW in 2017 (REN2018) and is expected to double by 2020 (IEA SolarPACES Technology Collaboration Programmes), while the global capacity of solar water heating collectors was estimated to be 472 GW thermal at the end of 2017 (REN2018).

To ensure the technological sustainability of the deep and growing penetration of solar power, a thorough knowledge and characterization of solar resources worldwide are needed. In addition, the operational use of solar power in tasks such as power grid management, dynamic electricity pricing, and solar energy adoption modeling requires further uncertainty reduction in the assessment of solar resources.

High-level expertise from several disciplines converges to assist in the deployment of solar power plants. Thermal and energy engineering and deep knowledge of optics contribute to plant design, marketing and finance help address issues of solar energy technology diffusion and bankability, and energy meteorology and geographic computing together enable the characterization of power output from solar energy systems.

Among all research and development endeavors, the mapping of solar resources is undoubtedly the fundamental activity in providing the information needed to establish the technical and regulatory basis for the diffusion of solar power technologies, especially in developing countries where financial resource mobilization for technological innovation is more challenging. Nevertheless, while the potential for solar power technology adoption in any country certainly depends on solar resource availability, it is ultimately determined by the permeability of the local social, economic, and political context to the introduction and diffusion of renewable energy solutions. In recognizing this dependency, the book aims at addressing both endogenous and exogenous aspects of solar resource mapping, including the ensuing operational applications, since both are involved in the decision-making processes that underlie the adoption of solar energy applications solutions.

The book starts with very basic information on solar radiation definitions and magnitudes in Chap. 1, where an overview on the interaction of solar radiation with the Earth's atmosphere and basic concepts of solar geometry are presented. Chapters 2–7 are dedicated to the background on measuring and modeling solar irradiance. Chapter 2 presents a thorough review of instruments to measure solar radiation. The working principles and types of radiometers (broadband and spectral) are widely described, and an overview of calibration and traceability is also presented. Chapter 3 continues by giving a detailed vision on all the aspects that must be considered when setting up a complete radiometric and meteorological station for monitoring the main involving variables. Recommendations on selecting the site, maintenance actions, sampling and data acquisition, shielding and safety and security are remarked in this chapter. The part focused on measuring solar radiation ends up with Chap. 4 which is dedicated to quality assurance of the measurements. Modeling solar radiation for clear sky and all sky conditions are covered through Chaps. 5–7. The state of the art on clear sky models is presented in Chap. 5. This chapter describes thoroughly the aspects from radiative transfer and atmospheric optics that influence in modeling the solar irradiance under cloudless conditions. It presents also the most updated and accurate clear sky models, their input needs, the impact of atmospheric aerosols and the validation and the sources of uncertainty. Chapter 6 is focused on reviewing the models for all sky conditions solar radiation derived from satellite imagery presenting the fundamental working principles of the models and lists the most widely used databases and products of solar irradiance retrieved from satellite information. The modeling part of the book ends with Chap. 7 dedicated to modeling solar radiation with numerical weather prediction models. Numerical weather models have been used mainly for meteorological forecasting, but they offer solar radiation among the main output variables, and they have been evolved recently to model accurately the solar irradiance. The chapter gives an overview of the background of numerical modeling the atmosphere and then focused on the use of weather models for solar resource assessment.

Chapters 8–10 constitute the part of the book focused on mapping and spatial analysis of solar radiation. Chapter 8 describes the different spatial interpolation techniques which can be applied to the grided output of the satellite and numerical models including examples with ArcGIS. It also presents the basis of solar radiation estimation with ArcGIS including practical examples of mapping solar radiation in the urban environment. Chapter 9 summarizes the basic steps in creating solar radiation maps with GIS software. It introduces the basic spatial data types and presents a very simple exercise of a solar radiation map with an open-source software GIS. Chapter 10 describes statistical techniques, clustering, for identifying specific regions according to solar radiation variability. It presents examples of using these statistical techniques with GIS tools for optimizing the selection of ground stations in a large spatial region. The short-term period ground measurements are important to correct the systematic bias of long-term modeled solar radiation from satellite images and numerical models.

Chapters 11–15 belong to the part of the book dedicated to specific applications of solar energy in a country level to improve environmental sustainability, solving problems such as global warming, lack of water, environmental pollution, and rapid consumption of natural resources. Chapter 11 provides a summary on modeling solar power plant performance (photovoltaic and concentrating solar power plants) for long-term characterization and yield performance analysis. Chapter 12 deals with the spatiotemporal analysis of solar radiation variability and the potential impact on the power grid. Spatial variability and smoothing effect are mentioned in this chapter, and some remarks for power grid management are also pointed out to improve the integration of solar energy. Chapter 13 continues with PV integration and presents a comprehensive overview of demand-side management and how this kind of analysis can foster PV integration. Chapter 14 presents a summary of the actual status of desalinization using concentrating solar systems. Finally, Chap. 15 presents another particular application of solar energy for water detoxification.

The contributors to this book are highly skilled experts in their knowledge areas. Many of them are among the most recognized experts worldwide in solar resource knowledge and assessment. We, as editors, feel the privilege and proud of having brought their participation to this book, and we wish to thank them extensively for their excellent work in each chapter. The editor board hopes that the reader enjoys the book and can find useful information to its professional activity.

Madrid, Spain   Jesús Polo
Doha, Qatar   Luis Martín-Pomares
Doha, Qatar   Antonio Sanfilippo
October 2018

Acknowledgements

The authors and editors thank the Solar Heating and Cooling Programme (SHC) and Photovoltaic Power Systems Programme (PVPS) of the International Energy Agency for facilitating Task 46 'Solar Resource Assessment and Forecasting' and Task 16 'Solar resource for high penetration and large scale applications,' without which this publication would not have been possible.

We thank also Energy Sector Management Assistance Program (ESMAP) from the World Bank Group (WB) whose data examples are presented here from several solar mapping projects executed in Vietnam and Tanzania.

Contents

Contributors

Diego C. Alarcón-Padilla Plataforma Solar de Almería (PSA-CIEMAT), Tabernas, Almería, Spain

José L. Balenzategui Photovoltaic Solar Energy Unit, Renewable Energy Division (Energy Department) of CIEMAT, Madrid, Spain

Islam Safak Bayram Qatar Environment and Energy Research Institute, Hamad Bin Khalifa University, Doha, Qatar

Alejandro Cabrera Escuela Universitaria de Ingeniería Mecánica, Universidad de Tarapacá, Arica, Chile

Javier Dominguez Renewable Energy Division (Energy Department) of CIEMAT, Madrid, Spain

Fernando Fabero Photovoltaic Solar Energy Unit, Renewable Energy Division (Energy Department) of CIEMAT, Madrid, Spain

Carlos Fernández-Peruchena National Renewable Energy Centre (CENER), Seville, Spain

Laura Frías-Paredes National Renewable Energy Centre (CENER), Sarriguren (Navarra), Spain

Martin Gastón Romeo National Renewable Energy Centre (CENER), Sarriguren (Navarra), Spain

Christian A. Gueymard Solar Consulting Services, Colebrook, NH, USA

Sue Ellen Haupt Research Applications Laboratory, National Center for Atmospheric Research, Boulder, CO, USA

Pedro A. Jiménez Research Applications Laboratory, National Center for Atmospheric Research, Boulder, CO, USA

Branko Kosovic Research Applications Laboratory, National Center for Atmospheric Research, Boulder, CO, USA

Jared A. Lee Research Applications Laboratory, National Center for Atmospheric Research, Boulder, CO, USA

Ana M. Martín Renewable Energy Division (Energy Department) of CIEMAT, Madrid, Spain

Luis Martín-Pomares Qatar Environment and Energy Research Institute, Hamad Bin Khalifa University, Doha, Qatar

Sara Miralles Solar Energy Research Centre (CIESOL), Almería, Spain; Chemical Engineering Department, University of Almería, Almería, Spain

Patricia Palenzuela Plataforma Solar de Almería (PSA-CIEMAT), Tabernas, Almería, Spain

Richard Perez Atmospheric Science Research Center, State University of New York, Albany, NY, USA

Jesús Polo Photovoltaic Solar Energy Unit, Renewable Energy Division (Energy Department) of CIEMAT, Madrid, Spain

Jan Remund Energy and Climate, Meteotest, Bern, Switzerland

Antonio Sanfilippo Qatar Environment and Energy Research Institute, Doha, Qatar

Lucas Santos-Juanes Departamento de Ingeniería Textil, Papelera de la Universitat Politécnica de Valencia, Valencia, Spain

José P. Silva Photovoltaic Solar Energy Unit, Renewable Energy Division (Energy Department) of CIEMAT, Madrid, Spain

Frank Vignola Solar Radiation Monitoring Laboratory, University of Oregon, Eugene, USA; Luminant LLC, Denver, USA

Stefan Wilbert Deutsches Zentrum für Luft- und Raumfahrt (DLR), Institute of Solar Research, Tabernas, Spain

Acronyms

ACR	Absolute Cavity Radiometer
AEMET	Agencia Estatal de Meteorología
AERONET	Aerosol Robotic Network
AM	Air mass
AOD	Aerosol Optical Depth
AOP	Advanced Oxidation Process
APOLLO	AVHRR Processing Scheme Over cLouds, Land and Ocean
ASE	Average standard error
AU	Astronomical Unit
AVHRR	Advanced Very High Resolution Radiometer
BDRF	Bidirectional reflectance distribution function
BHI	Beam Horizontal Irradiance
BIPM	International Bureau of Weights and Measures
BNI	Beam Normal Irradiance
BOM	Australian Bureau of Meteorology
BSRN	Baseline Surface Radiation Network
CAMS	Copernicus Atmosphere Monitoring Service
CDF	Cumulative distribution function
CIPM	International Committee for Weights and Measures
CIRA	Cooperative Institute for Research in the Atmosphere
CM SAF	Satellite Application Facility on Climate Monitoring
CMC	Calibration and Measurement Capability
CPC	Compound Parabolic Collector
CPP	Critical Peak Pricing
CSAR	Cryogenic Solar Absolute Radiometer
CSP	Concentrated Solar Power
CSRM	Clear Sky Radiation Model
DA	Data assimilation
DHI	Diffuse Horizontal Irradiance

DLR	Deutsches Zentrum für Luft- und Raumfahrt
DNI	Direct Normal Irradiance
DSM	Demand-Side Management
DSM	Digital Surface Model
DUT	Device under test
ECMWF	European Centre Mesoscale Weather Forecast
ERA	European ReAnalysis
ESRI	Environmental Systems Research Institute
ET	Equation of time
FS	Finkelstein-Schafer Statistic
GCOS	Global Climate Observing System
GEBA	Global Energy Balance Archive Data
GEOS	Global Earth Observing System
GHI	Global Horizontal Irradiance
GIS	Geographic Information System
GMS	Geostationary Meteorological Satellite
GOES	Geostationary Operational Environmental Satellite
GPS	Global Positioning System
GTI	Global Tilted Irradiance
GUM	Guide for Uncertainty Measurement
IEA	International Energy Agency
IMO	International Meteorological Organization
IODC	Indian Ocean Data Coverage
IPC	International Pyrheliometer Comparison
IrSOLaV	Investigaciones y Recursos Solares Avanzados
ITCZ	Intertropical Convergence Zone
JPL	Jet Propulsion Laboratory
LAT	Local Apparent Time
LEC	Levelized Electricity Cost
LED	Light-emitting diode
LST	Local Standard Time
LUT	Lookup table
LVRPA	Local volumetric rate of photon absorption
MAGIC	Mesoscale Atmospheric Global Irradiance Code
MBD	Mean bias difference
ME	Mean error
MED	Multi-Effect Distillation
MENA	Middle East and North Africa
MERRA	Modern-Era Retrospective Analysis for Research and Applications
MISR	Multi-angle Imaging SpectroRadiometer
MLWT	Multilayer-weighted transmittance
MODIS	MODerate Resolution Imager Spectroradiometer
MOS	Model Output Statistic
MPPT	Maximum Power Point Tracking

MSE	Mean standardized error
MSF	Multi-stage flash evaporation
MSG	Meteosat Second Generation
MTSAT	Multifunctional Transport Satellite
MVIRI	Meteosat Visible and Infrared Imager
NCAR	National Center for Atmospheric Research
NMI	National Metrology Institute
NOAA	US National Oceanic and Atmospheric Administration
NREL	National Renewable Energy Laboratory
NSRDB	National Solar Resource Data Bases
NWP	Numerical Weather Prediction
OPL	Optical path length
ORC	Organic Rankine Cycle
PACRAD	Primary Absolute Cavity Radiometer
PAR	Photosynthetically Active Radiation
PCD	Photoconductive device
PDF	Probability distribution function
POA	Plane Of Array
PR	Performance ratio
PV	Photovoltaic
PVGIS	Photovoltaic Geographical Information System
PVPS	Photovoltaic Power Systems
PW	Precipitable water
QMS	Quality Management System
RANS	Reynolds-averaged Navier–Stokes
RMSD	Root-mean-square difference
RMSE	Root-mean-square error
RMSSE	Root-mean-square standardized error
RO	Reverse osmosis
RPT	Real-time pricing
RTM	Radiative Transfer Model
SAM	System Advisor Model
SAPM	Sandia Array Performance Model
SC	Solar constant
SCH	Solar Heating and Cooling
SCIAMACHY	SCanning Imaging Absorption spectroMeter for Atmospheric CHartographY
SEVIRI	Spinning Enhanced Visible and Infrared Imager
SMARTS	Simple Model of the Atmospheric Radiative Transfer of Sunshine
SODA	Solar Radiation Data
SOLEMI	Solar Energy Mining
SRRL	Solar Radiation Research Laboratory
STE	Solar thermal energy
SUNY	State University of New York
TES	Thermal storage system

TMY	Typical Meteorological Year
TOA	Top Of Atmosphere
TOC	Total organic carbon
TOU	Time of Use
TSI	Total Solar Irradiance
TTY	Typical yield year
TZ	Time zone
UO SRML	University of Oregon Solar Radiation Monitoring Laboratory
UTC	Coordinated Universal Time
UV	Ultraviolet
VIS	Visible
VUF	Voltage unbalance factor
WMO	World Meteorological Organization
WRC	World Radiation Center
WRF	Weather Research and Forecasting Model
WRR	World Radiometric Reference
WSG	World Standard Group
WWTP	Wastewater Treatment Plant

Chapter 1
Fundamentals: Quantities, Definitions, and Units

**Jesús Polo, Luis Martín-Pomares, Christian A. Gueymard,
José L. Balenzategui, Fernando Fabero and José P. Silva**

Abstract Solar radiation is a generic term that refers to different magnitudes of the solar electromagnetic radiation. The quantification of solar radiation incident at the Earth's surface is of high interest in many disciplines (radiative transfer in the atmosphere, meteorology and climatology, remote sensing of the atmosphere, solar energy studies, etc.). This multidisciplinary aspect of solar radiation sometimes produces duplication of names, definitions, or units. Moreover, different application-specific conventions for variable naming or units exist, which can be confusing. The solar irradiance that reaches a point at the Earth's surface is basically dominated by (i) the geometric aspects of the Earth's orbit around the Sun, and the inclination of its rotation axis in the ecliptic plane that determines the incident angle of the Sun rays; and (ii) the interaction mechanisms of solar radiation with various types of atmospheric constituents. This chapter intends to give the reader an overview of the basic definitions of the main variables that are commonly found in solar energy, and hence in this book as well. In addition, some basic aspects of solar geometry are briefly presented, followed by a concise description of

J. Polo (✉) · J. L. Balenzategui · F. Fabero · J. P. Silva
Photovoltaic Solar Energy Unit, Renewable Energy Division (Energy Department)
of CIEMAT, Avda Complutense 40, 28040 Madrid, Spain
e-mail: jesus.polo@ciemat.es

J. L. Balenzategui
e-mail: jl.balenzategui@ciemat.es

F. Fabero
e-mail: fernando.fabero@ciemat.es

J. P. Silva
e-mail: josepedro.silva@ciemat.es

L. Martín-Pomares
Qatar Environment and Energy Research Institute, Hamad Bin Khalifa University,
P.O. Box 5825, Doha, Qatar
e-mail: lpomares@hbku.edu.qa

C. A. Gueymard
Solar Consulting Services, P.O. Box 392, Colebrook, NH 03576, USA
e-mail: Chris@SolarConsultingServices.com

© Springer Nature Switzerland AG 2019
J. Polo et al. (eds.), *Solar Resources Mapping*, Green Energy and Technology,
https://doi.org/10.1007/978-3-319-97484-2_1

1

the fundamentals of radiation-transfer modeling in the atmosphere. Detailed information on these topics, which is out of the scope of this book, can be found in many textbooks and the abundant literature on solar radiation, radiative transfer and atmospheric physics, to which the avid reader is referred for additional insight.

1 Basic Radiative Definitions

The Sun is a giant thermonuclear reactor: as a consequence of a chain of reactions of nuclear fusion type, helium is produced from hydrogen, releasing huge amounts of energy and charged particles into space. That influx of electromagnetic radiation is the primary source of energy on Earth. It is estimated that the Sun will continue emitting radiation in a *steady state* during about 5×10^9 more years. The electromagnetic radiation emitted by the Sun is distributed throughout space without interaction, following the inverse-square law. Therefore, the total power at the Sun's surface, 3.8×10^{26} W, is reduced to 1.7×10^{17} W at the mean Earth–Sun distance (1.496×10^{11} m), which defines the astronomical unit (AU). The spectral distribution of this radiation at the top of the Earth atmosphere (i.e., the variation of irradiance with wavelength λ or frequency v) roughly fits that of a blackbody emitting at a temperature of $\approx$5770 K. Each solar photon hitting the Earth's surface first travelled $\approx$8 min through free interstellar space.

The solar constant is defined as the mean radiant flux of energy at 1 AU. That radiant flux, called total solar irradiance (TSI) varies somewhat as a consequence of varying solar activity over time, which is characterized by the 11-year Sun cycle. Since 1978, spaceborne radiometers have been measuring TSI with high precision. The solar constant is finally determined by the average value of TSI over many Sun cycles. Various solar constant values have been used in the solar literature, usually in the range 1365–1370 W m^{-2} (Fröhlich and Brusa 1981; Iqbal 1983; Willson 1994). Recent measurements and analyses point out to a lower value, close to 1361 W m^{-2} (Kopp and Lean 2011; Coddington et al. 2016). This was confirmed by a new revision of the main existing databases, which resulted in a reconstituted 42-year TSI time series and an average value of the solar constant of 1361.1 W m^{-2} with an estimated standard uncertainty of 0.5 W m^{-2} (Gueymard 2018a).

The rate of radiant energy received by a surface per unit area is called *irradiance*, whose unit is the Watt per square meter (W m^{-2}). When irradiance (an instantaneous quantity) is accumulated over time, it becomes *irradiation* (sometimes also referred to as "radiant exposure"). In principle, it should be expressed in the proper SI unit, i.e., kJ m^{-2} or MJ m^{-2}. However, in most solar energy applications, the everyday kWh m^{-2} unit is more frequently used, for convenience (1 kWh = 3.6 MJ). In many cases, hourly or sub-hourly irradiations are reported in terms of *average irradiance*, hence expressed in W m^{-2}. At time scales longer than one day or more, irradiations are normally reported in MJ m^{-2} or kWh m^{-2}, but many climatological databases rather express irradiation in irradiance unit (W m^{-2}), which constitutes a source of confusion. What is implied here is that 1 W m^{-2} is assumed constant over 24 h, and if integrated over the whole day would actually represent 24 Wh m^{-2} or 86.4 kJ m^{-2}. For instance, the literature reports that, as a long-term average, the Earth receives ≈340 W m^{-2} at the top of its atmosphere and ≈240 W m^{-2} at the bottom. Note that the colloquial term *insolation* is vague and should be completely avoided since it is not a scientific term.

Generally speaking, the interaction between radiation and atmospheric constituents results in two basic types of attenuation: scattering and absorption. What is not either scattered or absorbed by the atmosphere is transmitted. (Note that what appears to be reflected by clouds is actually caused by scattering.) The transmitted solar radiation that reaches the surface is also partially reflected to space. The reflected fraction of the incident solar radiation is characterized by the *reflectance* (or *albedo*) of the surface. Similarly, the fraction absorbed is the *absorptance*, and finally, the fraction transmitted is called the *transmittance*. Globally, the sum of these three fractions must be equal to 1. Of that total, the long-term mean Earth albedo is ≈0.3. Ultimately, what is absorbed by the atmosphere and the surface is radiated back to space in the form of thermal (infrared) radiation.

In solar energy applications, the orientation and inclination of the receiver or collector (i.e., the observer's relative position) determine both the definition and the name of the incoming solar irradiance. Thus, the term *global horizontal irradiance* (GHI) refers to the total solar irradiance incident on a horizontal surface. Accordingly, the term *global tilted irradiance* (GTI) denotes the total solar irradiance that is captured by a surface tilted with respect to the horizontal plane. In the framework of photovoltaic (PV) solar systems, it is commonplace to rather find this component referred to as the *plane-of-array* (POA) irradiance.

In remote sensing and radiative transfer, the terms *intensity* and *radiance* are frequently used too. Intensity, expressed in W sr^{-1}, is defined as the power emitted by a point per solid angle. Radiance (expressed in W m^{-2} sr^{-1}) is the intensity emitted per unit of the projected surface. Lambertian surfaces are reflectors for which the emitted radiance is the same in all directions. Most natural surfaces are actually non-Lambertian to varying degrees. The irradiance incident on a surface can be obtained as the spatial integration of the radiance that it receives from all directions, e.g., from the whole sky or parts of it, and from surface reflections.

2 Solar Geometry

In all solar radiation studies, a key first step consists in precisely determining the apparent position of the solar disk relative to the observer. Since this needs to be done at any moment and for any observer's location, a general and accurate method is required. Various solar position algorithms have been proposed in the literature. One of the latest is called SG2 (Blanc and Wald 2012). Calculating the solar geometry involves the knowledge of several angles in the ecliptic and equatorial planes, inclination of the receiving surface, the plane of the Sun's apparent path, celestial dynamics, and the trigonometric relationships between all angles. For a horizontal surface, the Sun's position is determined by its zenith angle, Z, and azimuth, γ_{Sun}. In that case, the zenith angle is also the angle of incidence on the horizontal plane. Azimuths are normally measured clockwise from north, although some other conventions exist. Trigonometric functions exist between latitude, solar declination, and hour angle. Moreover, the solar constant must be corrected for the deterministic variation in Sun–Earth distance, related to the *eccentricity* of the planet's orbit. The necessary irradiance correction varies on a daily basis, within $\pm 3.34\%$ during the year. Basic expressions for all these processes can be found elsewhere (Iqbal 1983; Garg and Datta 1993).

For an arbitrarily oriented and inclined surface defined by its tilt, φ, and azimuth, γ, the direction vector of the rays coming from the solar disk are determined by the angle of incidence, θ_{in}, which is the angle between the Sun and the position of the observer's direction vector and the normal vector to the surface plane (Fig. 1). This angle of incidence can be estimated from the trigonometric relationship relating the Sun's azimuth and zenith angles to the surface's azimuth and tilt angles,

$$\cos \theta_{\text{in}} = \sin Z \cos(\gamma_{\text{Sun}} - \gamma) \sin \varphi + \cos \varphi \cos Z. \qquad (1)$$

Apart from its geographical latitude, the amount of radiation received by a surface depends, first, on the date and time, and second, on the relative position

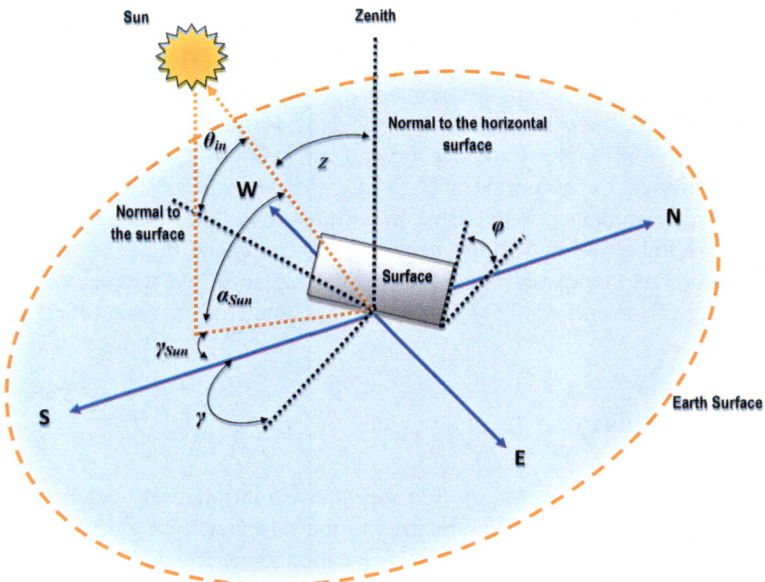

Fig. 1 Basic solar geometry for the angle of incidence relative to an arbitrary oriented and tilted surface

between the Sun and the perpendicular to the plane (with orientation γ and tilt φ). The direct irradiance is maximum at normal incidence ($\theta_{in} = 0$). Figure 2 shows an example of the variation of the Sun's position (γ_{Sun}, Z) along the year in the specific case of an observer located at Madrid, Spain. For any other direction or angle θ_{in} out from the normal to the surface, the direct irradiance is reduced as a function of $\cos(\theta_{in})$—the so-called Lambert's cosine law (McCluney 1994). Moreover, the optical reflectance $\rho(\lambda, \theta_{in})$ of the absorbing surface of detectors or collectors usually increases with incidence angle (Martin and Ruiz 2001; Balenzategui and Chenlo 2005).

Since the Sun's position is a strong function of time, it is important to define the temporal reference correctly. Measured and modeled databases report solar radiation data relative to a *timestamp*, which characterizes how the date and time information is related to the digital data. For an instantaneous value, the timestamp corresponds to that specific moment, or "snapshot," at which the event is recorded by a sensor, computer, or datalogger. In the case of irradiations, or irradiances averaged over short (hourly or sub-hourly) periods, such as one-min data in many observational databases, the timestamp may correspond to the start of the period (forward reference), to its end (backward), or (rarely) to its mid-time (middle reference). Moreover, the reported radiation quantities can have various meanings:

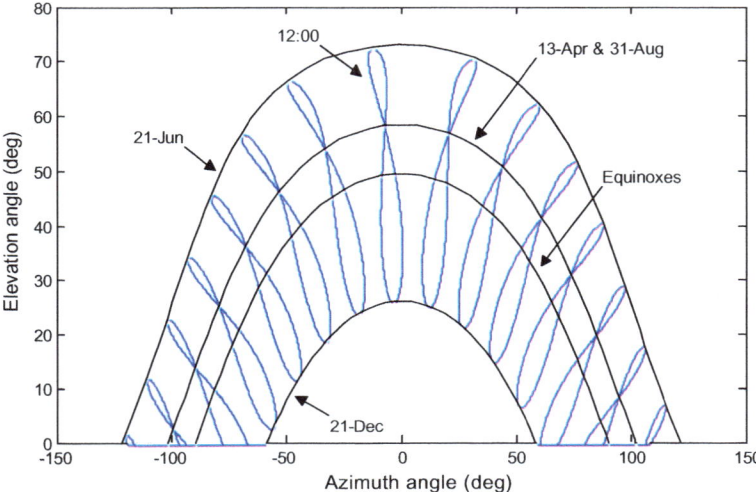

Fig. 2 Sun's position along the year for Madrid, Spain (located 40.45°N, 3.73°W). The dotted curves represent the *analemma*, a diagram formed by the Sun's position at a fixed time of the day along the year. Each analemma in the figure differs one hour to the next (given in UTC). The pattern represented by the analemma is caused by a combined effect of the Sun's declination, the tilt of the Earth's axis, and the Earth's orbital eccentricity. The Sun's position coincides at a certain hour for only two days in the year (13-Apr and 31-Aug) in the Northern hemisphere

- *Instantaneous* values, usually for time steps of 1 s ("true irradiances"), expressed in terms of W m^{-2};
- *Averaged* irradiance values, usually for sub-hourly to monthly periods (W m^{-2});
- *Integrated* irradiation values for hourly, daily, monthly, or annual periods, expressed as kWh m^{-2} or MJ m^{-2} over the appropriate period.

The recording time reference is shown in Fig. 3 for three types of hourly records. For instance, a radiation quantity reported with a timestamp of 01:00 h could mean:

- A value referred to the time interval from 1:00 to 1:59 h (case A in the figure).
- A value referred to the time interval from 0:01 to 1:00 h (case B).
- A value assigned to the middle of the hour, hence 1:30 h would refer to the time interval from 1:00 to 2:00 h (case C).

This possible ambiguity of recording time reference is eliminated in instantaneous records, which are typical of spectral radiation measurements with spectroradiometers or sunphotometers, for instance.

Additionally, the timestamp may be referenced in terms of *local standard time* (LST), coordinated universal time (UTC), or local apparent time (LAT, also called solar time). Those three temporal references are related by

$$\text{LST} = \text{UTC} + \text{TZ} = \text{LAT} + \text{TZ} - \text{ET} - \text{LL}/15 \tag{2}$$

where ET is the equation of time, TZ is the time zone (both expressed in hours), and LL is the local longitude (°). Both the latter and TZ are evaluated positively eastward of the Greenwich meridian and negatively westward. Note, however, that LST sometimes stands for local *solar* time, and that ET is sometimes defined with the opposite sign, which can create confusion. For years between 1900 and 2100, ET (in hours) can be approximated with

$$\text{ET} = 0.16450 \sin(2B) - 0.12783 \sin(B + 78.7) \tag{3}$$

where $B = 360(N - 81)/365$, N is the day number of the year (1–366), and all angles are in degrees.

In summary, the solar geometry affects the solar irradiance reaching the Earth's surface in two ways: (i) modification of the solar constant value due to the Sun–Earth astronomical distance, resulting in what is referred to as *extraterrestrial*

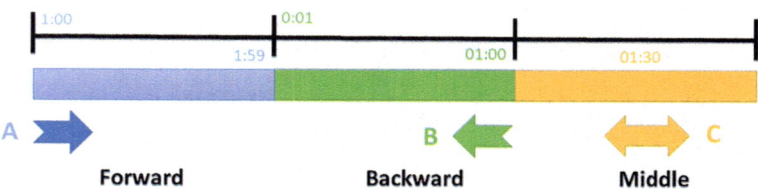

Fig. 3 Recording time reference

irradiance; and (ii) actual apparent position of the Sun center relatively to the observer or receiver. The latter effect is the most important because it also conditions the atmospheric attenuation. The latter is discussed in the next section.

3 Components of Solar Radiation and Atmospheric Interactions

The combination of all the absorption and scattering processes that take place between solar photons and atmospheric constituents (air molecules, water vapor, ozone, carbon dioxide, aerosols, etc.) is referred to as atmospheric *extinction* or *attenuation*. Compared to the unattenuated irradiance at the top of the atmosphere (or extraterrestrial irradiance, as defined above), the amount of transmitted energy reaching the surface is highly variable, depending on conditions. Depending on geographical location, atmospheric conditions, and surface orientation, as much as ≈75–80% of the extraterrestrial irradiance can be received at the Earth's surface on an hourly average. The sky dome illuminates the receiver with diffuse irradiance, while the direct and circumsolar irradiances are received from the solar disk and its aureole, respectively. The fundamentals of these physical processes are briefly described next. For more precise and extended information, the reader may consult specific references dealing with atmospheric radiation and optics (Lenoble 1993; Thomas and Stamnes 1999; Liou 2002; Petty 2006).

Scattering is a physical process by which photons of electromagnetic radiation hit a particle that redirects the energy in all directions following a specific angular distribution. In the atmosphere, scattering particles range in size from air molecules ($\sim 10^{-4}$ μm) and aerosols (~ 1 μm) to water droplets and ice crystals (~ 100 μm). The angular distribution of the scattered intensity is closely related to the relative size of the scattering particle compared to the wavelength of the incident radiation. Assuming a spherical particle, it is common to define a *size parameter*, $a = 2\pi r/\lambda$, where r is the radius of the particle. For very small values of the size parameter ($a \ll 1$), such as with air molecules, the process is called Rayleigh scattering. Its intensity is proportional to λ^{-4}, so that blue wavelengths are more intensely scattered than the red part of the spectrum, in turn creating the blue color of the sky.

For larger particle sizes, comparable to solar radiation wavelengths ($a \approx 1$), the process is called Mie scattering. This applies to the scattering caused by atmospheric aerosols. The angular distribution of the scattered intensity in Mie scattering is much larger in the forward direction and the wavelength dependence is weaker than in Rayleigh scattering. Finally, non-selective scattering occurs when the particles are much larger than the wavelength of radiation ($a \gg 1$). Non-selective scattering—a particular case of Mie scattering—is primarily caused by water droplets in the atmosphere. Its wavelength dependence is virtually non-existent, which makes fog and clouds appear white or gray.

The interaction of solar radiation with the atmosphere and the surface (land or water phases) produces the different components of the solar irradiance incident at the Earth's surface. In particular, the part of the incoming irradiance that is received by a plane normal to the direction of propagation and that comes directly from the solar disk without undergoing any attenuation is called direct normal irradiance (DNI). Its projection on the horizontal plane, more frequently used in atmospheric sciences, is called direct horizontal irradiance (DHI). Likewise, the solar irradiance component that is received from the whole sky as a result of the scattering process constitutes the diffuse irradiance. The global horizontal irradiance (GHI) is defined as the sum of DHI and the diffuse horizontal irradiance (DIF), such that

$$ \text{GHI} = \text{DHI} + \text{DIF} = \text{DNI} \cos Z + \text{DIF} \tag{4} $$

(Note that the solar component terminology can be confusing because some authors associate DHI with *diffuse* horizontal irradiance; additionally, the acronym DHI is sometimes replaced by BHI—beam horizontal irradiance—and, similarly, BNI is sometimes used as a replacement for DNI.) A part of GHI is reflected by the surface (more or less depending on its albedo) toward the sky. A fraction of this upward irradiance is scattered back to the surface, thus increasing GHI—a process called *backscattering*. This process is normally weak but can become intense in the case of a bright overcast sky over snow-covered ground. On a tilted surface, the global total irradiance (GTI) is defined as the sum of the direct, sky diffuse, and ground-reflected components incident on that surface.

The precise definition of DNI may be interpreted in different ways, depending on context. Consequently, slightly different meanings can be found in the literature, depending on whether the circumsolar irradiance emanating from the sun's aureole is accounted for or not, as reviewed by (Blanc et al. 2014). The circumsolar irradiance is the diffuse irradiance emanating from the sky region closely surrounding the solar disk, which is known as the solar aureole (Sengupta et al. 2017). The circumsolar irradiance is the result of Mie scattering in the forward direction of the Sun and thus depends on the amount and type of aerosols or thin clouds. It is, in essence, diffuse radiation that behaves like direct radiation.

The instruments used for measuring DNI, called pyrheliometers, have a field of view that includes the circumsolar irradiance, within ≈2.5° from the sun center. Thus, the strict definition referring to the photons that do not interact with the atmosphere is conceptually useful for atmospheric physics and radiative transfer but can be confusing for ground observations and for the manipulation of multiple sources of data. For solar energy systems, the most useful definition of DNI is the one that includes the circumsolar radiation since it is effectively measured by pyrheliometers, and can also be collected by planar solar systems (Blanc et al. 2014). When using concentrators with high concentration ratios and small opening angles (<1°), however, the measured DNI is slightly overestimated since a part of the circumsolar irradiance is not intercepted.

4 Spectral Solar Radiation and Conversion Applications

Atmospheric absorption is the process whereby an incoming photon is captured by a molecule or atom, thus producing an electronic, vibrational or rotational transition, and ultimately heat. Some gases in the atmosphere, like CH_4, CO, CO_2, N_2, N_2O, NO_2, O_2, O_3, or water vapor, absorb the incoming solar radiation more or less strongly in various wavebands, which creates recognizable patterns in the spectral irradiance distribution at the surface. Aerosols and clouds also absorb photons, but relatively much less than they scatter them. Overall, this absorption process is the main source of energy in the atmosphere and tends to increase its temperature in different layers.

The extinction of solar radiation passing through the atmosphere modifies the spectrum of the incoming solar radiation. Figure 4 shows the extraterrestrial solar spectrum (Gueymard 2018b) compared with direct normal spectral irradiance at sea level obtained with the SMARTS model (Gueymard 1995, 2001), assuming a zenith angle of 48.2°. The absorption processes are particularly intense in some wavebands (e.g., because of water vapor around 1400, 1850, and 2600 nm), resulting in the irradiance being partially or completely attenuated. Strong absorption due to ozone also exists in the UV, which protects biological organisms from excessive dangerous radiation.

Additionally, the available irradiance at the surface depends on the optical pathlength that sunlight has to cross through the atmosphere. This varies during the day as a consequence of Earth's rotation. The air mass, AM or m, is the

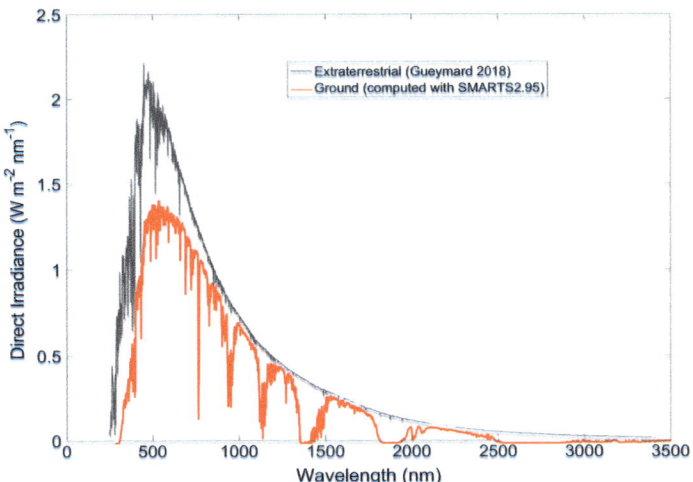

Fig. 4 Spectral irradiance at sea level under typical atmospheric conditions and a zenith angle of 48.2°, compared to its extraterrestrial counterpart. Both quantities are evaluated at normal incidence

conventional variable used to estimate the ratio between the slant optical pathlength through the atmosphere, L, and its zenith (or vertical) counterpart, L_0, according to:

$$m = \frac{L}{L_0} \approx \frac{1}{\cos Z} \qquad (5)$$

By definition, AM1 (or $m = 1$) is the air mass for a zenith sun. The simple expression in Eq. 5 is only an approximation, which starts to diverge at high zenith angles ($Z > 75°$). More elaborate expressions have been developed (e.g., Kasten and Young 1989; Young 1994), as further discussed in Chap. 5.

The air mass is an essential variable that conditions the magnitude of each broadband irradiance component, as well as the distribution and magnitude of its spectral counterpart. Conventionally, the spectral distribution outside the atmosphere is referred to as the AM0 spectrum (i.e., the spectrum for "zero atmosphere"). A common type of reference spectrum is used in solar applications such as photovoltaic or thermal systems, and is referred to as AM1.5 (corresponding to a zenith angle of 48.2°) because that air mass is representative of the mean annual sun's position at mid-latitudes. AM1.5 was historically selected as a reference for the development of standards such as ASTM G173 (ASTM 2012) or IEC 60904–3 (IEC 2016), in combination with specific atmospheric conditions derived from an analysis of solar irradiance data over the Southwestern USA (Gueymard et al. 2002). In these standards, the spectral distributions of both DNI and GTI are synthetically generated with the SMARTS code (Gueymard 1995, 2001).

From a broader spectral standpoint, solar radiation can be classified into three main wavebands:

- Ultraviolet (UV) radiation, for wavelengths below 400 nm (photons with energy larger than 3.1 eV). According to the International Electrotechnical Committee (IEC 1987), UV is further divided into three bands: UVA or A-type ($\lambda \in [315, 400]$ nm), UVB or B-type ($\lambda \in [280, 315]$ nm), and UVC or C-type ($\lambda \in [100, 280]$ nm). Fortunately, essentially all the dangerous UVC is absorbed in the stratosphere (mainly by ozone and oxygen). UVB is also strongly attenuated by ozone and is very low at the surface.
- Visible (VIS) radiation, for wavelengths between 400 and 760 nm (photon energy between 1.6 and 3.1 eV). This range corresponds to that of a typical human eye (the *photopic* range), though limits of sensitivity vary on an individual basis. Following the Commission Internationale de l'Éclairage (CIE 1987), the lower limit of the VIS range is sometimes taken between 360 and 400 nm, and the upper limit is sometimes extended up to 830 nm.
- Infrared (IR) radiation, for wavelengths larger than 760 nm (photon energy below 1.6 eV). The near-infrared (NIR) extends to ≈ 4 μm. Beyond that limit, solar radiation is extremely small at the surface. The extraterrestrial spectrum has only 0.8% of its total energy at wavelengths beyond 4 μm, and less than 0.06% beyond 10 μm.

The relative energetic importance of each of the wavebands just mentioned is compared in Table 1. As seen, ≈45–50% of the total irradiance is contained in either the VIS or NIR range.

Another general classification of importance here opposes solar radiation—also referred to as shortwave (SW) radiation—to terrestrial radiation—also referred to as longwave (LW), infrared, or thermal radiation. The Earth's radiation budget (ERB) analyzes the balance between the incoming radiation (SW) and the outgoing radiation (partly reflected SW, and partly emitted LW). The SW and LW wavebands overlap somewhat between 3 and 10 μm, and the limit between them is not clearly defined. The WMO–CIMO guide to meteorological instruments and methods of observation (CIMO 2017) limits SW radiation to the range 300–3000 nm and LW to the range 3–100 μm. Another common limit used in practice is 4 μm because the quartz window of pyrheliometers transmits radiation up to that wavelength. This is also why the standard spectra discussed above are defined up to that limit.

What Earth receives from the Sun in terms of electromagnetic radiation is perceived by humans in two ways: light and heat. Light commonly refers to the visible range of spectral irradiance, while heat is associated with any source of radiation producing a rise in the temperature of, e.g., a sensor or collector. This disambiguation can be directly applied to the field of energy conversion. Figure 5 shows different ways of harnessing solar energy. The plot's left side applies to the conversion of solar radiation into heat by thermal processes, whereas the right side describes the direct conversion of radiation into electricity through photonic processes.

Thermal systems can be divided into passive systems (without mechanical systems, such as in bioclimatic architecture, greenhouses, or thermosyphon hot water collectors) and active systems (if the produced heat energy is moved away forcibly). Without optical concentration, active solar collectors can just produce low temperatures, referred to as "low-grade" heat. With optical concentration, high temperatures can be achieved for industrial process heat (concentrated solar thermal systems, CST), or to produce electricity with turbines, as in thermal power plants

Table 1 Relative content of irradiance in selected wavelength ranges for different solar radiation spectral distributions according to the ASTM G173 Standard for the global spectrum (AM1.5G) on a 37° tilt and the direct normal spectrum (AM1.5D), as well as the corresponding extraterrestrial spectrum used by SMARTS to obtain these spectra

Waveband	Range (nm)	AM0	AM1.5G	AM1.5D
UV	280–400	7.6%	4.6%	3.4%
VIS	400–760	45.2%	50.1%	48.7%
NIR	760–4000	47.1%	45.2%	47.6%
	280–4000			
Integrated irradiance (W m^{-2})		1347.9	1000.4	900.1

Percentage values refer to the total irradiance over the whole spectral range (280–4000 nm) in each spectral distribution. At the top of the atmosphere, the irradiance between 280 and 4000 nm represents ≈98.6% of the solar constant

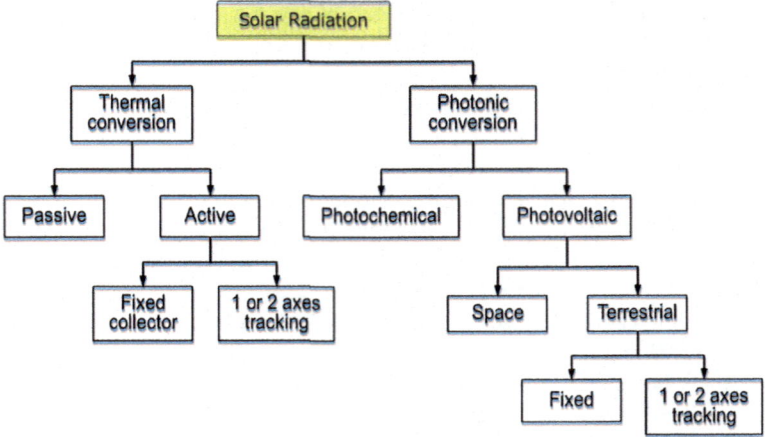

Fig. 5 Different forms of conversion of solar radiation into usable energy

(concentrated solar power systems, CSP). Concentrators require 1-axis or 2-axis tracking. Flat-plate collectors can also be installed on 1-axis or 2-axis tracking systems to increase their resource potential.

In parallel, photonic systems (such as photovoltaic panels) take advantage of the quantum energy of photons. Their absorption in a semiconductor material promotes electronic excitations or transitions, which are used to produce an electrical current. Many semiconductor materials, technologies, and structural designs are developed to improve PV systems for both space and terrestrial applications. In the future, it can be expected that artificial photosynthetic systems will imitate natural plants. In that case, the spectral range of interest would be the photosynthetically active radiation (PAR), between 360 and 760 nm.

The broad range of conversion processes just reviewed provides some important keys to understand the importance of the correct evaluation of the solar resource, beyond the basic needs of energy conversion, meteorology, environment, or climate change.

Acknowledgements This work has been partially supported by the Spanish National Funding Program for Scientific and Technical Research of Excellence, Generation of Knowledge Subprogram, 2017 call, DEPRISACR project (reference CGL2017-87299-P). The authors wish to thank Dr Stefan Wilbert from DLR for sharing several useful comments and remarks on this chapter.

References

ASTM (2012) Standard tables for reference solar spectral irradiances: direct normal and hemispherical on 37° tilted surface. Standard G173-03(2012). ASTM International. http://www.astm.org/Standards/G173.htm

Balenzategui JL, Chenlo F (2005) Measurement and analysis of angular response of bare and encapsulated silicon solar cells. Sol Energy Mater Sol Cells. https://doi.org/10.1016/j.solmat.2004.06.007

Blanc P, Wald L (2012) The SG2 algorithm for a fast and accurate computation of the position of the Sun for multi-decadal time period. Sol Energy 86:3072–3083. https://doi.org/10.1016/j.solener.2012.07.018

Blanc P, Espinar B, Geuder N et al (2014) Direct normal irradiance related definitions and applications: the circumsolar issue. Sol Energy 110:561–577. https://doi.org/10.1016/j.solener.2014.10.001

CIE (Commission Internationale de l'Éclairage) (1987) Methods of characterizing illuminance meters and luminance meters. CIE 69-1987. Vienna, Austria

CIMO (2017) Part I, Chapter 7 for solar radiation measurements. Part I, Chapter 16 for aerosols measurements. In: The CIMO guide. World Meteorological Organization. WMO guide to meteorological instruments and methods of observation (WMO-No. 8, 2014 edition, updated in 2017)

Coddington O, Lean JL, Pilewskie P et al (2016) A solar irradiance climate data record. Bull Am Meteorol Soc 97:1265–1282. https://doi.org/10.1175/BAMS-D-14-00265.1

Fröhlich C, Brusa RW (1981) Solar radiation and its variation in time. Sol Phys 74:209–215

Garg HP, Datta G (1993) Fundamentals and characteristics of solar radiation. Renew Energy 3:305–319. https://doi.org/10.1016/0960-1481(93)90098-2

Gueymard C (1995) SMARTS2: a simple model of the atmospheric radiative transfer of sunshine: algorithms and performance assessment. Rep. FSEC-PF-270-95. Florida Solar Energy Center, Cocoa, FL

Gueymard CA (2001) Parameterized transmittance model for direct beam and circumsolar spectral irradiance. Sol Energy 71:325–346. https://doi.org/10.1016/S0038-092X(01)00054-8

Gueymard CA (2018a) A reevaluation of the solar constant based on a 42-year total solar irradiance time series and a reconciliation of spaceborne observations. Sol Energy 168:2–9. https://doi.org/10.1016/j.solener.2018.04.001

Gueymard CA (2018b) Revised composite extraterrestrial spectrum based on recent solar irradiance observations. Sol Energy 169:434–440. https://doi.org/10.1016/j.solener.2018.04.067

Gueymard CA, Myers D, Emery K (2002) Proposed reference irradiance spectra for solar energy systems testing. Sol Energy 73:443–467. https://doi.org/10.1016/S0038-092X(03)00005-7

IEC (International Electrotechnical Commision) (1987) International electrotechnical vocabulary, Chapter 845: lighting. IEC 60050-845. Geneva, Switzerland

IEC (2016) Photovoltaic devices—Part 3: measurement principles for terrestrial photovoltaic (PV) solar devices with reference spectral irradiance data. IEC 60904-3:2016 RLV Standard. International Electrotechnical Commission, Geneva, Switzerland

Iqbal M (1983) An introduction to solar radiation. Academic Press, Canada

Kasten F, Young AT (1989) Revised optical air mass tables and approximation formula. Appl Opt 28:4735–4738

Kopp G, Lean JL (2011) A new, lower value of total solar irradiance: evidence and climate significance. Geophys Res Lett 38:L01706. https://doi.org/10.1029/2010gl045777

Lenoble J (1993) Atmospheric radiative transfer. Deepak Publishing, Hampton

Liou K-N (2002) An introduction to atmospheric radiation. Academic Press, San Diego, USA

Martin N, Ruiz JM (2001) Calculation of the PV modules angular losses under field conditions by means of an analytical model. Sol Energy Mater Sol Cells 70:25–38. https://doi.org/10.1016/S0927-0248(00)00408-6

McCluney R (1994) Introduction to radiometry and photometry, 2nd edn. Artech House, Boston, London

Petty GW (2006) A first course in atmospheric radiation, 2nd edn. Sundog Publishing, Madison

Sengupta M, Habte A, Gueymard C et al (2017) Best practices handbook for the collection and use of solar resource data for solar energy applications. pp 1–233. https://doi.org/10.18777/ieashc-task46-2015-0001

Thomas GE, Stamnes K (1999) Radiative transfer in the atmosphere and ocean. Cambridge University Press, Cambridge

Willson RC (1994) Atlas of satellite observations related to global change. In: Turner JL, Foster CLP (eds) Weather. Wiley-Blackwell, Hoboken, pp 5–18

Young AT (1994) Air mass and refraction. Appl Opt 33:1108–1110. https://doi.org/10.1364/AO.33.001108

Chapter 2
Solar Radiation Measurement and Solar Radiometers

José L. Balenzategui, Fernando Fabero and José P. Silva

> ...*too little interest is devoted to the calibration of instruments and quality of data. Measurements which are not reliable are useless.*
>
> WMO Technical note 172 (1981)

Abstract An instrument able to measure electromagnetic radiation, in its different forms and spectral ranges, is called a radiometer. This chapter focuses on the radiometers used for sensing solar radiation and on the measurements of different components and types of solar irradiance. As a simple classification, we will distinguish between broadband and spectral (narrowband) sensors. First, the fundamentals of physical sensors used to measure solar radiation are briefly described. Then, importance about calibration methods and uncertainty, as well as the structure of the traceability chain in the magnitude of solar irradiance, are presented. Next, solar radiometers and measurement techniques are described, starting from direct radiation in Earth's surface, global irradiance in horizontal and tilted surfaces, diffuse irradiance, and finally another kind of radiation sensor. This structure is not casual but follows a path similar to that of the traceability chain, starting from the more accurate to the less accurate instruments. There are two additional sections devoted to the measurement of the spectral distribution of irradiance and to the measurement of aerosol contents in the atmosphere by using filter radiometers.

J. L. Balenzategui (✉) · F. Fabero · J. P. Silva
CIEMAT—Photovoltaic Solar Energy Unit (Energy Department),
Avda. Complutense 40, 28040 Madrid, Spain
e-mail: jl.balenzategui@ciemat.es

F. Fabero
e-mail: fernando.fabero@ciemat.es

J. P. Silva
e-mail: josepedro.silva@ciemat.es

© Springer Nature Switzerland AG 2019
J. Polo et al. (eds.), *Solar Resources Mapping*, Green Energy and Technology,
https://doi.org/10.1007/978-3-319-97484-2_2

15

1 Sensing Solar Radiation

Instruments measuring solar irradiance are based on the shift of a certain physical property (e.g., an increase in the temperature) in a material or device when solar radiation is impinging in and being absorbed by it. Measurement of this shift allows quantifying the amount of solar irradiance. Therefore, there is no way for a direct measurement of solar radiation, and it is always estimated by an indirect or a two-step method, based well on thermal, or well on photonic effects.

In many cases, thermal detectors of solar radiation have also been used as detectors of infrared radiation (and vice versa). As well, photonic detectors of sunlight have also been used as general optoelectronic sensors of different radiation sources (VIS and UV lamps, laser systems, LEDs, etc.). In many cases, it is simply the shape, driving circuit, embodiment, structure or supporting case used, what differences a solar sensor from a conventional thermal, IR or photonic sensor used in other scientific areas. In other cases, however, sensibility, spectral range, or output signal levels are somewhat different.

Good historical reviews and descriptions of solar radiation instruments (in more detail than in this chapter) can be found elsewhere (Marchgraber 1970; Coulson 1975; Thekaekara 1976; Frohlich and London 1986; Zerlaut 1989; Fröhlich 1991; Vignola et al. 2012; CIMO 2017; Stanhill and Achiman 2017). Here, a brief about sensors and instruments is given.

To date, physical phenomena and practical devices used for sensing solar radiation include:

- *Thermoelectric sensors.* They are based on a thermoelectric effect: A temperature difference between two junctions of two different metals creates an electromotive force, as in the case of a thermocouple (TC). This thermoelectric effect, discovered by Seebeck in 1815, was first used for optical radiation measurements by Nobili and Melloni in 1835 (Palmer and Grant 2010). The sensing element is a thermopile, in which tens of these TC junctions are combined in series to increase the output signal, as shown in Fig. 1. Half of the TC junctions are in contact with a black absorbing plate exposed to the Sun. The other half, in the backside, is in contact with a second plate (ideally, a heat sink or thermal block) which gives a reference temperature (a lower temperature, can be ambient temperature). Thus, the temperature difference between both plates produces a voltage difference proportional to the irradiance absorbed on the front plate. First modern designs of this kind of thermopile effectively used in commercial solar instruments were developed in the 1920s (Moll 1922; Kimball and Hobbs 1923; Gorczyński 1924). State-of-the-art pyranometers and pyrheliometers (even secondary standards) are based on thermopiles of different designs and configurations. In general, they show good levels or responsivity (around some mV at 1000 W m^{-2}), good linearity over the range of terrestrial solar irradiances, relatively fast response to changes in irradiance (time constants of the order of seconds), and small influence of ambient temperature (Vignola et al. 2012). Specific details about the operation and characteristics of thermopile-based instruments will be discussed below.

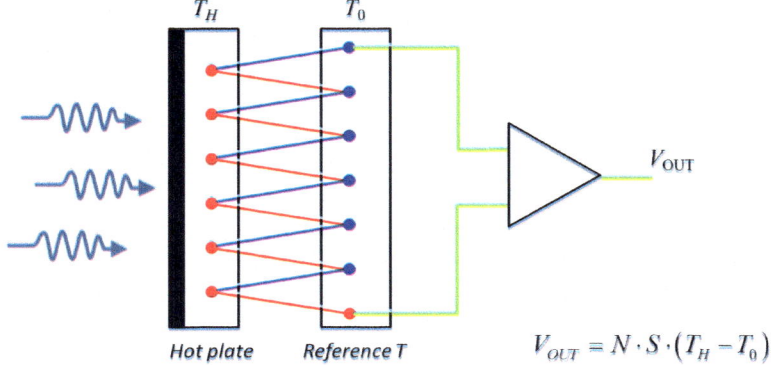

Fig. 1 Sketch of a thermopile: output voltage is proportional to the temperature difference $(T_H - T_0)$; S = Seebeck coefficient; N = number of thermocouples

Fig. 2 Some examples of black-and-white pyranometers: (left) a Yanishevsky pyranometer used for albedo measurement, (right) an Eppley model 8-48 diffuse pyranometer

- *Differential absorbing surfaces (black and white).* The idea of estimating solar irradiance as being proportional to a temperature difference between the two surfaces is extended in these sensors by allocating both surfaces exposed to the Sun. One of the surfaces is black (absorbing shortwave and longwave radiation), and the other is white/reflecting/metallic (only absorbing longwave radiation). Usually, several black-and-white areas (in the form of a chessboard or circular pie portions) are combined, as in the examples of Fig. 2. The temperature difference is measured with an electrical resistance thermometer or by a thermopile structure (hot junctions beneath the black surface, cold junctions under white one). With the development of higher-precision instruments, these sensors have been classified as of lower accuracy and used in many cases only for diffuse irradiance measurements. Some examples are: the Callendar pyranometer (1898, Callendar and Fowler 1906), the Angström compensation pyranometer (1919, Coulson 1975), the Eppley model 50 "light bulb" pyranometer (1930), the Yanishevsky pyranometer (1957), and the Eppley model 8-48 (1969) (Stewart et al. 1985; Vignola et al. 2012).

- *Calorimeter-like sensors.* A metal disk or cylindrical vessel (silver, brass) with a blackened absorbing surface, and filled with water or mercury (a liquid medium), is exposed to the Sun in the normal direction. Changes in the temperature of the liquid due to the absorption of radiation can be tracked by a conventional thermometer (a mercury thermometer, a resistance wire) in direct contact with the liquid or the disk, as depicted in Fig. 3. The disk was alternately exposed to sunlight (direct normal irradiance) and then shaded (or rotated) in periods of 2–5 min, in a sequential run. Examples of these types of instruments were the Pouillet's pyrheliometer (Pouillet 1838), the Abbot silver-disk pyrheliometer (Abbot and Fowle 1908), and the Marvin pyrheliometer (1910, Foote 1919). Abbot and Marvin's devices used a collimating tube defining the field of view. The difference between temperatures during the shaded and unshaded periods, together with characteristics of the sensor (heat capacity, area, etc.), allowed to estimate the irradiance. At that moment, an improved version of Abbot pyrheliometer (Abbot 1913) was one of the best state-of-the-art accurate instruments for direct solar irradiance and was adopted by the Smithsonian Institution to be the reference instrument to base its irradiance scale (Abbot and Aldrich 1913).

- *Electrical substitution radiometers.* Based on the principle of electrical substitution (and/or electrical compensation), first applied by Angström in 1893 (Ångström 1894; Angström 1899), these instruments are self-calibrated and considered as primary absolute radiometers. The principle of substitution assumes that heating produced by the absorption of solar radiation in a black metallic strip (or in a cavity) and heating produced by an electrical current

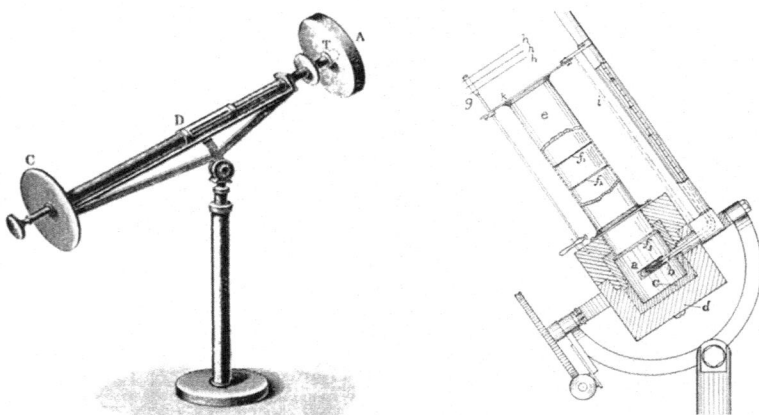

Fig. 3 (left) Pouillet's and (right) Abbot's pyrheliometers, two examples of calorimeter-based sensors of direct solar irradiance

circulating through the strip (or through a wire intimately endorsed to the cavity) are equivalent, as to produce the same temperature rise in a thermopile or resistance thermometer. As voltage and current can be measured with high accuracy, the irradiance is estimated from measurement of electrical power supplied to the sensing element. In the case of Angström electrical compensation pyrheliometer (1893), two strips of black-painted manganin foils are exposed alternatively to sunlight (one shaded, the other unshaded in every run) by means of a reversible shutter at the front of the collimator tube. The shaded strip is electrically heated to reach the same temperature as the exposed strip. In the case of the primary absolute cavity radiometer (PACRAD) developed by Kendall in 1969 (Kendall 1968), the front cavity is alternatively exposed to and shaded from sunlight, while a second twin cavity (compensation cavity) is kept in the dark at ambient temperature. In the closed period, the electrical current heats the front cavity until the same temperature difference with reference to the rear cavity, as in the open period when the front cavity is radiatively heated, is reached. Active- and passive-type absolute cavity radiometers (ACRs) were later developed based on Kendall's PACRAD and allowed WMO for the definition of the World Radiometric Reference (WRR) in 1979. Both Angström and ACR pyrheliometers are currently the primary reference instruments for the magnitude of solar irradiance in many national radiometric laboratories. Due to the key importance of these sensors, further details are later given in other sections of this chapter.

- *Photoelectric devices.* While previous instruments were thermal-type sensors, these are photon-type sensors, mainly based on the property of semiconductor materials and alloys of experiencing electronic transitions and excitations to different energy levels as a consequence of absorption of radiation photons. Thus, this description mainly refers to photovoltaic-type (PV) devices, as photodiodes (PD) and solar cells (SC), and not to photomultiplier tubes (PMT). PMT is really based on the photoelectric effect (the emission of electrons out from the surface of a material being illuminated) and then can be considered as photoemissive detectors. However, a PMT has a very different structure than PV devices and is constructed to detect very low levels of radiation (even photon-count devices) and not for solar irradiance levels. PV-type devices are designed to provide an electrical current (and to produce a difference of potential) proportional to the irradiance absorbed. Although both types of devices have a similar structure (usually based on one or more p–n junctions created by differential doping of the semiconductor), in PD the main focus is put on sensing radiation (linearity, speed, low noise, high sensitivity, etc.), while SC is devoted to converting solar radiation power into electrical power (high efficiency, low series resistance, low thermal coefficients, etc.). For the same reason, PD are prepared for working under low to moderate levels of irradiance (except if a diffuser/attenuator is added), while SC can be used at normal solar irradiance levels and can especially be designed to work under high levels of optical concentration (up to $\sim 2 \times 10^6$ W/m^2). However, in both cases, spectral range is limited by the bandgap of the material, and thermal/noise effects reduce

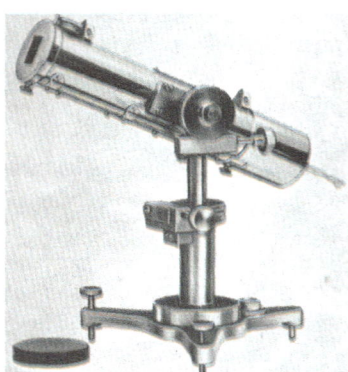

Fig. 4 (left) Angström strip-type and (right) PMO6, an ACR-type primary reference by PMOD/ Davos Instruments, both pyrheliometers based on electrical substitution or compensation principles

the performance of IR sensing devices except if cooled. Some commercial examples are shown later in Sect. 5.

- *Photoresistive and photoconductive sensors.* Except for some special devices (photoresistances used as penumbra/sunrise/sunset detectors), this type of sensors is not generally used in solar radiation sensing but as IR or thermal detectors. However, it is worth to mention them because of their close relationship with some of the previous instruments. In both cases, radiation received in the sensor promotes an increase in the temperature and, as consequence, in its electrical resistance (photoresistances) or its conductivity (photoconductive devices). A bolometer is a special kind of photoresistance with a high-temperature coefficient of resistance, made of a thin film of metal or semiconductor, and was invented by S. P. Langley in 1880 (Langley 1880; Callendar and Fowler 1906). Photoconductive (PC) sensors are made of semiconductor films and are based on the changes in the conductivity of the material (as a consequence of a change in the population of free electrons in the conduction band) produced by a temperature variation or by absorption of radiation (Palmer and Grant 2010). In the end, a PC device or a bolometer, connected as part of a circuit (see Fig. 5), functions as a resistor whose resistance depends (linearly or exponentially) on the light intensity.

- *Thermomechanical devices.* In these sensors, absorption of solar radiation energy produces some kind of appreciable mechanical perturbation. In the case of bimetallic radiation sensors (as in bimetallic thermometers), a couple of identical size strips of two metals with different expansion coefficient are joined (forming a bimetallic strip) and are connected to a gauge. Heating caused by solar radiation produces a differential thermal expansion on these metals, which can be sensed by the gauge. It is not used in practice for monitoring or recording of solar radiation by its lack of accuracy, and it is valid only for a rough naked-eye estimation of incident radiation. Another type of thermomechanical

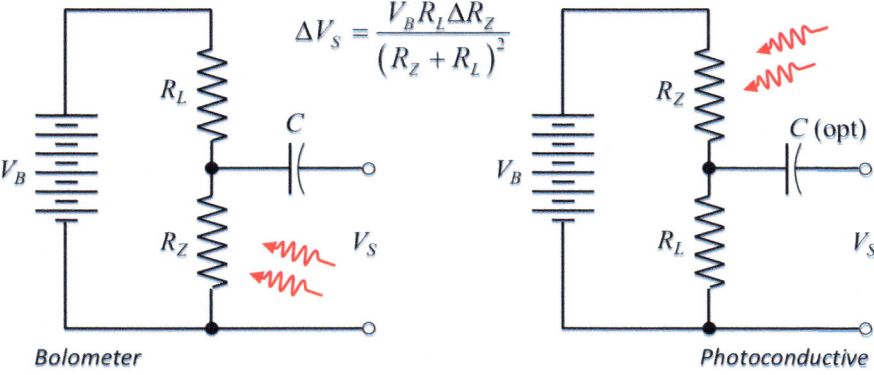

Fig. 5 Polarization circuits for a bolometer (half-bridge circuit) and for a photoconductive device (PCD). The capacitor is placed to block DC component when using a modulated signal/beam (required for bolometer, optional for PCD). R_Z represents the variable resistance of the bolometer or the PCD, while R_L is a load resistance, after Palmer and Grant (2010). In both cases, maximum power transfer is obtained when $R_Z = R_L$

instrument is the Crookes' radiometer or light mill (invented 1873) (Crookes 1874), based on the different absorbing properties of black-and-white/metallic surfaces. In this, a set of vanes which are mounted on a spindle inside an airtight glass bulb, containing a partial vacuum, can rotate with a speed proportional to the light intensity. The motion of the vanes is in fact promoted by the movement of gas molecules between both faces of the vanes. Nowadays, it is only used for demonstrative or academic purposes and not as a sensor itself. Modern versions of this instrument have been developed, as a monocolored curved-vane micromotor (Han et al. 2011) and a 100-nm gold light mill (Liu et al. 2010).

Specific details about the practical use of devices and sensing elements above presented are discussed later, in different sections of the chapter. Additionally, there are other IR and thermal sensors not mentioned here because, to our knowledge, they are not used in solar radiation applications, as pyroelectric devices (see, e.g., Putley 1977), phototransistors, and Golay cells (Golay 1947a, b).

2 Calibration and Traceability

Gaining accuracy in the determination of solar irradiance has been of the major concern since the early days of solar radiometry (Fröhlich 1991). The description of the various instruments and physical phenomena applied in this field, given in the previous section, together with the history and evolution of the irradiance scales along the preceding decades (see below), is a demonstration of the huge effort employed by many researchers worldwide along the time in the consecution of this objective.

As in all the scientific areas, every instrument and sensor devoted to estimate a physical magnitude (as solar irradiance) has to be referenced to universally accepted standards, scales, and units. It is the only way the results given by laboratories and researchers in different locations (even in different points of solar system and beyond) can be compared. The reference frame, internationally agreed since 1960, is the International System of Units (SI).

The irradiance of a radiation source is a derived magnitude in SI, defined as the surface density of radiant flux or power (McCluney 1994), the radiant flux per unit area in a specified surface that is incident on, passing through or emerging from a point in the specified surface (considering all directions in the hemispherical solid angle above or below the point in the surface), and has units of Watts per square Meter (W m^{-2}) in SI (see Fig. 6).

The organism in charge of defining, establishing, reviewing, and maintaining the SI is the Bureau International des Poids and Measures (BIPM, International Bureau of Weights and Measures) through the International Committee for Weights and Measures (CIPM). To guarantee the universal equivalence of measurements, the CIPM signs Mutual Recognition Arrangements (CIPM MRA) with National Metrology Institutes (NMI) and some international organisms, once they demonstrate the correspondence of their measurement standards and the calibration and measurement certificates they issue with the rest of NMI (see BIPM 2018), based on the results of international intercomparisons (called Key Comparisons, KC). Additionally, an NMI can give the responsibility of materializing units and scales, carrying out primary calibrations and the preservation of national standards, for one or some particular magnitudes, to a Designated Institute (DI) in its country.

The outcomes of the CIPM MRA are the Calibration and Measurement Capabilities (CMCs), that have to be individually recognized for every participating institute (NMI or DI), which list the different magnitudes or quantities for which calibration and measurements certificates are recognized by the rest of NMIs.

The current list of magnitudes and scientific areas covered under the possible CMCs (list of services, see KCDB 2018) includes three quantities directly

Fig. 6 Sketch for the definition of the irradiance, after McCluney (1994)

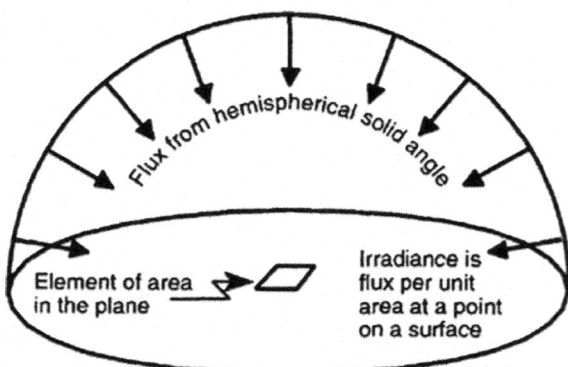

associated to solar radiation, although many others under the branch of photometry and radiometry also apply to detectors used in this field:

(a) Responsivity, solar, power.
(b) Responsivity, solar, irradiance.
(c) Responsivity, solar, spectral, irradiance.

Figure 7 shows these quantities and their current position in the table of recognized magnitudes in the KC—CMC.

Nevertheless, the current status of these solar radiation magnitudes within the SI is quite new, and solar irradiance scales are in fact pending of a better foundation within SI due to some discrepancies found in the past when compared to SI irradiance scales at NMI laboratories (see Appendix). A short review of the preceding history of solar irradiance scales is convenient at this point.

2.1 Solar Irradiance Scales and Reference Standards

Historically, solar irradiance was mainly considered as a meteorological variable, and thus, its natural place was under the cover of the International Meteorological Organization (IMO), created in 1873, superseded in 1950 by the World Meteorological Organization (WMO). The influence of solar radiation and its possible changes in weather and climate were one of the fields being researched into since the late nineteenth century. A complete review of the history of solar radiometry can be found elsewhere (Fröhlich 1991).

However, radiometry and photometry NMI laboratories used to base their measurement scales and standards in artificial sources (tungsten halogen lamps, UV sources, blackbodies, laser systems, etc.) and/or in related detectors adapted to these sources. Thus, as Sun was a natural (seasonally variable, unpredictable, unstable) source of irradiance, it was not usually considered under the scope of these NMI laboratories. At the same time, measurement capabilities of NMI laboratories were restricted to intensity levels relatively low and, as a consequence, not adapted to solar irradiance. Then, solar irradiance, taken as a physical magnitude, was not under the consideration of the CIPM and not specifically included in the CMCs, as irradiance, as a general quantity, did.

Inevitably, this promoted the independent evolution along the time of primary standards and scales for solar irradiance under the wings of IMO and WMO, out from the scope of CIPM and, in a certain form, out from the SI. Let us say *in a parallel way*, to express it in a less dramatic form. Figure 8 shows schematically the evolution of these scales and standards.

As shown, two irradiance scales were defined at the beginning of the twentieth century, almost at the same time: the Ångström scale (1905), based on the Ångström compensation pyrheliometer, and the Smithsonian scale (1913), based on the Abbot and Fowle stirred silver-disk pyrheliometer, already mentioned in

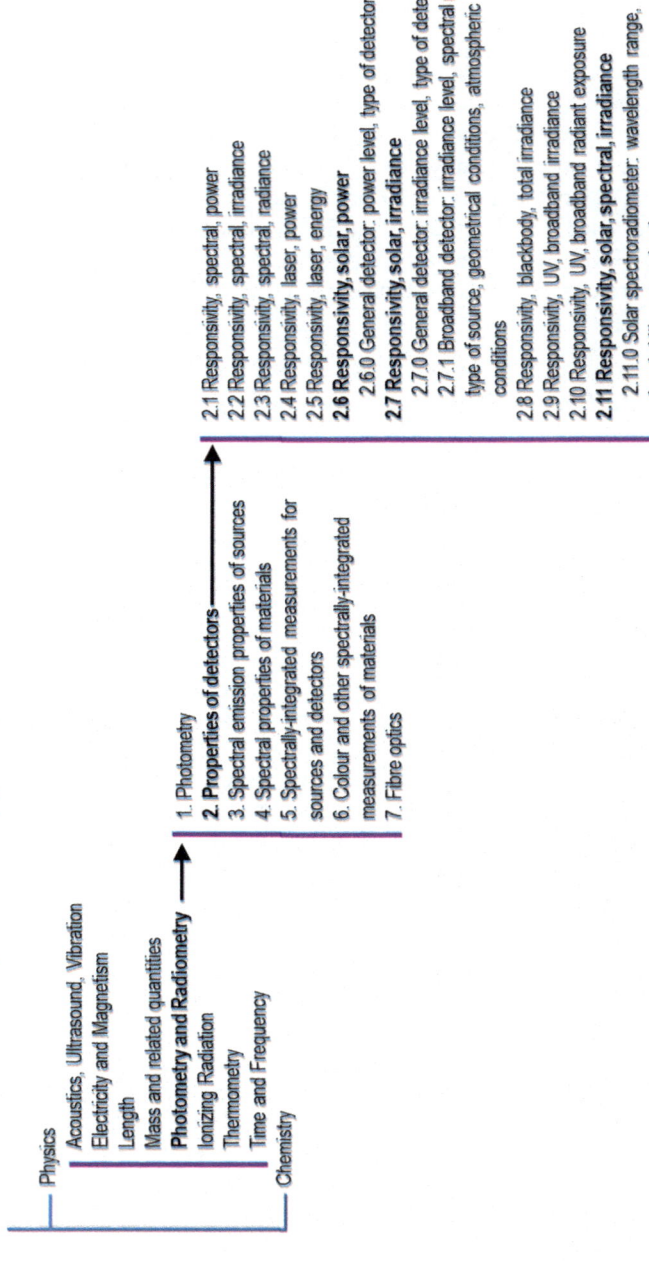

Fig. 7 Position of the three magnitudes associated to solar radiation in the SI, as given in the list of services of Calibration and Measurement Capabilities (CMC) in the Key Comparison Data Base (KCDB) of the BIPM (see text)

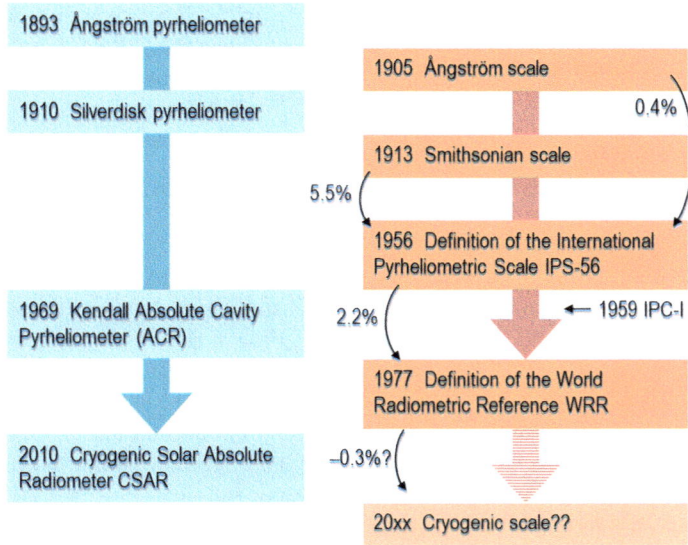

Fig. 8 Evolution of primary reference standards and scales in solar irradiance radiometry. After Finsterle (2015)

Sect. 1. However, several comparisons revealed important differences between both solar irradiance scales (Fröhlich 1973; Latimer 1973). With the objective of getting homogeneous irradiance measurements worldwide, the International Pyrheliometric Scale (IPS-56) was defined in 1956. In 1959, WMO organized the First International Pyrheliometer Comparison (IPC-I) in the Physikalisch Meteorologisches Observatorium Davos (PMOD, Davos, Switzerland). However, this first and the subsequent IPCs did not solve the discrepancies between both scales and neither the new IPS-56 was able to promote a clearer and more stable frame.

At the end of the decade of 1960, a new type of electrical substitution radiometers was developed in the Jet Propulsion Laboratories (JPL), in the framework of the US space race R&D efforts: the absolute cavity radiometers (Haley et al. 1965; Kendall 1968). Due to their better performance, similar instruments were readily developed in a few years, with different denominations: PACRAD, ACR, PMO, CROM, HF, TMI (Geist 1972; Crommelynck 1973; Willson 1973; Brusa and Fröhlich 1975; Hickey et al. 1977). The high accuracy and stability of these new radiometers led to the selection of new reference standards and the definition of a new irradiance scale. Results from the IPC-IV (1975) allowed the settlement of the World Radiometric Reference (WRR), defined as the mean value of 15 absolute radiometers of 9 different models with an estimated accuracy of 0.3% (Fröhlich 1978).

WRR was taken by WMO as the reference primary standard for solar irradiance measurements, and PMOD (Davos) was designed as World Radiation Center (WRC), in charge of maintaining the World Standard Group (WSG) of reference radiometers used to materialize the WRR. Every radiometer of the WSG is a practical realization of the unit of irradiance W m^{-2}. Since then, IPC has been organized every 5 years by WMO/PMOD to disseminate the WRR reference and to validate the stability of the WSG. E.g. IPC-XII was celebrated in 2015.

2.2 Next Steps to Join SI

As any other reference standard in the SI, the WRR has to show long-term stability and has to allow accurate and homogeneous worldwide measurements of its magnitude (the solar irradiance). This way, measurements done at different points of the Earth and in different moments along time have to be comparable and equivalent. For example, only a stable irradiance reference is able to detect subtle changes at a climatic, environmental, and Sun emission level. For weather, environment, or climatic applications, long-term stability can be more important than absolute precision. For solar energy applications, instead, absolute accuracy and small uncertainty can be as important as long-term stability (Finsterle 2015).

WRR can be considered as a primary reference which realizes the unit of solar irradiance, the W m^{-2}, and as such, it defines a scale based on a physical artifact or prototype (as in the past with the unit of mass, the kilogram).

However, according to BIPM principles, the primary standards must be guarded and maintained by NMIs. Then, METAS (NMI in Switzerland) named PMOD as Designated Institute and it signed the CIPM Mutual Recognition Arrangement (CIPM MRA) with BIPM in 2008 (Rüedi and Finsterle 2005). PMOD recognized capabilities include 2 CMCs for solar irradiance (for pyranometers and pyrheliometers) and 4 CMCs for UV.

On the other hand, WMO also signed a CIPM MRA in 2010 (after a working agreement previously signed between WMO and CIPM in 2001), being the second international organism in getting a CIPM MRA. WMO is then equivalent to an NMI and can define its own Designated Institutes to preserve primary standards and to carry out primary calibrations. WMO coincidently chose WRC/PMOD as DI for solar irradiance (2 CMCs) and UV (4 CMCs) within its CIPM MRA. Then, acting as WMO DI, the solar irradiance scale based on WRR is disseminated from PMOD/WRC to standard sensors in WMO regional and national radiometric centers, and to those responsible of BSRN stations, as before, but now until the formal structure of BIPM. IPC could also be used to carry out Key Comparisons between WRR and other standard ACRs of international NMI/DI.

Therefore, the circle is closed in the sense that WRR can be understood as the primary reference in a solar irradiance scale, both through the designation of PMOD by METAS (at Switzerland national scale) and through its designation by WMO (international scope). WRR-based (or any other future alternative) solar irradiance scales are no longer out of the scope of CIPM, and the magnitude of solar irradiance is fully integrated into SI.

However, there is only a little detail to be solved in the near future concerning the practical realization of the solar irradiance scale based on WRR. As a result of several intercomparison carried out between WRR and SI reference irradiance standards (absolute cryogenic radiometers), the WRR has currently no equivalence or compatibility to the SI laboratory irradiance scales implemented by NMIs (Romero et al. 1991, 1995; Finsterle et al. 2008; Fehlmann et al. 2012), and it is temporary out from SI (in terms of traceability). The difference in the determination of irradiance is larger than 0.3% between both scales, and their uncertainty ranges do not overlap this difference. In the Appendix of this chapter, there is a more detailed explanation of this issue.

Despite this transitory affair, the fundamental aspect to highlight here is that, whether a particular solar irradiance scale is based on WRR or another prototype realizing, the unit of W m^{-2} (that might supersede WRR in the future, as the cryogenic solar absolute radiometer CSAR could do, or perhaps a new, particular realization of the unit by an NMI), the integration of the magnitude "responsivity, solar, irradiance" into SI structure is solved.

2.3 Traceability of Solar Irradiance Detectors

In metrology, *calibration* of a specimen or device is the comparison of some of its measurands (its properties, characteristics or the output or signal delivered by it) to those of a reference standard of known accuracy (and, therefore, of known uncertainty) under specified working conditions. The calibration allows the determining of the bias, accuracy, and uncertainty of the measurable property of the device under test (DUT) under these conditions.

The *traceability* refers to an unbroken chain of documented comparisons (calibrations) by which the measurand of the DUT can be related to that of a national standard maintained by an NMI. Every calibration step performed between the national standard and the DUT, by means of intermediate standards in a hierarchical sequence (if any), contributes to (increases) the final measurement uncertainty. Then, less the intermediate calibration steps, (ideally) less the final uncertainty.

Figure 9 shows a sketch of the traceability chain for the different instruments used in a solar irradiance scale. The set at the bottom of this structure (the ultimate recipients of the work done in upper stages) are the working standards and field instruments which are used in (industry and laboratory) testing, in the monitoring of solar plants of different nature, and of weather and BSRN stations. As higher the accuracy and lower the uncertainty required for these applications, as stronger the requirements for the calibration procedures, for the metrological level of measuring instrumentation and for the quality of the standard sensors.

In this sense, some values of reference for these requirements are to be given. For example, the Global Climate Observing System (GCOS) points out the importance of a continuous recording of solar radiation through the BSRN grid (BSRN 2018) and sets a limit of 1 W/m^2 for absolute accuracy and 0.3 W/m^2 per decade in terms of stability (GCOS 2011, 2016). Requirements pointed out by the National Institute of Standards and Technology (NIST) for future space radiometers

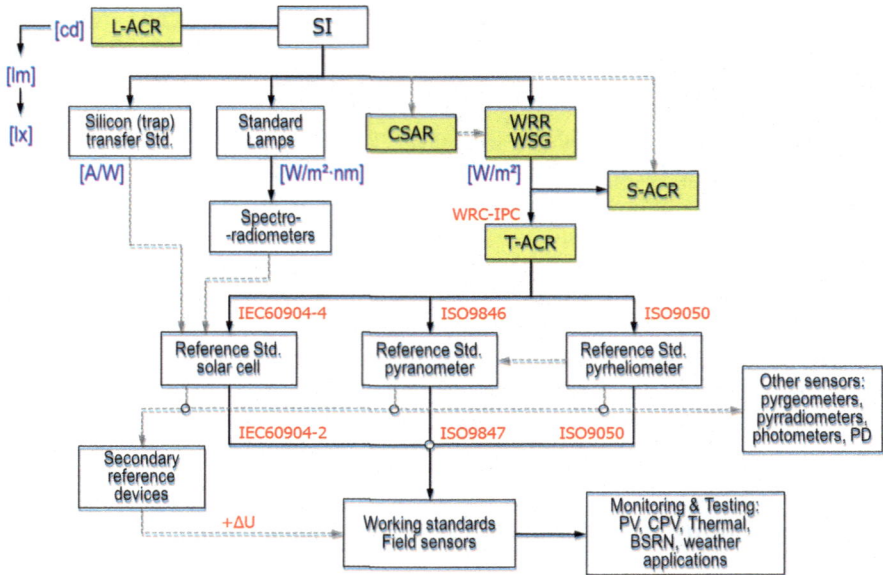

Fig. 9 Traceability chain for the calibration of solar sensors, from SI downwards. Yellow boxes refer to absolute radiometers. Blue brackets stand for SI units implemented in every branch. Some ISO and IEC Standards with calibration methods are indicated in red, although several alternative procedures exist

are even more demanding: Spectral radiation reflected by Earth surface needs a precision of a 0.2%, spectral solar radiance of 0.1%, and TSI up to 0.01% (Murdock and Pollock 1998; Pollock et al. 2000).

Finally, the accuracy of sensing devices for monitoring of solar PV plants is regulated by IEC 61724-1 Standard (IEC 2017). IEC 61724-1 classifies monitoring systems in three levels of complexity and required accuracy: Class A (high accuracy), Class B (medium accuracy), and Class C (basic accuracy). Class A is recommended for large PV systems, while Class C is recommended for small-size installations. Class A systems require the on-field measurement of several irradiance components (see Table 1), while Class B can either measure or estimate the magnitudes from meteorological stations or satellite data. All the classes must measure these quantities with a resolution ≤ 1 W m^{-2}.

Returning back to Fig. 9, it is interesting to remark some details. Procedures for calibration of reference pyrheliometers, pyranometers, and solar cells are described in different international standards (ISO, IEC, ASTM, etc.) and in the literature. According to ISO Standards, calibrations of pyrheliometer and pyranometer reference standards are obtained through comparison against a cavity radiometer, traceable to WRR, usually participating in the IPCs. Details about different calibration procedures are given later.

Table 1 Summary of requirements for monitoring systems in PV plants according to IEC 61724-1 Standard

Characteristic	Class A	Class B	Class C
Accuracy	High	Medium	Basic
Irradiance measurements			
In plane irradiance (POA) Global Horizontal (GHI) Direct Normal (DNI) Diffuse Horizontal (DIF)	Measured Measured Measured Measured	Measur./Estim. Measur./Estim. Measur./Estim. Measur./Estim.	Measur./Estim. – – –
Thermopile pyranometer class	Secondary (ISO) High Q (WMO)	First class (ISO) Good Q (WMO)	Any
Pyranometer uncertainty	$\leq 3\%$ for hourly totals	$\leq 8\%$ for hourly totals	–
Pyranometer ventilation	Required	Optional	–
Reference sensor heating	Required if ≥ 7 days affected	Required if ≥ 14 days affected	–
PV reference cell uncertainty	$\leq 3\%$	$\leq 8\%$	Any

ISO = ISO9060 Standard; *WMO* = WMO/CIMO Guide

The position of WRR in Fig. 9, as well as that of other standards (silicon trap standards, standard lamps), is below the main SI box to represent they are realizing derived units instead of fundamental ones. Several boxes are especially highlighted (in yellow) to indicate the position of cavity radiometers. Absolute cryogenic radiometers (L-ACR) are used at NMI laboratories to realize the fundamental unit of candela (luminous intensity) and derived units (lumen, lux). These are the kind of instruments WRR has been compared to in the WRR/SI comparisons already mentioned. As explained, WRR/WSG is the practical realization of the unit of W m^{-2}. To date, all the absolute cavity radiometers of terrestrial use (T-ACR), of active or passive type, mainly used by WMO national and regional centers and BSRN stations, or used by solar sensor manufacturers or by specialized calibration laboratories, are characterized by comparison against WRR during the IPCs. Special calibration procedures are applied for absolute cavity radiometers for use in space (S-ACR), against WRR or against SI laboratory-scale L-ACRs. Finally, the CSAR is introduced in this sketch occupying a position near WRR because of its potential inclusion in or substitution of WRR in the future.

The hierarchical level of pyrheliometers and pyranometers in the irradiance scale depends on the quality and metrological characteristics of the sensor. Two main classifications are recognized in this field: the ISO 9060 Standard (ISO 1990a) and the CIMO/WMO Guide (CIMO 2017). ISO 9060, for example, distinguishes (in its 1997 edition) three classes both for pyranometers and pyrheliometers: secondary standard, first class, and second class. This classification is made attending to different performance and physical aspects: response time, zero offsets, resolution, non-stability, temperature response, nonlinearity with irradiance, spectral sensitivity, tilt response, and directional response for beam radiation (only in the case of pyranometers). All these magnitudes of influence and technical aspects are later discussed (Sects. 3 and 4), and specific ranges for instrument classification are collected (see Tables 4 and 5). However, ISO 9060 is suffering a revision and its new edition (foreseen for fall of 2018) introduces some changes (see below).

Reference solar cells are special kind of irradiance sensors whose inclusion, in this context, makes sense for (a) the calibration of secondary cells and photodiode-based sensors, and (b) its role in the monitoring of PV plants (of matching technologies), though their spectral sensitivity ranges are narrower than those of thermopile-based instruments. This is also referred to in IEC 61724-1 Standard. Their primary calibration can be obtained by several methods and traceability chains, as indicated in Fig. 9, many of them covered in the IEC 60904-4 Standard (IEC 2009). One of these methods is based on the use of absolute cavity radiometers, traceable to WRR, as done with pyrheliometers and pyranometers (Emery et al. 1988; Osterwald et al. 1990). Secondary calibration of solar cells by comparison against reference solar cell is covered, for example, by the IEC 60904-2 Standard (IEC 2015).

3 Measurement of Direct Normal Irradiance

In a clear day, up to $\sim 90\%$ of irradiance reaching a surface on the ground, with adequate orientation and tilt, can come from the Sun disk and aureole/circumsolar regions of the sky. These, as a whole, form the direct normal irradiance (DNI). The sun disk subtends an angle of $\sim 0.535°$ (McCluney 1994) to an observer on the Earth, almost the same angle as the Moon. This is why, during a solar eclipse, the Moon perfectly covers the Sun although its mean orbit is about 385,000 km (0.00257 AU), much closer to Earth than the Sun.

However, the amount and the character of the circumsolar radiation vary widely with geographic location, climate, season, time of day, and the observing wavelength (Buie and Monger 2004). There is also some uncertainty in the edge limit of the solar disk (Blanc et al. 2014) and the solid angle that must be considered to account for DNI, although it has lately been determined as of a radial displacement or half angle of 4.65 mrad or 0.266° (Puliaev et al. 2000). The profile of the radiance, or of the radiation intensity, decreases from the center of the solar disk to the edges, and circumsolar radiation influence extends up to about ±2.5° with a linear-like dependence in a log–log plot (see Fig. 10). The relative intensity of the circumsolar region to that of the sun disk is obviously also dependent on the atmospheric scattering and contents on particles, aerosols, etc.

Out from this narrow solid angle subtended by ±2.5°, the rest of the skydome emits diffuse irradiance over the receptor. This is, therefore, the reference aperture for the opening angle θ_O used to define the field of view (FOV) of an instrument

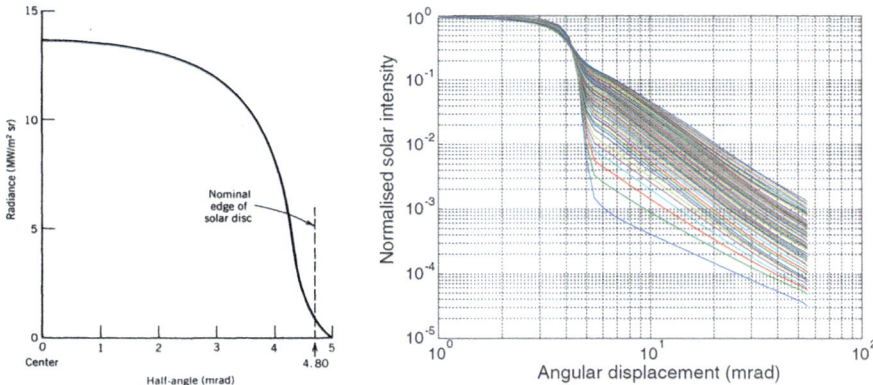

Fig. 10 Radiance distribution of the solar disk and the circumsolar radiation. Nominal edge angle of the solar disk corresponds to 0.266°. Images after Stine and Geyer (2001), Rabl and Bendt (1982), Buie and Monger (2004)

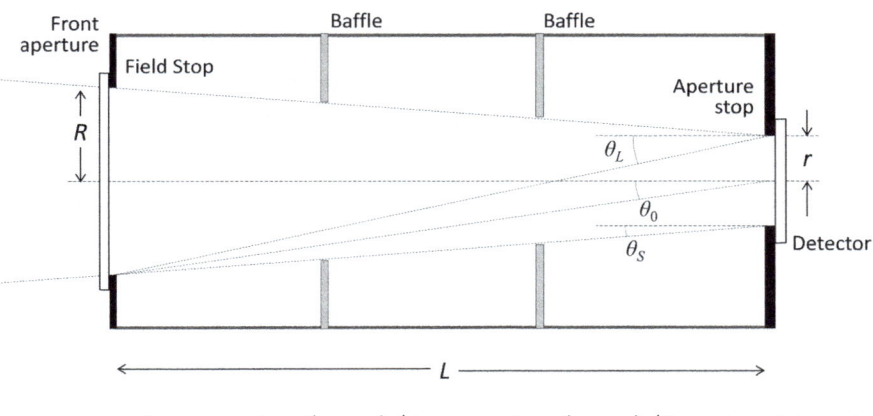

$$\tan\theta_O = R/L \quad \tan\theta_S = (R-r)/L \quad \tan\theta_L = (R+r)/L \quad \theta_L = 2\theta_O - \theta_S$$

Fig. 11 View-limiting geometry of an instrument measuring DNI. Opening θ_O, slope θ_S and limit θ_L angles are dependent on the size or radius r of the sensor aperture stop, that of the front aperture R (field stop) and the distance L between them

designed to measure DNI over the Earth's surface, as a pyrheliometer. These apertures are usually arranged inside a pyrheliometer in front of the detector, concentric around a common optical axis with rotational symmetry, in the form of a collimating tube, and thus defining a view-limiting geometry as that shown in Fig. 11. WMO CIMO currently recommends values of $\theta_O = 2.5°$ and $\theta_S = 1°$ (CIMO 2017), although there are other accepted criteria with wider ranges (e.g., ASTM E1125-99). As an example, Table 2 includes the angles and dimensions for some cavity- and common thermopile-based pyrheliometers.

Table 2 Overview of limiting geometries of cavity radiometers and instruments used to measure the DNI

Instrument	R (mm)	r (mm)	L (mm)	$2\theta_O$ (°)	θ_S (°)	θ_L (°)
CIMO spec.				5.00	1.00	4.00
PMO2	3.6	2.5	85	4.84	0.74	4.10
PMO5	3.7	2.5	95.4	4.44	0.72	3.72
CROM 2L	6.29	4.999	144.05	4.99	0.51	4.48
PAC 3	8.18	5.64	190.5	4.91	0.76	4.15
Eppley HF—AHF	5.86	3.98	134.3	5.00	0.8	4.20
PMOD PMO6	4.2	2.5	98.5	4.88	1.00	3.89
Eppley NIP	10.3	4	203	5.81	1.78	4.03
Kipp Zonen CHP1	13	5	168	5.00	1.00	4.00
Eppley sNIP	5.82	3.99	133.6	4.99	0.79	4.20

"CIMO spec." refers to WMO CIMO guide. After Rodríguez-Outón et al. (2012), Gueymard (1998), Gueymard and Wilcox (2011) and Dutton (2002). Data for sNIP and CHP1 kindly provided by Eppley Labs and Kipp Zonen respectively. Manufacturing tolerances are not included and could produce some variation in real angles

Far from the atmosphere limits, solar irradiance is only direct, and there are no diffuse components (interstellar space is in practice a no dispersing medium, despite the existing small amounts of gas, dust, and particles). Irradiance on the top of the outer atmosphere varies around ~7% along the year (between aphelion and perihelion) only because of the movement of the Earth along its orbit around the Sun, and the small eccentricity of this orbit. This does not affect the value of the solar constant, which is determined for a fixed distance of 1 AU and whose small periodic variations (of the order of 0.1%) have no relationship with this orbital displacement but with the apparition of dark spots and faculae in the Sun's surface. Some additional details on the measurement of total solar irradiance (TSI) in space are described in Appendix.

3.1 Solar Absolute Cavity Radiometers (ACR)

As introduced before, pyrheliometers on the top metrology level for on-ground measurement of DNI are absolute cavity radiometers (ACR) working under the principle of electrical substitution. These are open air, unencumbered sensors (neither windows nor domes are usually used), so they are sensitive to both SW and LW radiation. Solar ACRs are constructed having two twin cavities, one intermittently exposed to the sunlight and a second (reference or compensating) cavity operating at an ambient temperature in the dark. Figure 12 shows a diagram of one of these cavity radiometers. The realization of the unit of irradiance W m^{-2} in one ACR is based on the materialization of two other magnitudes: area (m^2) and power (W). While the area is obtained by the accurate measurement of the diameter of the

Ring of thermocouples   Precision aperture   View limiting aperture   Shutter

Primary cavity

Reference cavity

Heat sink

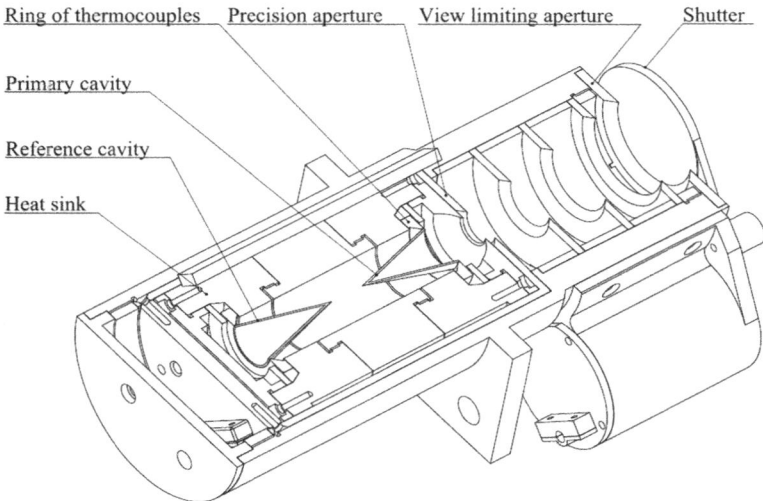

Fig. 12 Cross section of an absolute cavity radiometer, with a double cavity arrangement. Only the frontal aperture (sunlight entering by the shutter side) is illuminated while rear (compensating) cavity is in the dark and acts as reference cavity working at ambient temperature. After Fang et al. (2014)

precision aperture at the entrance of the cavity, the power is calculated in an indirect way by measuring the temperature (or the heat flux) reached in the cavity when it is illuminated by sunlight and when it is heated by an electrical current. It is then based on the assumption of the equivalence between heat produced radiatively and heat produced electrically (Joule effect). An extremely careful design of the cavity, the heating electrical circuit, the temperature sensors, and their measurement circuits try to ensure that this equivalence principle holds.

Although with differences in operation modes among different instruments (mainly between passive and active types), they operate in two steps, phases, or stages: the open phase and the closed phase, in reference to the position of the frontal shutter. During the open phase, DNI sunlight enters through the view-limiting aperture and heats the cavity, with a radiant power of up to ~ 45 mW (for a typical aperture area of 0.5 cm^2). During the closed phase, a small current of tens of milliamps circulates through a heating circuit whose wires are intimately adhered to the cavity by its rear side. The voltage/current injected during the closed phase is regulated until the temperature difference between both cavities (or the heat flux toward the heat sink) is the same as in the open phase. Therefore, equal heat flux (equal temperature difference) must imply equality between electrical power ($P = V \times I$) and radiant power.

In practice, real absolute cavities are affected by slight deviations from the ideal behavior, contributed by several optical, radiative, thermal, and electrical effects, some of them grouped under the term non-equivalence. Therefore, solar ACRs are to be characterized, by two different ways: (a) through the calibration and assessment of every of the magnitudes of influence in the equation governing its operation, or (b) by direct comparison against WSG/WRR (or any future standard reference embodying the unit W m^{-2} and realizing a scale of solar irradiance). However, and due to the relative ease of the characterization, by comparison, WMO suggests the second way for the cavity instruments giving support to WMO national and regional centers and to BSRN stations. Every radiometer measuring solar irradiance must be traceable to WRR, according to WMO guidelines. Transference is performed in the International Pyrheliometer Comparison (IPC) celebrated in the PMOD/WRC every 5 years. After the IPC, a deviation factor with respect to WRR (and the corresponding 1σ uncertainty) is calculated for every participating instrument. This deviation factor must be included after in the calibrations performed by every particular ACR and the uncertainty included in its final uncertainty budget. Details about the IPC operation, participants, data acquisition and validation, results for instruments, and the procedures applied for calculating the reference irradiance based on the measurements from the WSG, can be obtained from the IPC reports (Finsterle 2011, 2016).

3.2 Field Pyrheliometers

Pyrheliometers for on-field operation, solar (thermal, PV) power plants, and weather/BSRN network stations, while accomplishing for the view-limiting geometry already described, are much simpler than cavity radiometers. Correspondingly, they are less accurate. Classic design pyrheliometers, still in operation and in the market, are based on a thermopile as radiation sensing element, being therefore analog, passive instruments not requiring a power supply to operate. Typical sensibility, depending on thermopile configuration, is between 5 and 20 μV/W m^{-2}. The internal structure of these instruments is depicted in Fig. 13, also showing the placement of sensors at the end of the collimating tube.

Some examples of pyrheliometers commercially available in the market are collected in Fig. 14. Almost all of these instruments have several common features: a long cylindrical-shaped architecture, associated to a sunlight collimating tube; a metallic body (aluminum, stainless steel) for long-term durability; an alignment

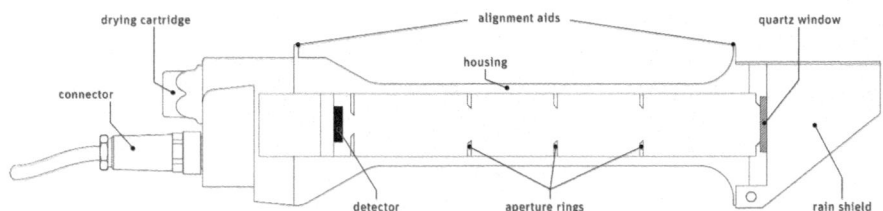

Fig. 13 Internal structure of a (secondary standard, first class) pyrheliometer. After Kipp and Zonen CHP1 manual

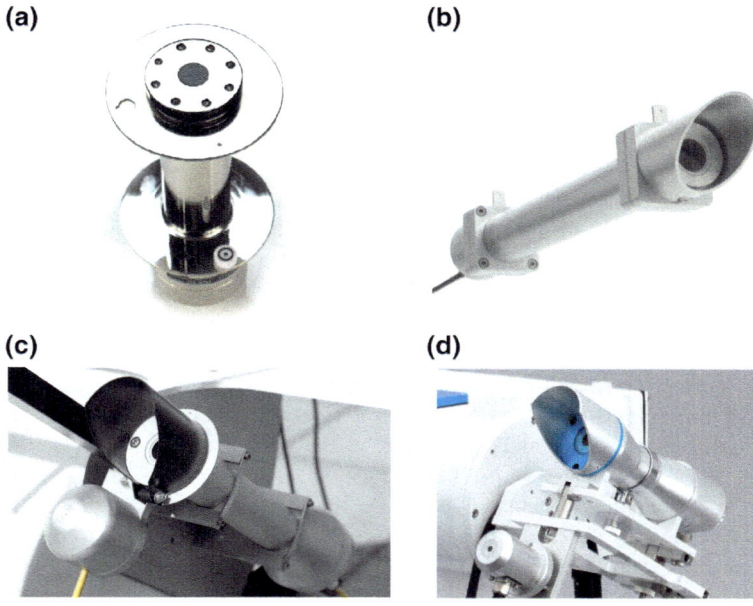

Fig. 14 Examples of some commercial type (secondary standard, first class) pyrheliometers: **a** sNIP by Eppley, **b** DR01 by Hukseflux, **c** CHP1 by Kipp and Zonen, **d** MS-56 by EKO

sight, sunspot indicator, or pointing aid; a frontal window to protect the sensor and create a watertight enclosure; and, on the rear side, an external connector for output signals and a small removable plastic tube or cartridge with desiccant. As they use frontal windows, LW radiation is filtered and only SW is measured. They are usually fixed to the supporting structure (in a Sun tracker, see below) by clamps or braces wrap around the cylindrical body.

Modern versions of these instruments, recently introduced in the market, incorporate one or some of the following features: a faster response, amplified voltage signal (e.g., 1 mV/W m^{-2}), output in 4–20 mA current loop, digital outputs (e.g., Modbus RS485), microprocessor, heating elements, tilt angle sensor, temperature sensor, and temperature compensation circuit. While the T sensor is still an analog signal, the rest of the characteristics require to feed the pyrheliometer with an external power supply. It is then a change in the philosophy of sensing solar radiation: simple-passive-analog devices versus complex-active-digital ones. This change does not necessarily imply an improvement in accuracy or uncertainty, neither in the internal structure nor in the external shape. It is more an extension of equivalent solutions applied for other sensors to attend the demand of large, extended, and accurate monitoring systems in industry and power plants. A similar technical development is being experienced in pyranometers' market.

On the other hand, due to the need of continuously pointing to the Sun's disk, pyrheliometers require the use of an automated/motorized platform which follows the Sun trajectory in the sky along the day and along the year. These platforms are mounted in two-axis Sun trackers or sun trackers, which use motors and gear trains driven by microprocessors and location-, date- and time-based algorithms to calculate the position of the Sun. These systems are preferred to trackers only based on sensors, although a pointing sensor is usually added to the tracker operative in order to check the accuracy of the algorithm, the performance of the tracker and to introduce on-the-fly corrections. Pointing errors up to 0.1° can be admissible for sensing DNI (CIMO 2017) while other applications could demand better accuracy. Sun trackers of diverse designs are also used to move solar panels in PV power plants, to carry parabolic troughs, Fresnel reflectors, lenses, or the mirrors of a heliostat. Figure 15 shows an example of sun trackers used for solar radiation measurements. The cost of a two-axis tracker is considerably higher than that of a pyrheliometer, even a secondary standard, by the way.

Finally, it is worth to mention a new type of sensor for assessing circumsolar radiation and to account for its influence in the calibration and in the DNI measurements done with different pyrheliometers. It was developed by Black Photon and successfully used during last IPC-XII (2015), to the point of being used as an additional criterion for acceptance and validation of data. Figure 16 shows an image of these new sensors, BPI-CSR460. The calculation of circumsolar radiation for a given solid angle was obtained by the difference between output signals of partially shaded and unshaded electronic sensors. It was used as an indication of varying and unstable irradiance conditions (Finsterle 2016). It had also been tested previously for evaluating the influence of circumsolar variations on concentrating solar collectors (Wilbert et al. 2013).

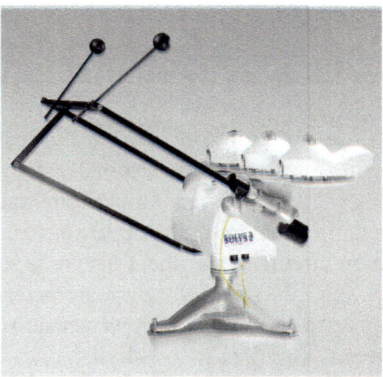 

Fig. 15 Examples of two-axis sun trackers used for solar radiation instruments, by Kipp Zonen (left) and Hukseflux (right)

Fig. 16 Measurement of circumsolar radiation from the difference between DNI signal of partially shaded and unshaded sensors during IPC-XII in Davos, performed by Black Photon Instruments

3.3   Classification of Pyrheliometers

The performance of pyrheliometers is dependent both on technical capabilities (and/or limitations) and on external working influences (of environmental nature). The difference in the quality and accuracy of different instruments, excluding primary absolute radiometers, made necessary the classification of pyrheliometers to state and to compare their metrological level. Although quite similar in some aspects, the classifications of pyrheliometers according to both WMO CIMO Guide and ISO 9060 Standard (ISO 1990a) specifications are collected in Table 4. Two categories are recognized by WMO, while three are distinguished by ISO in the case of pyrheliometers, and both establish three (near identical) categories for pyranometers. Better class implies higher metrological level and instrument quality. The characteristics included and categorized in Table 4 (and similarly in Table 5 for pyranometers) are mainly related to the behavior of thermopiles as sensors and not all of these characteristics are directly portable to another kind of instrument. To our knowledge, the new version of the ISO 9060 Standard seems to keep near identical the requirements for each class included in Table 4. The name of the categories changes to Classes A, B, and C (substituting respectively to secondary standard and first a second class), and a new AA class, of higher requirements, is introduced in the standard, although it seems to be only applied to standard reference instruments (as primary ones). Specific details about other minor changes are to be confirmed in this new edition.

The motivation of these classifications is, in the end, to define which kind of sensor is able to guarantee which level of confidence in the measurement of solar irradiance without the need of measuring additional working parameters (e.g., sensor temperature and ambient temperature). Classifications were conceived having in mind classic analog-passive type of solar sensors. A linear device for DNI or GHI irradiances, and independent of all of the rest of influences, would be the ideal sensor as it avoids performing corrections in the recorded values. Table 3 shows the values reflected by CIMO Guide of achievable uncertainty both for pyrheliometers and pyranometers. Therefore, according to the needs of the accuracy of a particular application (and according to the affordable budget), a given quality of the sensor has to be chosen.

Calibration of pyrheliometers is always performed under natural sunlight by comparison against a standard pyrheliometer, with equal or higher metrological level (equal or better class), and with a known sensibility ($\mu V/W\ m^{-2}$) traceable to WRR. ISO 9059 Standard (ISO 1990b) describes the procedure for calibration, by comparison, valid ranges for minimum irradiance and maximum turbidity, and an indication of uncertainty determination. Usually, calibration during clear and stable days is recommendable and this limits the availability of acceptable days. Different view-limiting geometries between DUT and standard, as well as different time constants and temperature coefficients, can result in calibration errors. Characteristics included in Table 4, common with some of Table 5, are indicative of the relative influence of these parameters for different classes.

Response time of thermopiles is dependent on its size, number of thermojunctions, and the thermal capacity of the structure. Under varying irradiance conditions, the thermopile changes the output voltage following an exponential function with a given time constant. The 95% level is referred to the final value the sensor output would reach under a stable irradiance condition after the change.

With reference to "zero offset," there are several thermal effects that can be analyzed. On the one hand, the term can be associated with the signal measured on the sensor for a null irradiance condition. Although it should ideally be zero, a thermopile-based radiometer can show a nonzero output value or even a negative one, depending on the amount and the direction of the thermal flux across the thermopile and the temperatures on both sides (at the end, a thermopile can be assimilated to a thermal flux meter). This effect has been extensively analyzed in the case of pyranometers and much less for pyrheliometers, because of the lower

Table 3 Achievable uncertainty for every class of thermopile-based solar radiometer according to WMO CIMO guide

	Pyrheliometers		Pyranometers		
Achievable uncertainty (95% confidence level)	High quality	Good quality	High quality	Good quality	Moderate quality
1 min totals (%)	0.9	1.8	–	–	–
Hourly totals (%)	0.7	1.5	3	8	20
Daily totals (%)	0.5	1.0	2	5	10

Table 4 WMO CIMO and ISO 9060:1997 specifications for pyrheliometers (PH)

Specification →	WMO CIMO		ISO 9060:1997		
Characteristic	High quality	Good quality	Secondary standard	First class	Second class
Response time (for 95% response) (s)	<15	<30	<15	<20	<30
Zero offset: response to 5 K/h change in ambient temperature (W/m²)	2	4	±1	±3	±6
Resolution: smallest detectable change in irradiance (W/m²)	0.51	1	–	–	–
Non-stability: percentage change in responsivity per year (%)	0.1	0.5	±0.5	±1	±2
Temperature response (%)	1	2	±1	±2	±10
Nonlinearity (%)	0.2	0.5	±0.2	±0.5	±2
Spectral sensitivity (WMO) or selectivity (ISO) (%)	0.5	1.0	±0.5	±1	±5
Tilt response (%)	0.2	0.5	±0.2	±0.5	±2
Traceability (maintained by periodic comparison)	–	–	With primary standard PH	With secondary or better PH	With first class or better PH

- Spectral sensitivity (WMO) or selectivity (ISO): percentage deviation of the product of spectral absorptance and spectral transmittance from the corresponding mean within the range 300–3000 nm (WMO) or within 0.35 µ and 1.5 µm (ISO)
- Nonlinearity: deviation from the responsivity at 500 W/m² due to changes in irradiance within the range of 100–1100 W/m²
- Temperature response: percentage maximum error (WMO) or deviation (ISO) caused by changes in ambient temperature within an interval of 50 K
- Tilt response: deviation from the responsivity at 0° tilt (horizontal) due to changes in tilt from 0°–90° at 1000 W/m² irradiance

impact in the latter (see paragraph 4). On the other hand, "zero offset" refers, in the case of Table 4, to the possible change on the output of the pyrheliometer due to a slow change on the ambient temperature (5 K/h) under constant irradiance (sometimes referred to as zero offset type B). Again, it should ideally be zero because the output of the thermopile should be independent of these slow changes. In the end, both effects have a thermal origin and are intimately related one to each other, but the first one is referred to null irradiance while the second is determined under illumination.

Similarly, there is a characteristic related to a thermal coefficient of the output ('Temperature response'). In this case, the sensitivity of the sensor is obtained indoors after stabilization of output signal under constant irradiance, but with a wide temperature range covering the normal operation of field pyrheliometers

Table 5 WMO CIMO and ISO 9060:1997 specifications for pyranometers (PN)

Specification →	WMO CIMO			ISO 9060:1997		
Characteristic	High quality	Good quality	Moderate quality	Secondary standard	First class	Second class
Response time (95% response) (s)	<15	<30	<60	<15	<30	<60
Zero offset: response to 200 W/m² net thermal radiation (ventilated); response to 5 K/h change in ambient temperature (W/m²)	7 2	15 4	30 8	±7 ±2	±15 ±4	±30 ±8
Resolution: smallest detectable change in irradiance (W/m²)	1	5	10	–	–	–
Non-stability: percentage change in responsivity per year (%)	0.8	1.5	3.0	±0.8	±1.5	±3
Temperature response (%)	2	4	8	2	4	8
Directional response for beam radiation (W/m²)	10	20	30	±10	±20	±30
Nonlinearity (%)	0.5	1	3	±0.5	±1	±3
Spectral sensitivity (WMO) or selectivity (ISO) (%)	2	5	10	±3	±5	±10
Tilt response (%)	0.5	2	5	±0.5	±2	±5

- Spectral sensitivity: deviation of the product of spectral absorptance and spectral transmittance from the corresponding mean within the range 300–3000 nm (WMO) or within 0.35 μ and 1.5 μm (ISO)
- Nonlinearity: deviation from the responsivity at 500 W/m² due to changes in irradiance within the range of 100–1100 W/m²
- Tilt response: deviation from the responsivity at 0° tilt (horizontal) due to changes in tilt from 0°–90° at 1000 W/m² irradiance
- Temperature response: percentage maximum error (WMO) or total percentage deviation (ISO) caused by any change of ambient temperature within an interval of 50 K
- Directional response for beam radiation: range of errors caused by assuming that the normal incidence responsivity is valid for all directions when measuring, from any direction, a beam radiation with a normal incidence irradiance of 1000 W/m².

(50 K), usually in steps of 10 K. The resulting data points can be fitted to a curve or straight line to carry out corrections on temperature variations if required.

The rest of the parameters listed in Table 4 are easily understood and will not require further analysis or discussion. Tilt response is of importance for an instrument subjected to a shift in orientation and slope when arranged in a sun tracker. The more sophisticated test required for pyrheliometer classification refers to optical characteristics (absorptance of the sensor and transmittance of frontal windows), limited to the range of shortwave radiation, and which can only be carried out by specialized laboratories. A number of other magnitudes of influence have also been investigated (Thacher et al. 2000).

4 Measurement of Global Irradiance on Horizontal and Tilted Surfaces

Global irradiance is a wider term associated with radiation received in all the directions of space. A round-shaped sensor (a sensing ball) could potentially measure radiation in a solid angle of 4π, while sensors with flat surfaces can receive hemispherical irradiance in a 2π solid angle. When applied to solar sensors, global irradiance stands for radiation composed of BHI (sun disk and circumsolar radiation), diffuse DIF sky radiation, and even reflected (ground, albedo) radiation.

In practice, sensors commonly used to measure solar global irradiance have flat surfaces (thermopiles and thermal flux sensors, solar cells, photodiodes, etc.) and therefore are able to measure hemispherical irradiance. The most extended device for global irradiance measurements is a pyranometer. A pyranometer is a thermopile-based instrument, covered by one or two hemispherical glass domes, and therefore able to measure SW radiation in a 2π solid angle. Except for the different field of view (FOV), the working principle is therefore equal to a pyrheliometer. When placed in a horizontal position, it measures GHI, only composed of direct horizontal irradiance DHI and DIF sky radiation. Pyranometers can also be placed on inclined surfaces, e.g., to measure plane of array (POA) irradiance parallel to a photovoltaic array, or can be placed downwards, in inverted orientation, to only measure ground reflected global radiation (such instruments are called albedometers if they are combined with a horizontal pyranometer). Being placed horizontal, they can also be partially shaded or screened to avoid DNI contribution and thus exclusively measuring diffuse horizontal irradiance (DIF). Although the usual sensing element of a pyranometer is a thermopile, there also exist some models of pyranometers based on solar cells and photodiodes, at the expense of measuring limited spectral ranges of solar irradiance.

Figure 17 shows some examples of commercially available pyranometers of classical design (passive, analog type), while Fig. 18 includes a cross section of these instruments. The sensing thermopile is intimately bonded to a ceramic disk painted black or a plate with an anodized surface, with round shape, and covered by the glass domes. The main body or housing is usually made of aluminum, with three levelling feet for tilt adjustment and a bubble level as an aid for getting the horizontal position, an external connector and a removable cartridge with desiccant. They also incorporate a white sun shield (plastic, metal) in the form of a truncated cone. Some models have a temperature sensor inside (Pt100, thermistor) available through the external connector. Many manufacturers also offer as an option a ventilated unit (fan based) and even an external heating element that can be added to the pyranometers body to avoid or reduce effects of dust, dew, frost, snow, ice, etc., that affect the performance of the instrument and the availability of valid data.

However, as in the case of pyrheliometers, modern versions of pyranometers can offer many other features for adapting to requirements of International Standards and monitoring networks: heating and ventilating elements already embodied in the pyranometer structure, microprocessor control, analog and digital outputs,

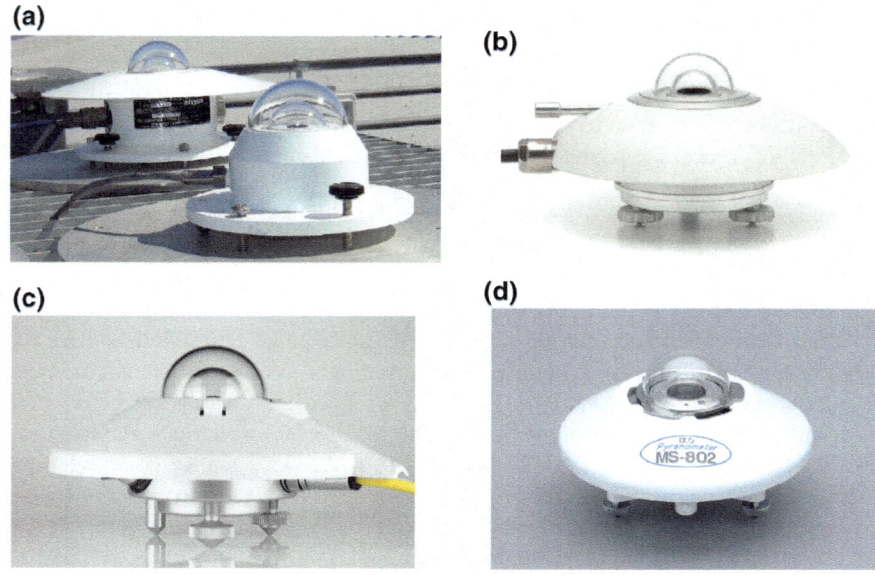

Fig. 17 Examples of commercial type standard pyranometers (secondary standard): **a** GPP and PSP (behind) by Eppley, **b** SR20 by Hukseflux, **c** CMP22 by Kipp and Zonen, **d** MS-802 by EKO. New and improved versions are shown in Fig. 19

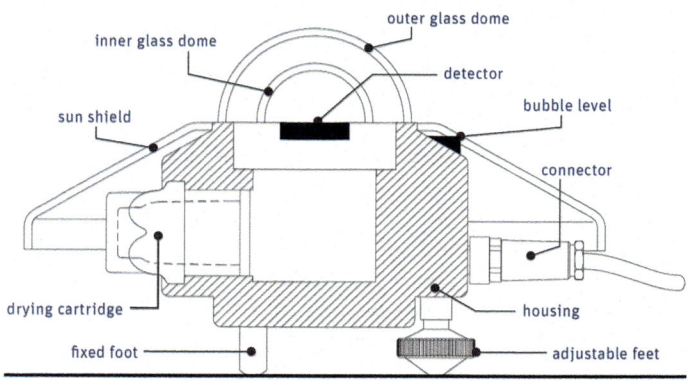

Fig. 18 Cross section of a double dome pyranometer. After Kipp Zonen CMP22 manual

amplified voltage and current loop outputs, compensating temperature circuit, lower offsets, and temperature and tilt angle sensors. Another of the most interesting improvements refers to the sensing element: reduced in size thermopile placed on a cavity and covered by a quartz diffuser results in a much faster and sensitive device.

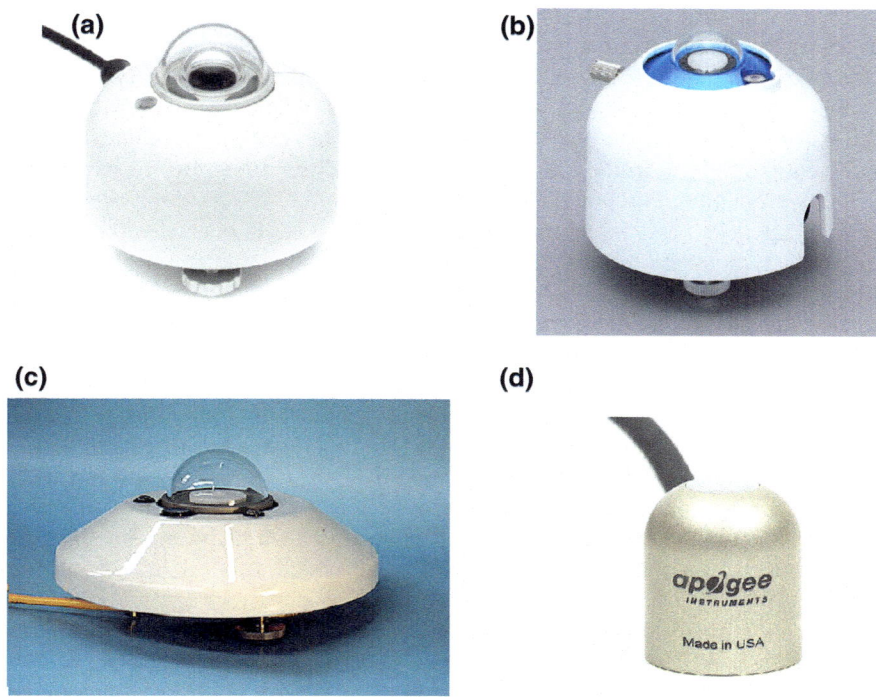

Fig. 19 Examples of new generation of thermopile pyranometers, with improved features (see text). **a** SR30-D1 by Hukseflux; **b** MS-80 by EKO; **c** ER08-SE by Middleton Solar; **d** SP-510 by Apogee Instruments

Some examples of modern design pyranometers are shown in Fig. 19. Again, there is a shift from passive-analog instruments to active-digital ones.

The characteristics and classification of pyranometers, according to current ISO 9060 and WMO CIMO Guide, are summarized in Table 5. Remember that the classification for a particular characteristic is conceived as an acceptance criterion of accomplishment of the indicated range. Many of the working operational issues already commented for pyrheliometers are common to pyranometers. In general, requirements are more restrictive for pyrheliometers than for pyranometers for a given class. Due to their 2π FOV and the short distance between domes and thermopile, pyrheliometers are prone to experience larger influences originated by a tilt angle, orientation, directional response (not applicable to pyrheliometers), and zero irradiance offsets.

Zero offsets are separate in two different contributions: the zero offset type A, caused by the longwave radiation emitted inside and outer the instrument and by the different temperatures of thermopile and domes; and the zero offset type B, the possible deviation in the output produced by drifts in ambient temperature.

In particular, zero offset type A in pyranometers have been the subject of dedicated research efforts in the literature (Reda and Myers 1999; Bush et al. 2000; Haeffelin et al. 2001; Philipona 2002; Hernandez et al. 2015). As they are made of different materials, have different thermal capacity, and are in contact with different parts of the radiometer, there are differences in operating temperatures of the black disk, the inner dome, and the external one. Temperature differences promote radiative transfer among these components (due to their different emissivities), and between outer dome and atmosphere or sky (sky can have effective temperatures up to 50 °C cooler in a clear day). This transfer leads to the apparition of a small negative signal which reduces to the output signal of the thermopile. As a result, the true irradiance can be underestimated. However, although it is identified as a source of error in irradiance measurements, these thermal offsets are still not well accounted for, and no clear methodologies have been defined for their assessment. There are also discrepancies among the differences between nighttime and daytime offsets, and between the offsets obtained when measuring global or diffuse irradiance.

On the other hand, with reference to Table 5, it is important to remark that ISO 9060 Standard is being currently under revision (2018) and that new edition includes some noticeable changes with respect to the previous one, especially in the case of pyranometers (Hukseflux 2018). The instrument is to be classified under accuracy classes now labeled as A, B, and C. In the case of pyranometers, Class A devices require the individual testing (and reporting) of temperature response and directional response for every instrument. There is also an extension of every class for "spectrally flat" devices, recommended for POA, diffuse, albedo, and reflected solar measurements. This "spectrally flat" category will apply to instruments not installed in horizontal and exposed to spectral distributions different than that of GHI. Any case, it is necessary to wait until the issue of the new edition to better know all the changes.

With respect to calibration of pyranometers, there are several methods reported in the literature and in International Standards, such as ISO 9847 (ISO 1992) and ISO 9846 (ISO 1993). CIMO Guide also briefly describes some of the most important procedures, including those of ISO 9847. These can be summarized as follows:

(A) Outdoor methods:

- Comparison of a DUT against a standard pyrheliometer (DNI) and a calibrated pyranometer (diffuse sky irradiance).
- Comparison of a DUT against a standard pyrheliometer (DNI) by using a removable shading disk for pyranometer (sun and shade method).
- Comparison of two DUT against a standard pyrheliometer (DNI) by alternatively measuring GHI and DIF with every pyranometer.

- Comparison of a DUT against a standard pyranometer, even under cloudy and partly cloudy conditions.
- Comparison of a DUT against a standard pyrheliometer (DNI) by using a collimating tube in the pyranometer.

(B) Indoor methods:

- Comparison of a DUT against a similar pyranometer (previously calibrated outdoors) on an optical bench with an artificial source. This can be carried out at normal incidence or at another angle of incidence.
- Comparison of a DUT against a similar pyranometer (previously calibrated outdoors) inside an integrating chamber simulating diffuse sky radiation.

Finally, it is important to include in this paragraph another two families of sensors used in many applications for measuring of global irradiance (mainly for GHI and POA). These are photoelectric-based devices: solar cells and photodiodes, which are encapsulated or embedded in suitable structures and cable connections as to guarantee long-term stability and performance, and to make easy their direct installation and use on the field. Figure 20 shows some examples of photodiode-based pyranometers, while Fig. 21 includes various types of reference solar cells used in PV power plants and smaller PV systems.

While at a lower cost than pyranometers, both kinds of devices measure irradiance only in a limited range of solar spectral distribution (e.g., silicon between 300 and 1150 nm) and therefore are subject to some spectral errors in different moments of the day and along the seasons. Corrections for temperature and spectral sensitivity can improve the measurement results. However, they have a very fast response to varying irradiance. As a whole, they can be an adequate solution for monitoring PV plants of the same or equivalent technology, or for applications only requiring accuracies equivalent to first class or good quality in Table 5.

5 Measuring Diffuse Irradiance

For some applications, ad hoc assessment of diffuse (DIF) irradiance can be advisable or mandatory. Classical solutions for measuring diffuse irradiance are based on the same type of sensors used for measuring global irradiance, mainly thermopile pyranometers. But we can classify approaches usually applied in two categories:

1) *By computing the difference between GHI and DNI.* The basic idea is quite simple: After simultaneous measurement of global horizontal (GHI) and direct normal (DNI) irradiances with a horizontal pyranometer and a pyrheliometer, the DIF can be computed by the known relation:

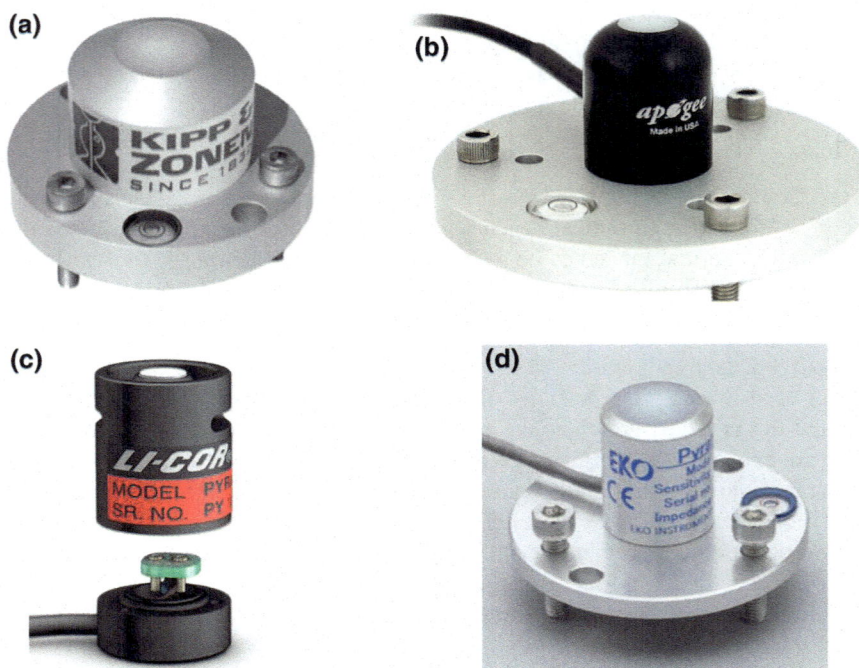

Fig. 20 Examples of Si photodiode-based pyranometers: **a** SP Lite2 by Kipp and Zonen; **b** Apogee SP-212; **c** a LI-200R photometric sensor by LI-COR, detached from removable base; **d** ML-01 Si-Pyranometer by EKO

$$\text{GHI} = \text{DNI} \cdot \cos\theta_{Sun} + \text{DHI} \qquad \rightarrow \qquad \text{DHI} = \text{GHI} - \text{DNI} \cdot \cos\theta_{Sun}$$

The resulting uncertainty of this computation will be affected for those of the individual measurement of global and direct components, having in mind all the characteristic parameters of influence already commented in previous sections. On the other hand, this approach would not be admissible for Class A (high accuracy) monitoring systems for PV power plants regulated by IEC 61724-1 Standard, because direct measurement of DIF is required.

2) *Applying a static or sun-tracking shadow over a horizontal pyranometer.* The easier way of evaluating DIF is to block up or occlude the DNI on a pyra-nometer measuring GHI by using an opaque shading gadget. More accurate results are obtained when the solid angle subtended by the shading device over the pyranometer sensing element equals that of the pyrheliometer measuring DNI. Otherwise, some corrections should be applied to account for the differ-ence in the FOV. Figure 22 shows the traditional solutions developed to shadow horizontal pyranometers for measuring DIF. These comprise, first, static ele-ments as shadow rings or shadow bands tilted in such a way that is coincident

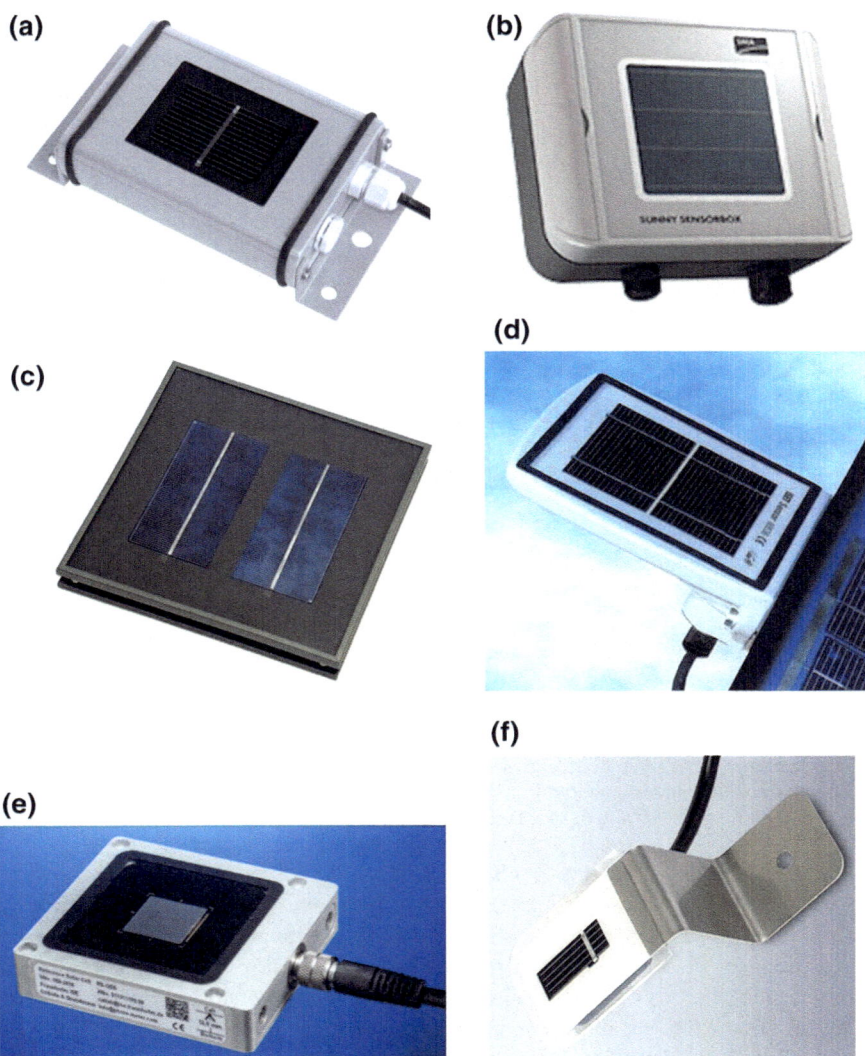

Fig. 21 Examples of commercial reference solar cells used as irradiance sensors: **a** Si sensor by Mencke and Tegtmeyer; **b** Sunny Boy sensor by SMA; **c** Temperature compensated MET solar cell by ATERSA; **d** same idea in the ISET sensor by IKS Photovoltaik; **e** Fraunhofer ISE's outdoor reference solar cell; **f** Fronius irradiation sensor

with the Sun's path (ecliptic) along the day. As the apparent motion of the Sun varies between maximum trajectories occurring at solstices, it is necessary to correct the position of the shadow element from time to time along the year. Besides, the shadow band or shadow ring is screening the sensor from a portion of the diffuse radiation coming in from the sky, and corrections have to be

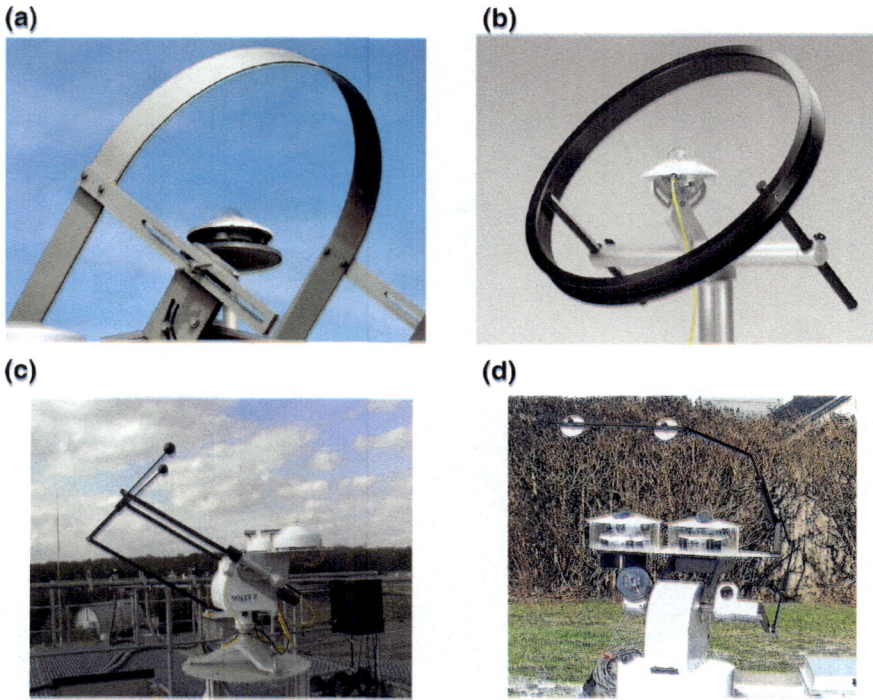

Fig. 22 Examples of traditional arrangements for measuring diffuse irradiance based on standard pyranometers: **a, b** shadow band and shadow ring manually adjusted; **c, d** shadow balls and shadow disks arranged in two-axis sun trackers

applied (Batlles et al. 1995; CIMO 2017). The second type of shadowing solution is based on two-axis tracking systems to which light articulated arms or structures with shadow balls or shadow disks are arranged. The tracking system continuously displaces the disk's or ball's arm to follow Sun's position at all times, and therefore, the pyranometer is permanently shadowed from DNI. Additionally, zero irradiance signals can become an important source of errors and methods to minimize its influence are to be introduced (Hegner et al. 1998).

An alternative to such conventional arrangements is a motorized rotating shadow band which intermittently occludes from direct irradiance a photodiode-based pyranometer (see examples of Fig. 23). The arm is moved around the sensor head which measures GHI when unshaded and measures DIF when shaded. Most of the systems available in the market work with a continuous rotation (constant angular velocity), while a few move the band back and forth at periodic intervals. These instruments can also compute the DNI by using the recorded values of GHI and DIF. Operational and performance details can be found elsewhere (Wilbert et al. 2015, 2016).

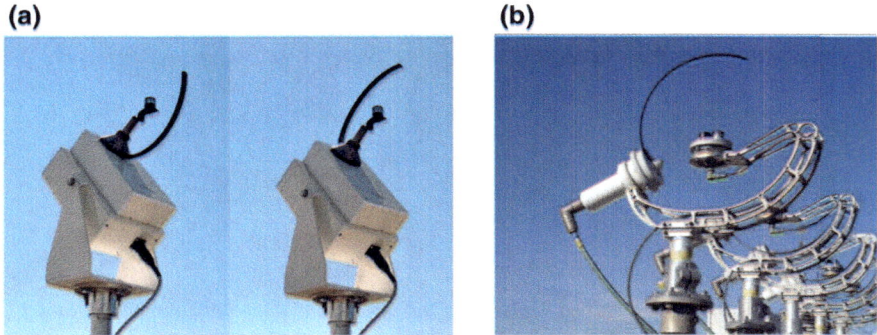

Fig. 23 Two equivalent concepts of rotating shadow band based on a fast response Si photodiode as irradiance sensor

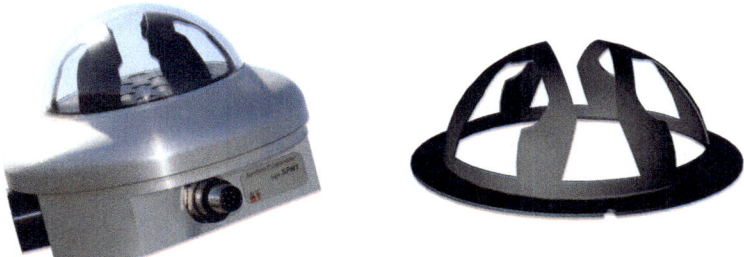

Fig. 24 A new concept for the measurement of global and diffuse irradiance with a shadow mask and without any moveable parts: SPN1 pyranometer by Delta-T. There is always at least one shaded and one unshaded sensor

Another alternative to measure DIF and GHI with a single instrument and without any moveable components is the SPN1 developed by Delta-T Devices (see Fig. 24). The instrument is based in a set of seven fast thermopile detectors, distributed in the same plane in a hexagonal pattern, and covered by diffuser disks. A specially designed shadow mask, created from a hemispherical surface, is placed over the devices and under a glass dome. With this mask, there is always at least one sensor shaded and at least one sensor unshaded for any position of the Sun in the sky. Both sensors (rotating shadow band and masked shadow) have demonstrated a similar accuracy for measuring GHI, but measurements of DIF with SPN1 have higher errors than those obtained with rotating shadow band.

(a)
(b)
(c)
(d)

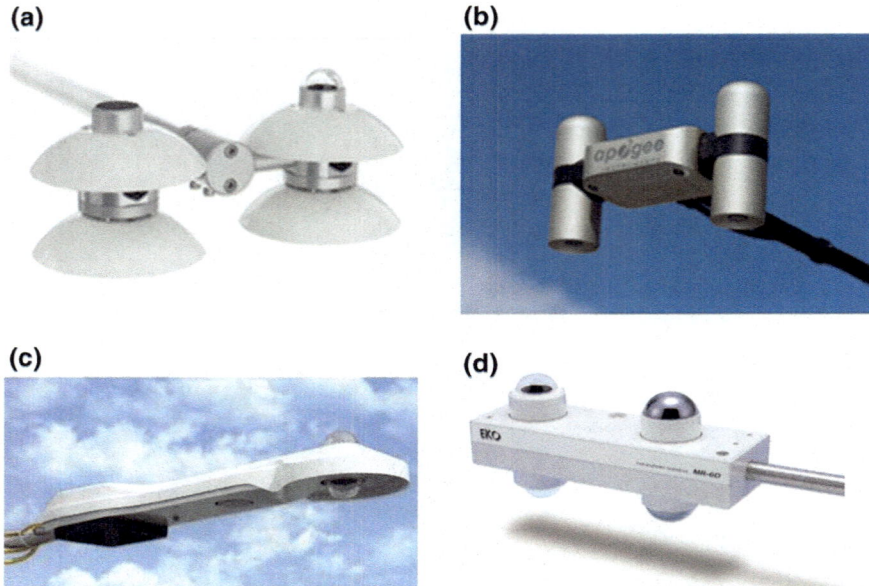

Fig. 25 Examples of four-component net (pyr)radiometers: **a** NR01 net radiometer by Hukseflux; **b** Apogee SN-500 net radiometer; **c** CNR4 net radiometer by Kipp and Zonen; **d** MR-60 net radiometer by EKO

6 Other Broadband Solar Sensors: Total and Longwave Radiation

Preceding instruments are devoted to measure and characterize SW irradiance (λ from 0.3 to 3 µm), but, as commented in the first chapter, there is a great interest in the determination of LW irradiance (λ from 3 to 100 µm) that has a terrestrial and atmospheric origin (thermal radiation). The measurement of these components at the ground level is very important to compare them with those being measured in the outer atmosphere by radiometers in spacecrafts.

Two kinds of instruments are to be commented to this respect: pyrradiometers and pyrgeometers. Put together in pairs to measure downward and upward radiation components, or in association with a couple of pyranometers measuring GHI and albedo in the SW range, they conform a set of four-component net radiometers which are the basis for evaluating total radiation budget at terrestrial level. Figure 25 includes some examples of these four-component net radiometers available commercially.

A pyrradiometer is a thermopile-based instrument, able to measure total radiation, including SW and LW, in a hemispherical 2π solid angle. They must have a constant sensitivity in the entire spectral range SW + LW (λ from 0.3 to 100 µ). Computing the difference of two of these instruments arranged for measuring downwards and upwards, the net radiation can be obtained.

A pyrgeometer is designed for measuring LW radiation, also sensing thermal radiation with a thermopile. In most cases, the shorter λ range is eliminated by means of high- or long-pass filters (e.g., domes or disks made of silicon with additional solar blind filters, or directly deposited over the thermopile sensor) to make them opaque to SW while keeping constant transmittance in the LW range. In some cases, when SW range is not fully filtered by the instrument optics, they have to be used only at night.

When measuring with these instruments having filters added, it is important to note that the own domes, covers, and filters emit radiation themselves because of their operating temperature (blackbody radiation) and temperature sensors have to be included to account for this contribution. Internal heating elements to keep the instrument above dew point and to avoid water vapor condensation is also important, because water filters LW radiation and can alter the measurement results. Additional sources of error, operational characteristics, and classification of instruments can be found in the WMO CIMO Guide (CIMO 2017) and are not included here for completion.

7 Solar Spectral Measurements

The knowledge of the spectral distribution of solar radiation (as well as of other artificial light sources) is of major interest for many scientifical areas: biology, agriculture, human health, weather, air quality, etc. The measurement of spectral distribution is performed by instruments named spectroradiometers. It is again one of the magnitudes of basic knowledge to assess several essential climate variables under WMO GCOS. The specific spectral range every technical area is interested in, the intensity of the irradiance received by a particular object or substance, and the technical capabilities of different types of sensors, have resulted in a wide variety of spectroradiometers available in the market. However, many of these instruments are conceived for its use in laboratory environments, for working under low-intensity light levels, with indirect or reflected light beams, or in relatively narrow spectral ranges, and therefore can become not suitable for solar applications.

The basic element in a spectroradiometer is a spectrally dispersive device, like a prism, a ruled diffraction grating or a holographic diffraction grating. Chromatic dispersion of light is a natural phenomenon that people are familiar with, because the formation of the rainbow is due to the same physical process (light scattered by raindrops). Rainbow-like dispersive effects in the surface of a CD or a DVD, or in the border of a curved glass lens, are also result of the same phenomenon.

The grating in a spectroradiometer separates (diffracts) the incoming white (broadband) light into its component wavelengths λ in a continuous spatial distribution, because every wavelength interval $\Delta\lambda$ is progressively dispersed from the surface of the grating in sequential adjacent angle $\Delta\theta$. Diffraction gratings can be manufactured to work by transmission or by reflection, but reflection is the usual election for optical instruments. Reflection gratings are manufactured by "sculping" a set of closely and uniformly spaced parallel grooves, in a mirror coating with a flat glass substrate. The angular width of the scattered light, the spectral resolution, and the spectral range in which the grating disperses the light are design parameters dependent on the physical dimensions of these grooves (density of lines/mm, angle of the grooves, sinusoidal or sawtooth shape, etc.). Besides, an optoelectronic sensor sensitive to the spectral range matching that of the dispersive grating is required to measure the light intensity in every $\Delta\lambda$ interval. Usually, semiconductor photodiodes or thermopiles are used for this purpose or PMT tubes for very low-intensity light sources. Examples of common semiconductor detectors are silicon (300–1150 nm), InGaAs (900–1850 nm), or PbS (1–4 μm).

But in order to resolve very narrow $\Delta\lambda$ intervals (very narrow $\Delta\theta$), two different solutions have been applied: first, to use only one detector and a very thin slit the dispersed light crosses, and to turn the diffraction grating with a step motor to select every $\Delta\lambda$. These are scanning spectroradiometers (SSR), and their architecture is based in an instrument called monochromator. Second, to use a grating in a fixed position and to multiply the number of sensors, arranged in the form of a linear array (as in a linear CCD). These are called array spectroradiometers (ASR). First ones were more accurate at the expense of the time required for scanning the spectral range of interest, and good λ resolution was at the end dependent on the size of the monochromator. Additionally, straight light dispersion in the UV range requires the use of a double monochromator system. Its size and optical quality components are better prepared for a laboratory environment. Second ones are fast instruments, but the λ resolution is dependent on the number of array elements; they can also be affected by straight light and by second-order dispersion effects. However, these latter have become very popular for many applications due to their small size, portability, ease of use, and lower cost.

In the case of solar spectral distribution, main difficulties arise by the high intensity of solar irradiance (making necessary to attenuate the sunlight with neutral density filters of with integrating spheres) and by the use in outdoor conditions, where temperature, dust particles, and wind can affect the performance of these instruments. On the other hand, the desired spectral range for many solar applications would be between 280 and 2500 nm, and this range is not covered by a single instrument. Scanning SSR, being sensitive and delicate instruments, require protective and rugged cases, watertight, adequate thermal insulation, and robust construction for being used outdoors. Long scanning times can also be an issue when fast measurements are required. Figure 26 shows a couple of examples of these scanning spectroradiometers for solar applications. Nowadays, it is difficult to find this SSR prepared for solar applications in the market and those available are quite expensive.

Small, optical fiber-based ASRs are easy-to-find instruments in the current market (almost all the companies manufacturing optical equipment have some in their product catalog) but are neither well prepared for solar applications, and their accuracy is directly related to their cost (higher the accuracy, higher the cost). Optical fibers are also affected by transmittance issues, and CCD arrays are very sensitive to temperature (overall in the IR range), so well-insulated temperature stable cases and cooled sensors (e.g., with Peltier stages) would be required to

(a)  **(b)**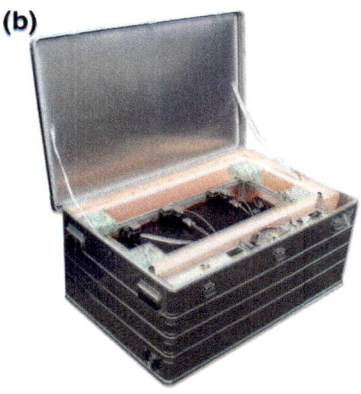

Fig. 26 Scanning spectroradiometers designed for solar irradiance applications: **a** LI-COR 1800, one of the most popular instruments during decades of solar radiation research, now discontinued (after Estellés et al. 2006); **b** Enviro300 wide spectral range solar spectroradiometer, by Bentham, enclosed into a rugged case for outdoor operation

ensure reproducibility and stability in an outdoor environment. Additionally, spectral resolution in the IR range is usually low [typically 10–20 nm spectral full width at half maximum (FWHM)].

However, in the last few years, some new ASRs, specially designed for solar applications, have been developed. Some examples are collected in Fig. 27. Solar ASRs are still limited in spectral range and IR resolution. There are a few models available, mainly based on silicon arrays (300–1050 nm), but with good accuracy (±2%) and with adequate spectral resolution achievable (1.5–6 nm). However, there are hardly any instruments working in IR range (1100–2500 nm), these spectroradiometers are based on InGaAs arrays, and these have still a low number of detectors as to have a good spectral resolution.

8 Narrowband Filter Radiometry: Aerosol Optical Depth Measurements

The last group of instruments here compiled, due to the high interest developed in recent years at international level, are filter radiometers. These instruments are devoted to the measurement of on-ground irradiance in special and selected spectral narrowband and spectral lines in which absorption (mainly due to gases as H_2O vapor, O_3, O_2, CO_2, CO, N_2O, CH_4, etc.) and scattering processes in the atmosphere (mainly caused by suspended particles, the aerosols) produce a characteristic reduction of spectral irradiance (see Fig. 28). Scattering of light by particles (known as Mie scattering) occurs when their diameters are approximately equal to the wavelength of the incident light, and takes place in the lower 5 km of the atmosphere.

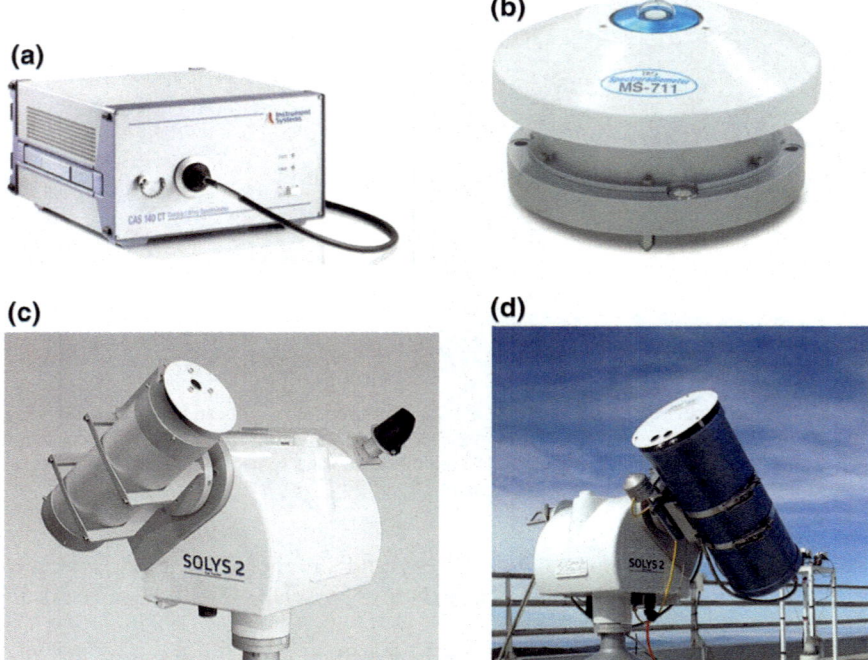

Fig. 27 Examples of CCD array-based spectroradiometers: **a** CAS 140CT by instrument systems; **b** MS-711 by EKO, also with a model for IR range; **c** Kipp and Zonen PGS-100 sun photometer; **d** precision spectral radiometer by PMOD. Last three are specially adapted for outdoor operation and sunlight irradiance levels

The origin and nature of the aerosols are wide-ranging, both from natural and anthropogenic sources: sea salts, mineral windblown dust, volcanic ash, smoke from wildfires; sulfates, nitrates, and organics from chemical reaction of gases in the atmosphere producing non-volatile products that condense to form particles; condensation of semivolatile substances such as certain herbicides and pesticides on existing particles; pollution from factories (WMO GAW 2005; CIMO 2017; Earth Observatory 2018). The influence of these aerosols in the climate change, in the acid rain, in the formation and annihilation of clouds, in favouring or impeding precipitations, or in the air quality and human health, are of the major concern.

The study of these influences, as well as of their dynamics, production, reactions, and interactions, has resulted in the creation of several international networks of ground measurement stations that are compiling and sharing data, trying to complement those obtained by satellite and upper atmosphere (aircraft and balloons) observations: GAW-PFR, AERONET, SKYNET, and SURFRAD are examples of these international networks.

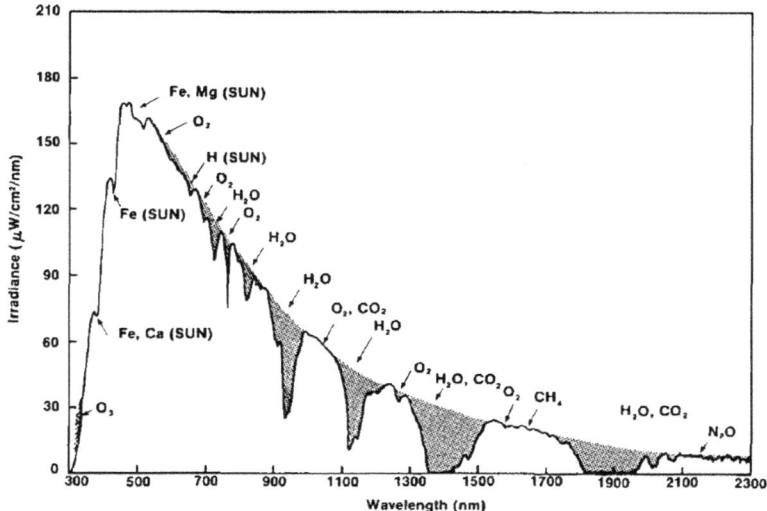

Fig. 28 Example of measured global spectral irradiance with spectral absorption features identified. After Bird et al. (1982)

The degree of the beam extinction by aerosols and gases is directly related with their density or concentration in air. Aerosol optical depth (AOD) is obtained from measurements of atmospheric spectral transmittance. The solar spectral irradiance E at a given wavelength λ can be expressed as (CIMO 2017):

$$E(\lambda) = E_0(\lambda) \exp(-m \cdot \delta(\lambda))$$

being E_0 the extraterrestrial irradiance, m the air mass, and δ the total optical depth. The value of $\delta(\lambda)$ includes terms associated to different effects: Rayleigh scattering δR (by gas molecules), absorption by trace gases δG, and extinction by aerosols δA. Then, AOD is obtained by subtracting these components from the total optical depth: $\delta A = \delta - \delta G - \delta R$. An AOD value of 0.01 corresponds to a very clean atmosphere, while a value >0.5 would correspond to a quite hazy ambient. An average aerosol optical depth for the USA is between 0.1 and 0.2.

In order to avoid large errors in the estimation of δA, several wavelengths and bandpasses out from the ranges in which attenuation is dominated by other components (extinction by water vapor, NO_x, and ozone) are usually selected. For example, WMO GAW-PFR recommends to measure at 3 or more channels among 368, 412, 500, 675, 778, and 862 nm with a bandwidth of 5 nm (OSCAR 2018), but other networks have selected different λ for their specific needs. Most networks coincide in using λ around 500 ± 3 and 865 ± 5 nm (CIMO 2017). Some instruments scan additional λ ranges to specifically account for optical depths of water vapor and ozone too.

Measurements of AOD are mainly performed with three families of instruments: LIDAR, sun photometers, and Brewer spectrophotometers. The latter, based on turning diffraction gratings and PMT sensors, are very sensitive instruments used for specialized spectral measurements in the UV range. With reference to this chapter, we are going to describe just the second group.

An alternative solution to obtain the different optical depths is the measurement of direct solar spectral irradiance with a spectroradiometer, in a wide wavelength range (e.g., 300–1100 nm) and with a fine spectral resolution, and then to select the bands or wavelengths of interest. This can be performed with the instruments described in the previous section of this chapter, by adding suitable collimators to receive DNI. However, as we have seen, spectroradiometers were usually sensitive and delicate instruments, not specially prepared for outdoors operation, and more robust and rugged solutions are convenient. In addition, the total (integrated) irradiance is also required to compute AOD, so spectroradiometers need the simultaneous reading from a broadband pyrheliometer.

Sun photometers were designed based on the use of one or several sensors and a set of interference narrowband filters to select discrete wavelengths or narrow bands to scan. First filter radiometers (see Fig. 29) were simple instruments based on a standard thermopile pyrheliometer and a filter wheel (manually interchanged).

Modern versions of these instruments use fully automatized filter wheels of several positions over one or two broadband sensors (usually, UV-enhanced Si, standard Si and InGaAs photodiodes, thermally stabilized), mounted in two-axis sun trackers and with the programmed operation to scan direct or diffuse components, by pointing to different sections of the sky. Perhaps a small disadvantage of using only one (two) sensor(s) is the extended time interval required to measure in every λ band and in interchanging filters while having the advantage of using simpler electronics and driven circuits.

Alternative designs to filter-wheel-based radiometers are multifilter multisensor-based sun photometers. The basic principle here is to use a dedicated "sensor + filter" couple for every wavelength or band to be measured, without moving parts inside the radiometer head. An additional, unfiltered sensor enclosed in the same instrument measures broadband, integrated irradiance. Examples of these instruments are shown in Fig. 30. Si and InGaAs photodiodes are also used for a fast response to varying sky conditions.

An additional application of some of these instruments, able to record in six and more wavelength channels, is the synthetic generation of solar spectra based on the attenuation measured at these bands (Tatsiankou et al. 2013) and by using simulation codes as SMARTS2. However, these "spectral" instruments are only valid to simulate the solar spectra and are not valid for any other light source.

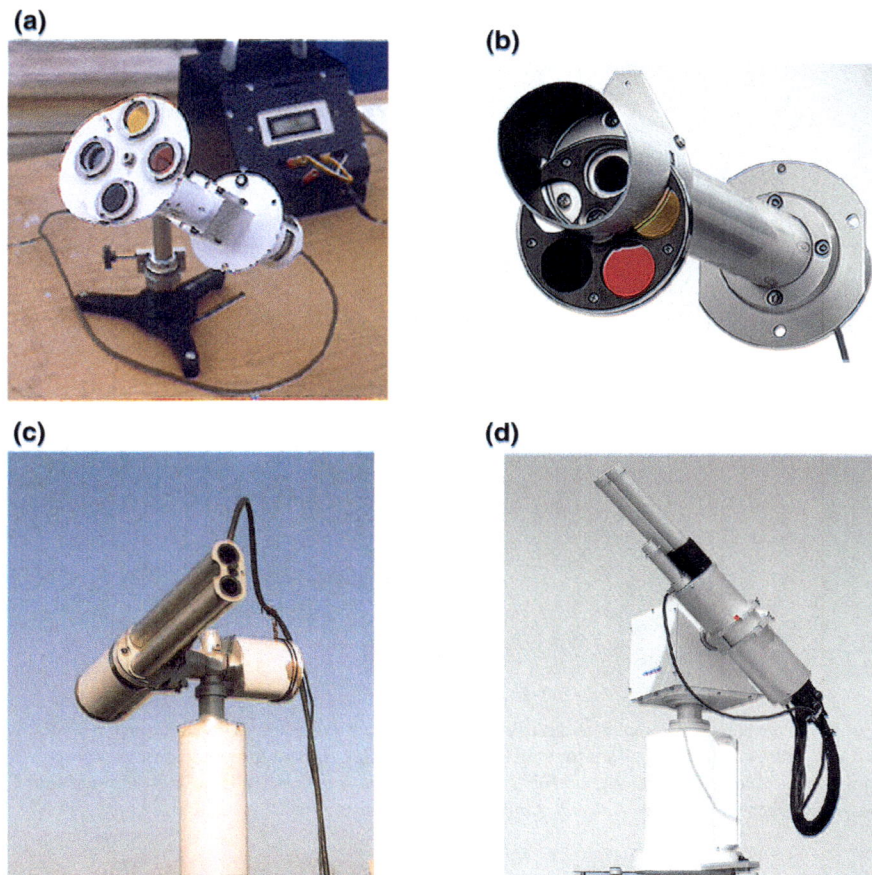

Fig. 29 Examples of filter radiometers based on rotating filter wheels and one (or two) sensors. **a** Eppley NIP pyrheliometer with a manual 4-position filter wheel and three narrowband filters; **b** delta ohm LP PYRHE 16 with a 5-position filter wheel; **c** CIMEL 318 sun photometer, with 9 and 12 λ versions available, fully automated filter wheel, assembled in a two-axis sun tracker; **d** Kipp and Zonen POM-01 and POM-02 sky radiometers, also with fully automated 7 and 11 filters versions

Finally, there is nowadays some limitations in the absolute traceability of these instruments, in part because of the large uncertainty associated with the calculations and estimations for the optical depths associated with different contributions. According to WMO, as traceability is not currently possible based on physical measurement systems, the initial form of traceability will be based on different criteria (WMO GAW 2005).

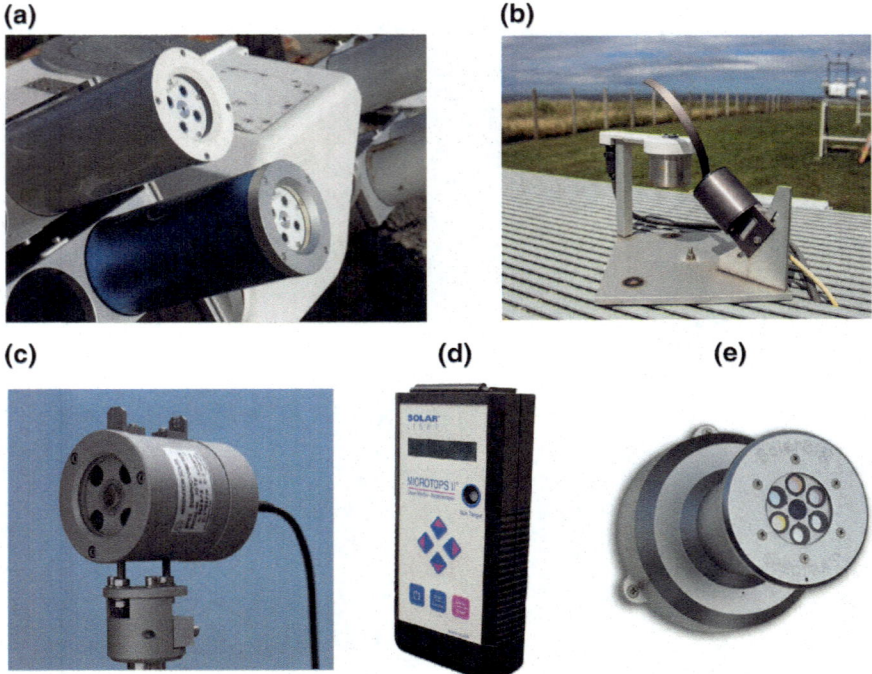

Fig. 30 Examples of narrowband multifilter radiometers: **a** Precision Filer Radiometer (PFR) by PMOD, with 4 λ filters and sensors; **b** Multifilter rotating shadowband radiometer MFR-7 by Yankee (6 λ channels); **c** Middleton SP02 sun photometer (4 channels); **d** Solar light microtops II sun photometer (5 channels with two configurations); and **e** Spectrafy solar SIM (6 channels)

The World Optical Depth Research Calibration Center (WORCC) was established in 1996 at the PMOD/WRC by WMO, to serve as an international reference in this field. WORCC designated a set of standard instruments, and the WORCC standard group of three precision filter radiometers (the so-called PFR triad) against the rest of field instruments are compared. Traceability is gained when 95% of the measurements performed by an instrument, during an intercomparison, are between specified limits (0.005 + 0.01/m optical depths) of the average value obtained from the "PFR triad". Development of new instruments and techniques are of high importance in this field for gaining in absolute accuracy and traceability.

Acknowledgements This work has been partially supported by the Spanish National Funding Program for Scientific and Technical Research of Excellence, Generation of Knowledge Subprogram, 2017 call, DEPRISACR project (reference CGL2017-87299-P). The authors also wish to thank Dr. Stefan Wilbert from DLR for sharing several useful comments and remarks on this chapter.

Appendix: Current Status of the WRR

1.1 The SI Laboratory Absolute Radiometers, the Space Radiometers, and the SI-WRR Conflict

As described in Sect. 2, the development of absolute cavity radiometers at the end of the decade of 1960, initially by the JPL and soon by other laboratories and commercial companies, provided at the end the foundation by the WMO of the WRR as the top reference standard in the solar irradiance scale, and the designation of PMOD (Davos) as the WRC where the group of WSG standards were conservated, maintained, and disseminated.

On the other hand, electrical substitution-based cavity radiometers continued evolving in the environment of the NMI laboratories, until the development of the cryogenic absolute radiometers (Quinn and Martin 1985). Unlike solar absolute radiometers, which work at ambient temperature, cryogenic radiometers work with reference temperatures in the cold reservoir between 2 and 20 K, by using liquid He and N. However, their application was first the determination of the Stephan–Boltzmann constant and of the thermodynamic temperature in radiation thermometry (Fox 2001).

Shortly after that, a primary standard radiometer was developed in the NPL (UK) for applications in optical radiometry (Martin et al. 1985). An improved design, by using a mechanical cooling engine to reach temperatures of 15 K, resulted in a compact instrument that became the standard to implement at NMI level the fundamental unit of candela (cd) and its derived units (lumen, lux) in the SI (Fox et al. 1995). Figure 31 shows a sketch of this kind of cryogenic radiometer. Additional reviews about the characteristics and operation of these absolute radiometers, their design, and their historical evolution can be found elsewhere (Hengstberger 1989; Fox and Rice 2005).

Due to the somewhat independent evolution of solar irradiance scale, based on the WRR, with respect to the SI optical radiometry scales at NMIs, based on cryogenic radiometers, intercomparisons between both scales were necessary and were done in a repeated form to determine their mutual transference and equivalence, and to check the stability of the results (Romero et al. 1991, 1995; Finsterle et al. 2008; Fehlmann et al. 2012). The intercomparison process is not immediate because of the differences between relative intensities of every scale and due to the different operation modes of the instruments, what forced the use of transfer standards (trap detectors) in some cases.

First, two comparisons (1991 and 1995) gave as result differences below 0.3% and within the respective uncertainty of each scale, what was considered as reasonable. The third comparison in 2005 produced an excessively low result and doubts about the linearity of the transfer detectors used were posed. In 2010, a new

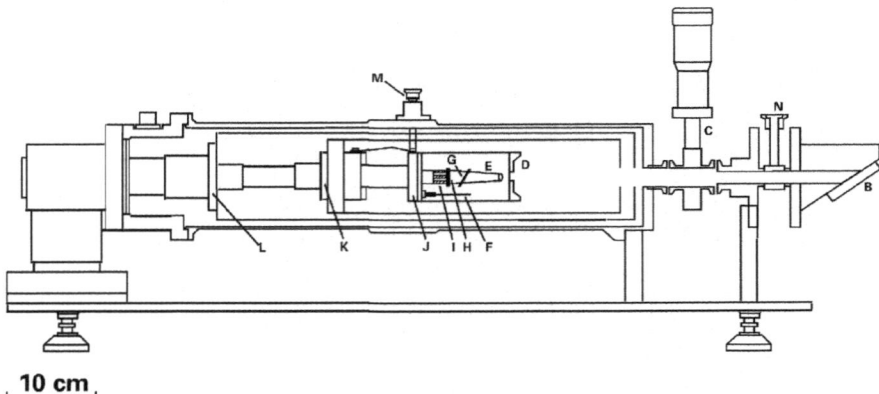

10 cm

Fig. 31 Basic structure of a cryogenic radiometer for measurements of radiant power of lasers in NMI laboratories. Taken from Fox et al. (1995)

intercomparison was carried out, with results at the same level of the two first comparisons for measurements in power mode, but with differences of $(0.34 \pm 0.18)\%$ between scales (WRR above SI) for measurements in irradiance mode. Discrepancies mainly arose due to the different modes of operation of the standards (irradiance versus power, light beam entirely covering or not the input port, see Fig. 32). Successive comparison along 2012 and 2014 confirmed differences of $(0.31 \pm 0.6)\%$ in the ratio WRR/SI (Suter et al. 2012; Finsterle 2015) but without overlapping the difference between respective uncertainties, what leaves in question the transfer between scales.

The third group of absolute cavity radiometers, of great importance for this exposition, is that formed by solar radiometers used for the determination of total solar irradiance (TSI) and related quantities in successive space satellite and shuttle missions since the 1970s. Though their fundamental structure is quite similar to that of terrestrial absolute radiometers (TSI level of the order of $\sim 1365 \text{ W m}^{-2} \pm 3.5\%$ of yearly oscillation), there are differences in two important working conditions in space: operation under vacuum (absence of air convection and atmospheric pressure effects) and operation at very low reference temperatures. Successive generations of instruments and space missions (e.g., NIMBUS7/ERB, SMM/ACRIM1, UARS/ACRIM2, SOHO/VIRGO, SORCE/TIM, ACRIMSAT/ACRIM3) have introduced progressive improvements in their design (Fröhlich 2013) and have contributed to the current recording of more than 35 years of TSI data. These space measurements allowed not only to determine the solar constant but also its natural variability in periodic 11-year cycles (corresponding to sunspots cycles), which is in the order of 0.1% (Yeo et al. 2014). However, data obtained from different experiments and instruments in space were not consistent

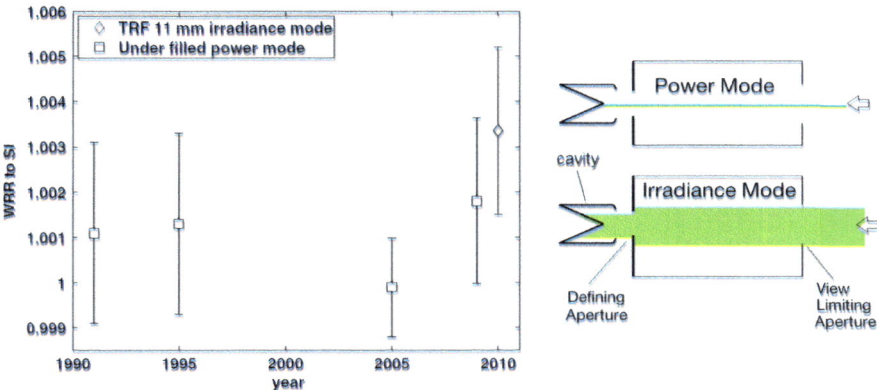

Fig. 32 (Left) Results of WRR to SI intercomparisons. Taken from Fehlmann et al. (2012). (Right) different operation modes for radiometers of the SI and WRR scales (power versus irradiance). Taken from Suter et al. (2012)

with reference to their absolute irradiance values (see Fig. 33). The value of 1361.1 W m^{-2} has been confirmed by a new revision with an estimated standard uncertainty of 0.5 W m^{-2} (Gueymard 2018).

Differences were particularly enhanced when total irradiance monitor (TIM) radiometer (Kopp et al. 2005) went into operation in 2003 and measured values 0.35% lower than those of the variability of the solar irradiance and gravity oscillations (VIRGO) mission (Fröhlich et al. 1997). Besides new research into the origin of these differences, a new laboratory able to compare the twin reserve instruments (kept on Earth), the total solar irradiance radiometer facility (TFR) in the Laboratory for Atmospheric and Space Physics (LASP, Univ. Colorado, USA) was created (Kopp et al. 2007). This advanced installation allows the absolute radiometers to work both in power and irradiance modes, under good vacuum and well normal atmospheric pressure conditions, and it is then suitable to compare different types of instruments and irradiance scales. Thanks to TFR, it was checked how part of the differences found in space was due to the respective traceabilities of radiometers to WRR and SI scales. A new intercomparison WRR/SI with the PMO/PREMOS radiometer was carried out in TRF (Schmutz et al. 2013), and equivalent differences were found in the ground as in space (Fehlmann et al. 2012).

These results with space absolute radiometers contributed to demonstrate how WRR irradiance scale was out of concordance or equivalence to SI irradiance scale due to operating and functional differences between instruments and reference standards.

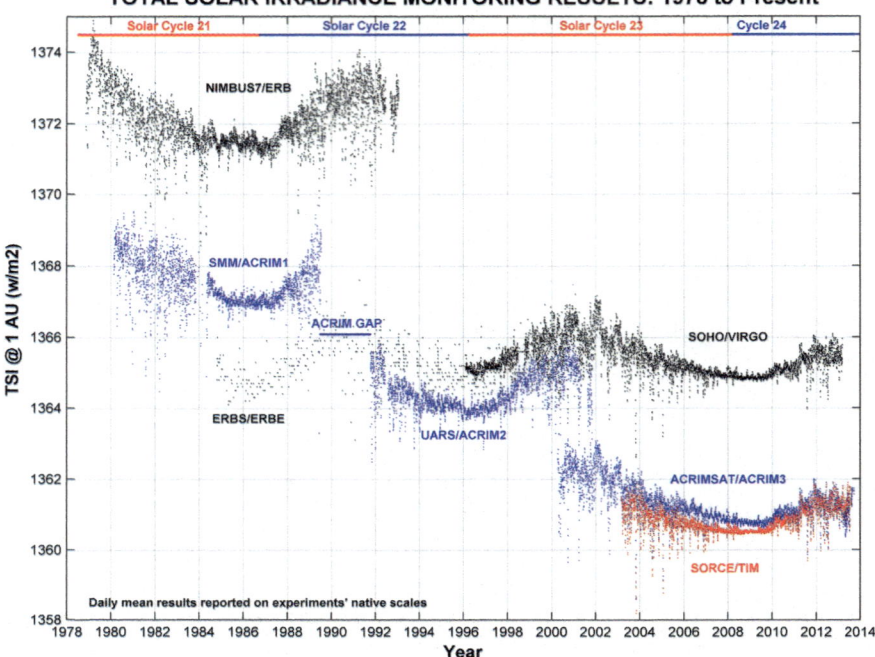

Fig. 33 Total solar irradiance recording by absolute radiometers in different space missions since 1978. Taken from www.acrim.com (as available in July/2018)

1.2   Current Status of WRR

Current lack of transference, equivalence, and/or compatibility between WRR and SI is being objected of an in-depth revision by WMO and CIPM, who agreed to cooperate to ensure that meteorological data could be adequately traced to SI. WRR is nowadays forming an "island of traceability" (Finsterle 2015), temporary out from SI, due to the WRR/SI ratio differences higher than 0.3% and the uncertainty associated to the comparison results.

Status of the WRR at a technical level is also delicate because many of the instruments originally integrating the WSG had to be ruled out of the group because of malfunctioning or drift. Currently, the WRR is implemented with at least four of the six surviving instruments, but some of them have cumulated more than 35 years of operation and can fail at any moment. Therefore, it is urgent the incorporation of new components to the WSG, or to search for new standard references, alternative to WSG, with enough precision, stability and low uncertainty, even by holding an irradiance scale based on artifacts.

Fig. 34 Picture of the WSG realizing the WRR, together with other cavity radiometers, and of the new CSAR and MITRA devices (left-lower side on the tracker). Taken at PMOD/WRR (Davos) in 2015

A possible solution to the problem could be the use as a reference of a new absolute cavity radiometer called cryogenic solar absolute radiometer (CSAR) (Martin and Fox 1993), developed in collaboration between PMOD/WRC, METAS (Switzerland) and NPL (UK). CSAR bases their outdoor measurements in a supplementary unit, monitor for integrated transmittance (MITRA) (Walter et al. 2014), which is responsible for detect changes on window transmittance. As a whole, CSAR and MITRA present an impressive accuracy (150 ppm) in the determination of solar irradiance.

The operation of a radiometer with cryogenic temperatures allows the use of larger cavities, with enhanced absorptivity, and thanks to a reduction in thermal gradients in the cavity, it ensures the equivalence between radiative/thermal heating and electrical heating. However, these low temperatures require the use of vacuum for operating the cavities and to add an optical window whose spectral transmittance can change due to ambient temperature and intensity of received radiation. MITRA allows introducing corrections due to these factors in a synchronous form with CSAR measurements. As radiometer, CSAR can also work at ambient temperature without requiring cooling (Fig. 34).

First functional probes of CSAR on the ground and first intercomparisons against cryogenic radiometers of SI scale laboratories seem to have given very promising results (Walter 2016) in terms of stability and traceability to SI. However, it is necessary to wait for the CIMO/WMO working group to decide what is the solution for the near future for the solar irradiance scale. The huge technical capabilities, complexity, and economical budget of an instrument like CSAR or of an installation as TFR-LASP do not seem to be easily expandable concepts to other NMIs in an extensive form.

References

Abbot CG (1913) Standard water-stir pyrheliometer no. 4. Ann Astro-physical Obs Smithson Inst 3:21–49

Abbot CG, Aldrich LB (1913) Smithsonian pyrheliometry revised. Smithson Misc Collect 60

Abbot CG, Fowle FE (1908) Chapter 2, Apparatus for solar constant determinations. Ann Astrophys Obs Smithson Inst 2:21–49

Ångström K (1894) The quantitative determination of radiant heat by the method of electrical compensation. Phys Rev (Series I) 1:365–372. https://doi.org/10.1103/physrevseriesi.1.365

Angström K (1899) The absolute determination of the radiation of heat with the electric compensation pyrheliometer, with examples of application of this instrument. Astrophys J 9:332–346

ASTM E1125-99 Standard (2009) Standard test method for calibration of primary non-concentrator terrestrial photovoltaic reference cells using a tabular spectrum. ASTM International, West Conshohocken, PA, 2016. www.astm.org. This Standard (1999) is currently superseded by a new versio

Batlles FJ, Olmo FJ, Alados-Arboledas L (1995) On shadowband correction methods for diffuse irradiance measurements. Sol Energy. https://doi.org/10.1016/0038-092x(94)00115-t

BIPM (2018) Information provided by Bureau International des Poids and measures in the webpage. https://www.bipm.org/en/cipm-mra/. Accessed 12 Jul 2018

Bird RE, Hulstrom RL, Kliman AW, Eldering HG (1982) Solar spectral measurements in the terrestrial environment. Appl Opt 21:1430–1436

Blanc P, Espinar B, Geuder N et al (2014) Direct normal irradiance related definitions and applications: the circumsolar issue. Sol Energy 110:561–577. https://doi.org/10.1016/j.solener.2014.10.001

Brusa RW, Fröhlich C (1975) Realization of the absolute scale of total irradiance. Scientific discussions. International pyrheliometer comparisons IPC-IV, Davos

BSRN (2018) World Radiation Monitoring Center (WRMC). Central archive of the BSRN. https://bsrn.awi.de/

Buie D, Monger AG (2004) The effect of circumsolar radiation on a solar concentrating system. Sol Energy. https://doi.org/10.1016/j.solener.2003.07.032

Bush BC, Valero FPJ, Simpson AS, Bignone L (2000) Characterization of thermal effects in pyranometers: a data correction algorithm for improved measurement of surface insolation. J Atmos Ocean Tech. https://doi.org/10.1175/1520-0426(2000)017%3c0165:coteip%3e2.0.co;2

Callendar HL, Fowler A (1906) The horizontal bolometer. In: Royal society series A. 77. pp 15–16

CIMO (2017) Part I, Chapter 7 for solar radiation measurements. Part I, Chapter 16 for aerosols measurements. In: The CIMO guide. World Meteorological Organization. WMO guide to meteorological instruments and methods of observation (WMO-No. 8, 2014 edition, updated in 2017)

Coulson KL (1975) Solar terrestrial radiations. Methods and measurements. Academic Press, New York

Crommelynck D (1973) Theorie instrumentale en radiometrie absolue. Pub Ser A, No. 81

Crookes W (1874) On attraction and repulsion resulting from radiation. Philos Trans R Soc Lond 164:501–527

Dutton EG (2002) Report on GEWES Baseline Surface Radiation Network (BSRN). Global energy and water cycle experiment

Earth Observatory (2018) EOS project science office at NASA Goddard Space Flight Center. Global maps. Aerosol optical depth. https://earthobservatory.nasa.gov/global-maps/MODAL2_M_AER_OD. Accessed 28 Aug 2018

Emery KA, Soterwald CR, Kazmerski LL, Hart RE (1988) Calibration of primary terrestrial reference cells when compared with primary AM0 reference cells. In: 8th European PVSEC. pp 64–68

Estellés V, Utrillas MP, Martínez-Lozano JA et al (2006) Intercomparison of spectroradiometers and sun photometers for the determination of the aerosol optical depth during the VELETA-2002 field campaign. J Geophys Res Atmos. https://doi.org/10.1029/2005jd006047

Fang W, Wang H, Li H, Wang Y (2014) Total solar irradiance monitor for Chinese FY-3A and FY-3B satellites—instrument design. Sol Phys. https://doi.org/10.1007/s11207-014-0595-6

Fehlmann A, Kopp G, Schmutz W et al (2012) Fourth world radiometric reference to SI radiometric scale comparison and implications for on-orbit measurements of the total solar irradiance. Metrologia. https://doi.org/10.1088/0026-1394/49/2/s34

Finsterle W (2011) WMO international pyrheliometer comparison (IPC-XI): final report. IOM report-no. 128

Finsterle W (2015) Status and future of the WRR in the SI. Presentation at IPC-XII seminar, PMOD/WRC (Davos). Available at: ftp://ftp.pmodwrc.ch/stealth/ipcxii/Seminar/status_of_the_WRR.pdf

Finsterle W (2016) International pyrheliometer comparison (IPC-XII): 28. Sep-16. Oct 2015. Final report. IOM report-no. 124. Davos, Switzerland

Finsterle W, Blattner P, Moebus S et al (2008) Third comparison of the world radiometric reference and the SI radiometric scale. Metrologia. https://doi.org/10.1088/0026-1394/45/4/001

Foote PD (1919) Some characteristics of the Marvin pyrheliometer. Sci Pap Bur Stand 605–635

Fox NP (2001) Developments in optical radiometry. In: Recent advances in metrology & fundamental constants, pp 537–571

Fox NP, Rice JP (2005) Absolute radiometers. In: Optical radiometry. Series: experimental methods in the physical sciences, vol 51, pp 35–96

Fox NP, Haycocks PR, Martin JE, Ul-Haq I (1995) A mechanically cooled portable cryogenic radiometer. Metrologia 32(6):581

Fröhlich C (1973) The relation between the IPS now in use and Smithsonian scale 1913, Angstrom scale and absolute scale. In: Symposium solar radiation measurements instrument, pp 61–77

Fröhlich C (1978) World radiometric reference. WMO/CIMO final report, Annex IV, WMO no. 490

Fröhlich C (1991) History of solar radiometry and the world radiometric reference. Metrologia. https://doi.org/10.1088/0026-1394/28/3/001

Fröhlich C (2013) Solar radiometry. Observing photons in space. Springer, New York, pp 565–581

Frohlich C, London J (1986) Revised instruction manual on radiation instruments and measurements. Chapter 4. World Climate Research Programme (WCRP) publications series no. 7. WMO/TD—No. 149

Fröhlich C, Crommelynck DA, Wehrli C et al (1997) In-flight performance of the VIRGO solar irradiance instruments on SOHO. The first results from SOHO. Springer, Netherlands, pp 267–286

GCOS (2011) Systematic observation requirements for satellite-based data products for climate (2011 Update). WMO/GCOS-154 report

GCOS (2016) The global observing system for climate: implementation needs. WMO/GCOS-200 report

Geist J (1972) Optical radiation measurements. Technical note 594-1. Washington, D.C., USA

Golay MJE (1947a) A pneumatic infra-red detector. Rev Sci Instrum. https://doi.org/10.1063/1. 1740949

Golay MJE (1947b) Theoretical consideration in heat and infra-red detection, with particular reference to the pneumatic detector. Rev Sci Instrum 18(5):347–356

Gorczyński L (1924) On a simple method of recording the total and partial intensities of solar radiation. J Opt Soc Am 9:455. https://doi.org/10.1364/josa.9.000455

Gueymard CA (1998) Turbidity determination from broadband irradiance measurements: a detailed multicoefficient approach. J Appl Meteorol. https://doi.org/10.1175/1520-0450(1998) 037%3c0414:tdfbim%3e2.0.co;2

Gueymard CA (2018) A reevaluation of the solar constant based on a 42-year total solar irradiance time series and a reconciliation of spaceborne observations. Sol Energy 168:2–9. https://doi. org/10.1016/j.solener.2018.04.001

Gueymard CA, Wilcox SM (2011) Assessment of spatial and temporal variability in the US solar resource from radiometric measurements and predictions from models using ground-based or satellite data. Sol Energy. https://doi.org/10.1016/j.solener.2011.02.030

Haeffelin M, Kato S, Smith AM et al (2001) Determination of the thermal offset of the Eppley precision spectral pyranometer. Appl Opt. https://doi.org/10.1364/ao.40.000472

Haley F, Kendall JM. S, Plamondon J (1965) Cavity type radiometer for absolute total intensity measurement of visible and IR radiation. 11th National Aerospace Instrumentation Symposium. ISA Preprint 1(11):3–65

Han LH, Wu S, Condit JC et al (2011) Light-powered micromotor: design, fabrication, and mathematical modeling. J Microelectromechanical Syst 20:487–496. https://doi.org/10.1109/ jmems.2011.2105249

Hegner H, Müller G, Nespor V et al (1998) Update of the technical plan for BSRN data, vol 3. pp 38

Hengstberger F (1989) Absolute radiometry: electrically calibrated thermal detectors of optical radiation. Elsevier Science, Netherlands

Hernandez GS, Serrano A, Cancillo ML, Garcia JA (2015) Pyranometer thermal offset: measurement and analysis. J Atmos Ocean Technol. https://doi.org/10.1175/jtech-d-14-00082.1

Hickey JR, Frieden RG, Griffin FJ et al (1977) The self-calibrating sensor of the eclectic satellite pyrheliometer/ESP/program. Thermopile sensor with cavity-type receiver

Hukseflux (2018) PV monitoring and meteorological industries prepare for revised pyranometers standard ISO 9060:2018

IEC 2009. IEC 60904-4 Standard (2009) Ed.2, Photovoltaic devices—Part 4: Reference solar devices—Procedures for establishing calibration traceability. Geneva, Switzerland

IEC 2015. IEC 60904-2 Standard (2015) Ed.3, Photovoltaic devices—Part 2: Requirements for photovoltaic reference devices. Geneva, Switzerland

IEC 2017. IEC 61724-1 Standard (2017) Ed.1, Photovoltaic system performance—Part 1: Monitoring. Geneva, Switzerland

ISO (1990a) ISO 1990. ISO 9060 Standard (1990) Solar energy—specification and classification of instruments for measuring hemispherical solar and direct solar radiation

ISO (1990b) ISO 9059:1990 Standard. Solar energy—calibration of field pyrheliometers by comparison to a reference pyrheliometer

ISO (1992) ISO 9847:1992: Solar energy—calibration of field pyranometers by comparison to a reference pyranometer

ISO (1993) ISO 9846:1993 Standard. Solar energy—calibration of a pyranometer using a pyrheliometer

KCDB (2018) Bureau International des Poids and Measures (BIPM), Key comparison data base, Appendix C, Calibration and measurement capabilities—CMCs. https://kcdb.bipm.org/AppendixC/ and https://kcdb.bipm.org/AppendixC/PR/PR_services.pdf. Accessed 12 Jul 2018

Kendall J.M. (1968) The JPL standard total-radiation absolute radiometer. JPL technical report 32-7263

Kimball HH, Hobbs HE (1923) A new form of thermoelectric recording pyrheliometer. Mon Weather Rev 51:239–242

Kopp G, Lawrence G, Rottman G (2005) The total irradiance monitor (TIM): science results. The solar radiation and climate experiment (SORCE). Springer, New York, pp 129–139

Kopp G, Heuerman K, Harber D, Drake G (2007) The TSI radiometer facility: absolute calibrations for total solar irradiance instruments. In: Butler JJ, Xiong J (eds). International society for optics and photonics, p 667709

Langley SP (1880) The bolometer. In: Proceedings of the American Metrological Society, vol 2. pp 184–190

Latimer JR (1973) On the Ångström and Smithsonian absolute pyrheliometric scales and the international pyrheliometric scale 1956. Tellus 25:586–592. https://doi.org/10.3402/tellusa.v25i6.9723

Liu M, Zentgraf T, Liu Y, et al (2010) Light-driven nanoscale plasmonic motors. Nat Nanotechnol. https://doi.org/10.1038/nnano.2010.128

Marchgraber RM (1970) The development of standard instruments for radiation measurements. Meteorological observations and instrumentation. American Meteorological Society, Boston, MA, pp 302–314

Martin JE, Fox NP (1993) Cryogenic solar absolute radiometer (CSAR). Metrologia. https://doi.org/10.1088/0026-1394/30/4/016

Martin JE, Fox NP, Key PJ (1985) A cryogenic radiometer for absolute radiometric measurements. Metrologia. https://doi.org/10.1088/0026-1394/21/3/007

McCluney R (1994) Introduction to radiometry and photometry, 2nd edn. Artech House, Boston, London

Moll WJH (1922) A thermopile for measuring radiation. Proc Phys Soc Lond 35:257–260. https://doi.org/10.1088/1478-7814/35/1/336

Murdock TL, Pollock DB (1998) High accuracy space based remote sensing requirements. National Institute of Standards and Technology, NIST GCR 98–748

OSCAR (2018) WMO observing systems capability analysis and review tool. https://www.wmo-sat.info/oscar/variables/view/6

Osterwald CR, Emery KA, Myers DR, Hart RE (1990) Primary reference cell calibrations, at SERI: history and methods. In: IEEE conference on photovoltaic specialists. pp 1062–1067

Palmer JM, Grant BG (2010) The art of radiometry. SPIE Press, USA

Philipona R (2002) Underestimation of solar global and diffuse radiation measured at Earth's surface. J Geophys Res Atmos. https://doi.org/10.1029/2002jd002396

Pollock DB, Murdock TL, Datla RU, Thompson A (2000) Radiometric standards in space: the next step. Metrologia. https://doi.org/10.1088/0026-1394/37/5/12

Pouillet C-S-M (1791–1868). A du texte (1838) Mémoire sur la chaleur solaire : sur les pouvoirs rayonnants et absorbants de l'air atmosphérique et sur la température de l'espace/par M. Pouillet

Puliaev S, Penna JL, Jilinski EG, Andrei AH (2000) Solar diameter observations at Observatório Nacional in 1998–1999. Astron Astrophys Suppl Ser 143:265–267. https://doi.org/10.1051/aas:2000180

Putley EH (1977) Chapter 3 InSb submillimeter photoconductive detectors. Semicond Semimetals 12:143–168. https://doi.org/10.1016/s0080-8784(08)60148-9

Quinn TJ, Martin JE (1985) A radiometric determination of the Stefan-Boltzmann constant and thermodynamic temperatures between −40 °C and +100 °C. Philos Trans R Soc Lond A 316:85–189

Rabl A, Bendt P (1982) Effect of circumsolar radiation on performance of focusing collectors. J Sol Energy Eng. https://doi.org/10.1115/1.3266308

Reda I, Myers D (1999) Calculating the diffuse responsivity of solar pyranometers. NREL report NREL/TP-560-26483

Rodríguez-Outón I, Balenzategui JL, Fabero F, Chenlo F (2012) Development of optical collimators for accurate calibration of reference solar cells. In: 27th European photovoltaic solar energy conference and exhibition

Romero J, Fox NP, Fröhlich C (1991) First comparison of the solar and an SI radiometric scale. Metrologia 28:125–128. https://doi.org/10.1088/0026-1394/28/3/004

Romero J, Fox NP, Fröhlich C (1995) Improved comparison of the world radiometric reference and the SI radiometric scale. Metrologia 32(6):523

Rüedi I, Finsterle W (2005) The world radiometric reference and its quality system. In: Technical conference on meteorology & environment instruments & methods of observation TECO, Session 3(15)

Schmutz W, Fehlmann A, Finsterle W et al (2013) Total solar irradiance measurements with PREMOS/PICARD. In: AIP conference proceedings. American Institute of Physics, pp 624–627

Stanhill G, Achiman O (2017) Early global radiation measurements: a review. Int J Climatol 37:1665–1671. https://doi.org/10.1002/joc.4826

Stewart R, Spencer DW, Perez R (1985) In: Böer KW, Duffie JA (eds) The measurement of solar radiation BT—advances in solar energy: an annual review of research and development, vol 2. Springer US, Boston, MA, pp 1–49

Stine WB, Geyer M (2001) Power from the sun. Chapter 2. In: Online book available at: http://www.powerfromthesun.net

Suter M, Finsterle W, Kopp G (2012) WRR to SI comparison with DARA. In: Technical conference on meteorology & environment instruments & methods of observation, TECO, Session 4(5)

Tatsiankou V, Hinzer K, Mohammed J et al (2013) Reconstruction of solar spectral resource using limited spectral sampling for concentrating photovoltaic systems. In: Cheben P, Schmid J, Boudoux C et al (eds) International society for optics and photonics, p 891506

Thacher PD, Boyson WE, King DL (2000) Investigation of factors influencing the accuracy of pyrheliometer calibrations. In: Conference record of the twenty-eighth IEEE photovoltaic specialists conference—2000 (Cat. No.00CH37036). pp 1395–1398

Thekaekara MP (1976) Solar radiation measurement: techniques and instrumentation. Sol Energy 18:309–325. https://doi.org/10.1016/0038-092x(76)90058-x

Vignola F, Michalsky J, Stoffel T (2012) Solar and infrared radiation measurements (Energy and the environment). CRC Press, USA

Walter B (2016) Direct solar irradiance measurements with a cryogenic solar absolute radiometer. In: Radiation processes in the atmosphere and ocean (IRS2016) AIP conference proceedings 1810, 080007-1/4

Walter B, Fehlmann A, Finsterle W et al (2014) Spectrally integrated window transmittance measurements for a cryogenic solar absolute radiometer. Metrologia 51:S344–S349. https://doi.org/10.1088/0026-1394/51/6/s344

Wilbert S, Pitz-Paal R, Jaus J (2013) Comparison of measurement techniques for the determination of circumsolar irradiance. In: AIP conference proceedings. American Institute of Physics, USA, pp 162–167

Wilbert S, Geuder N, Schwandt M et al (2015) Task 46: best practices for solar irradiance measurements with rotating shadowband irradiometers. IEA SHC Sol Updat Newsl 62:10–11

Wilbert S, Kleindiek S, Nouri B et al (2016) Uncertainty of rotating shadowband irradiometers and Si-pyranometers including the spectral irradiance error. In: AIP conference proceedings. AIP Publishing LLC, USA, p 150009

Willson RC (1973) New radiometric techniques and solar constant measurements. Sol Energy. https://doi.org/10.1016/0038-092x(73)90035-2

WMO GAW (2005) WMO/GAW experts workshop on a global surface-based network for long term observations of column aerosol optical properties. World Meteorological Organization—Global Atmosphere Watch. WMO TD no. 1287; GAW report-no. 162

Yeo KL, Krivova NA, Solanki SK (2014) Solar cycle variation in solar irradiance. https://doi.org/10.1007/s11214-014-0061-7

Zerlaut G. (1989) Solar radiation instrumentation. In: Hulstrom RL (ed) Solar resources. MIT Press, Cambridge, MA

Chapter 3
Establishing a Solar Monitoring Station with Auxiliary Measurements

Frank Vignola

Abstract A considerable amount of thought and planning should occur when considering the establishment of a solar monitoring station. One needs to understand the goals and limitations of the project and what can be accomplished with the instruments that fit within the budget. Other initial issues are: how the station will be maintained; how the data will be collected and analyzed; and how long the station will be operating. Of paramount importance is how the data are going to be used. For example, there is a considerable difference between monitoring to evaluate the long-term variability of the solar resource and using site data to provide the performance estimates to help obtain financing for a solar electric facility. This chapter consists of eleven sections. Section 1 will discuss overall considerations that set the basis for how the station should be configured. Section 2 will cover the instrumentation. The choice of location for the solar monitoring site is discussed next. Sections 5–7 cover the data logger and logistics such as maintenance and communications. Section 8 describes auxiliary measurements, and Sect. 9 will cover other useful instruments. Section 10 is on grounding, and Sect. 11 presents the physical layout of a hypothetical solar monitoring station.

1 Considerations for Setting up a Solar Monitoring Station

In order to optimally design a solar monitoring station, the purpose and goals for the solar monitoring station should be well defined. It is just as important to consider the accuracy desired for the measurements and the budget constraints for the initial capital cost and continued operation and maintenance of the station.

Three typical solar monitoring stations and their goals are:

F. Vignola (✉)
Solar Radiation Monitoring Laboratory, University of Oregon, Eugene, USA
e-mail: fev@uoregon.edu

F. Vignola
Luminant LLC, Denver, USA

© Springer Nature Switzerland AG 2019
J. Polo et al. (eds.), *Solar Resources Mapping*, Green Energy and Technology,
https://doi.org/10.1007/978-3-319-97484-2_3

1. A high-quality solar monitoring station that fully characterizes the solar resource and to serve as a reference facility for the development and testing of various solar resource models. A Baseline Surface Radiation Network (BSRN) station is typical of this type of facility.
2. A station to evaluate the solar resource at a potential solar facility and provide information necessary for optimum design and operation of the facility. In addition, this solar resource database will form the basis for performance estimates needed for funding of the project.
3. The third type of station is one that helps validate the facility performance and provides information necessary for forecasting system performance. This is an operational facility that requires real-time data and a high degree of reliability.

There are several initial steps that should be taken before designing the solar monitoring station. Record keeping is essential, independent of station type being considered. These records range from: the calibration history of the instruments; location of the equipment; and maintenance records and their storage. If records are maintained on a computer, regular backup is necessary to prevent loss of data. Plans should be made in advance on how these records are maintained and what position is responsible for maintaining and documenting these records.

Initial, ongoing, and final calibrations of the instruments should be performed. Factory calibrations should be validated with field calibrations because the factory calibrations are typically done indoors under controlled conditions. When instruments are used in the field, they are exposed to a combination of elements that are not easily replicated in the laboratory. Field calibrations are necessary to understand how the instrument performs in the outdoor conditions. The calibrations can either be done on-site or at facilities with capabilities necessary to provide accurate calibrations traceable to the world radiometric standard. The facilities should be structured to facilitate the calibrations. In other words, when field calibrations are conducted, convenient platforms should be available for side by side comparisons. If the instruments are exchanged out and sent to a testing facility, then the mounting for the instruments should be designed to facilitate the removal of the instruments and the substitution of replacement units. A better understanding of the changes in the solar environment can be maintained if the calibrated instruments are returned to their place in the station. Each instrument has its unique, albeit similar, performance characteristics. Keeping the same instrument in place can reduce one source of variance resulting in differences between instrument characteristics and calibration uncertainties.

Budgets for calibration of instruments and maintenance personnel should be established along with money for regular analysis of the data. As complete a record as possible should be one goal, and completeness of record is best obtained if the data are checked regularly. Software for data analysis is also a necessity. This software should include plotting routines and file-splitting features. Also, editing of the data or flagging questionable data is important. A database structure is helpful if multiple stations are established; however, this requires a programmer to create and maintain the database.

High-Quality Stations: High-quality stations, such as a BSRN station, require class A instruments, formally refered to as secondary standard instruments, an automatic tracker for aligning the pyrheliometer (direct normal irradiance [DNI] instrument) with the sun and shade disk for diffuse pyranometer (diffuse horizontal irradiance [DIF] sensor), along with a class A pyranometer for global horizontal irradiance (GHI) measurements. The station should also allow for testing of various sensors along with a comprehensive set of instruments for auxiliary measurements. The requirements for BSRN stations are available at BSRN Web site and provide useful information on the needs for a high-quality station.

Solar Resource Stations: Stations for solar resource assessment do not necessarily require the best instruments available, but high-quality instruments are recommended. The quality of instruments used at these stations is often dependent on the budget available. These stations are typically designed to last for a year or two, while data are being gathered for planning and financing. If the facility is established, more appropriate equipment can be purchased and the equipment at the site for initial analysis can be moved. If the plans call for using this equipment at the facility, then the instruments to be used at the facility should be used in the resource assessment stage.

Site Facility: The solar monitoring station at a solar electric facility is designed for real-time use and contains components that might not be considered with other types of facilities. These include sky imagers and forecasting ability.

Therefore, purpose and budget have considerable influence on the solar monitoring site. This information provides a perspective from which to view the following sections. The information in the following sections is recommendations and best practices, but the final decisions and designs are often modified by the budget and limitations at the available site.

2 Solar Monitoring Instrumentation

A good solar monitoring station should always include instruments that would enable the monitoring of the GHI, DNI, and DIF components of solar radiation. The details of the relationship between the three components are covered in Chap. 1:

$$GHI = DNI \cos Z + DIF \qquad (1)$$

where Z is the solar zenith angle. With the three components, the consistency of the measurements can be determined and any problems with collected data can be easier to identify and/or flagged for further examination.

The classification and their associated uncertainties for pyranometers that measure GHI and DIF are given in Table 2.5 (WMO 2017). The ISO 9066:2018 classification for pyranometer has classifications of A, B, and C. These are approximately equivalent to the WMO classification of high quality, good quality, and moderate quality, respectively. The ISO 9060:2018 and WMO (2017) specification are slightly different. The quality of pyranometers and pyrheliometers has improved in recent years as the demands for more accurate measurements have increased. The classification and associated uncertainties for the pyrheliometers that measure DNI are given in Table 2.4 (WMO 2017). Both the ISO and WMO specifications for solar sensors are being or have been updated, and new specification should be available by the end of 2018.

If instruments are classified as high quality, good quality, or moderate quality, their specifications should match or exceed the performance in each category. For example, a photodiode-based pyranometer has an extremely fast time constant that exceeds the high-quality specifications, but its spectral response is such as to classify the pyranometer as moderate quality. When choosing the appropriate instrument for the station, there is usually a trade-off between accuracy and funding available. In the long term, the best available equipment that fits within the budget increases the overall usefulness of the data. Table 1 provides an assessment of the pyranometer classification and the appropriate use of the instrument. If the data are to be used for evaluation and/or testing, the high-quality instruments are preferable.

The appropriate pyrheliometer classification is shown in Table 2. A new ISO 9066:2108 classification "AA" has been established, and cavity radiometers meet the specifications for this category and act as reference instruments. No such cavity radiometers have been certified for pyranometers. The achievable uncertainties in Table 2 are for extremely well-maintained instruments.

Table 1 Appropriate pyranometer use

WMO classification	High quality	Good quality	Moderate quality
ISO classification	A	B	C
Achievable uncertainty (WMO 2017)			
Hourly totals	±3%	±8%	±20%
Daily totals	±2%	±5%	±10%
Suitable applications	Working reference	Network operations	Low-cost network

Table 2 Appropriate pyrheliometer use

WMO classification	High quality	High quality	Good quality	–
ISO classification	AA	A	B	C
Achievable uncertainty (WMO 2017)				
Hourly totals	±0.4%	±0.7%	±1.5%	±3%
Daily totals	–	±0.5%	±1.0%	±2%
Suitable applications	Reference	Working reference	Network operations	Low-cost network

2.1 Solar Instrumentation

Besides the solar monitoring instruments, auxiliary equipment used in conjunction with the solar instruments needs to be included in station design. For example, if one is going to measure DNI with a pyrheliometer, a tracker is needed to aim the pyrheliometer at the sun. While manually adjusted trackers are still available, reliably automated trackers are now available and have the advantage that they do not need to be aligned two or three times a week. Some automatic trackers come equipped with a GPS or a sun-aiming device to more precisely align the tracker with the sun. Trackers equipped with GPS have very accurate information on time, latitude, and longitude. The sun-aiming devices ensure the pyrheliometers are pointed at the brightest point in the sky consistent with the calculated position of the solar disk. This provides accurate aiming even if there are minor errors in the original alignment of the tracker or a timing or a minor solar position algorithm issue. Another advantage of automated trackers is that most of them include a shade arm device that enables one to measure the DIF component of irradiance. Having DNI and DIF measurements enables one to calculate the GHI with a fair degree of accuracy. Values of GHI obtained by using Eq. 1 can often identify systematic biases in GHI values obtained from pyranometer measurements. The availability of all three solar radiation components enhances the quality control and analysis of the data. Pyranometers should be mounted in ventilators. This lessens the soiling of the pyranometer dome and can significantly reduce the amount of snow or ice that can accumulate on the pyranometer. Most of the automatic trackers use an AC power source, although some manufacturers have solar trackers that operate on DC with low power consumption (around 1–2 W). Ventilators can operate from AC or DC sources, and there is a debate as to which type of ventilator is best to use (Michalsky et al. 2017). Some ventilators also have small heaters that slightly warm the air that blows across the dome of the pyranometer. This reduces the periods when ice or frost can cover the dome.

Alternatively, there are devices such as the rotating shadow-band radiometer (RSR) that are powered by PV panels and produce GHI, DNI, and DIF data using one instrument that measures all three components. RSR requires algorithms to remove the systematic biases that are inherent in these instruments (Augustyn et al. 2002, 2004; Vignola 2006; Wilbert et al. 2015; Vignola et al. 2016). The algorithms are built into the data logger or data analysis programs and yield fairly reliable results. Since RSR needs only one photodiode-based pyranometer, a second pyranometer is often colocated to assist with quality control checks. RSRs are typically used in remote locations where maintenance is not as routinely performed. Some studies show that RSRs are less affected by soiling than pyrheliometers (Michalsky et al. 1988; Pape et al. 2009; Wilbert et al. 2015). Soiling degrades the performance of all pyrheliometers, but some may be less susceptible to some forms of soiling than others.

As understanding the performance of photodiode-based pyranometers increases, improved algorithms are being developed, although there are limits to the improvements because of the characteristics of how the RSR measurements are made (Vignola 2012).

Radiometer sensors can output digital or analog signals. The digital is recommended for big solar plants where the radiometers are distributed all along the plant, to avoid the attenuation of values over long cable runs. If the radiometers are close to the data loggers, radiometers with analog signals are recommended. The output of analog radiometers can be conditioned as voltage or current signals. For places where the distance is longer than several meters between the sensor and the data logger and where the electromagnetic interference can be a problem, using current signals from radiometers can reduce the uncertainty in the measurements.

Devices such as motors, relays, and "noisy" power supplies can induce voltages onto signal lines that can degrade the voltage sensor signal. Also, a voltage signal is susceptible to voltage drops caused by wire resistance, especially over long cable runs. An intensity signal, on the other hand, offers increased immunity to both electrical interference and signal loss over long cable runs. And most newer data loggers will accept current signals. Signals in the 4–20 mA range provide inherent error condition detection since the signal, even at its lowest value, is still active. Even at the extreme low end, or "zero" position, the sensor is still providing a 4 mA signal. If the value ever goes to 0 mA, something is wrong. The same cannot be said for an mV sensor. Zero volts could mean zero position, or it could mean that your sensor has ceased to function. In some cases, 4–20 mA sensors can be slightly costlier compared to mV sensors. But the cost difference is becoming increasingly smaller as more sensor types incorporate current output capability.

Thermopile-based irradiance sensors often produce negative voltage signals at night that are the result of the thermal offset. These negative signals can be used to evaluate the thermal offset of the sensor and should not be automatically set to zero. The nighttime values can also be used to identify noise that is picked up on the cable. If a current mode is used, one has to clearly identify the zero reading.

2.2 Circumsolar Measurement

Besides instruments used to measure GHI, DNI, and DIF, there has been increased interest in measuring circumsolar radiation, especially the circumsolar radiation that is included in DNI measurements. Instruments that measure DNI have a 2.5° field of view. However, many concentrating systems can only effectively concentrate the irradiance that comes from an area of about 0.5°, the size of the solar disk. In order to accurately estimate the performance of a CSP, the circumsolar contribution to the DNI needs to be estimated (Blanc et al. 2014).

Several different instruments are under analysis for circumsolar measurements. Any solar facility for which these circumsolar measurements are important needs to include these instruments in the facility design. Instruments that measure circumsolar irradiance need a clear field of view of the sun as it crosses the sky and should not block the field of view of other solar sensors.

The choice of instruments to use for the solar monitoring station boils down to a few basic considerations listed below:

1. The accuracy of data required
2. Type of data required
3. Budget available
4. The availability of AC power at the site
5. Level of maintenance that will be available at the site.

3 Selecting a Location

Several factors are important for the location of a solar monitoring station. Paramount is a clear field of view. An attempt should be made to find a location where there would be no obstructions that rise about 5° above the horizon. Sometimes, this condition is difficult to meet, especially if a tower for measuring wind speed is included. One should evaluate the site using an instrument or a map that plots the path of the sun across the sky for all the seasons of the year (see Fig. 1). The goal is to find the best location with minimal blockage of direct sunlight.

Other factors are also important, and sometimes choices have to be made between the best field of view and other factors. The security of the solar monitoring station is also important, especially if it is located in a remote area where there is little maintenance and vandals can damage or steal the instruments. Ease of access is important, especially if the station is to be well maintained. People who maintain the station will not be as motivated if the station takes a considerable time to visit or if it is located on a roof that is difficult to access.

Power to the site is another important consideration. If AC power is not available within a reasonable distance, then a photovoltaic-charged battery system could be used. Of course, the data logger and other equipment should be backed up when the inevitable power outage does occur. A good power backup system is important to get as complete a record as possible.

Locating the station along a dusty road is a problem. If at all possible, the location of the instruments should be in areas where dust and soiling are less of a problem. One problem that does occur at some locations is birds roosting or feeding on the instruments. Some thought should be given if this is likely to occur.

An evaluation of trade-offs is often necessary when locating a solar monitoring station:

• The proximity of a person who will maintain the station
• Availability of AC power
• Clear field of view

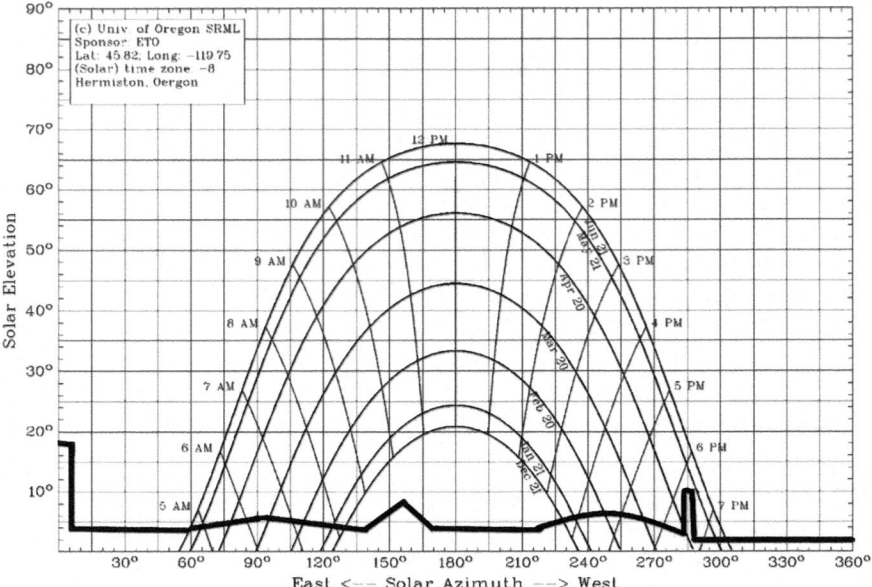

Fig. 1 Sun path chart of a site in Hermiston, Oregon. The black silhouette at the bottom of the chart is a pseudo-horizon diagram. A wind tower is represented by the tall silhouette in the north. A power pole is on the west side of the silhouette and does block the sun during part of the year. Sun path chart from the University of Oregon Solar Radiation Monitoring Laboratory's web page at http://solardata.uoregon.edu/SunChartProgram.php

- Security
- Distance from sources of airborne dust.

4 Security Recommendations for Ground Radiometric Stations

Security measures against severe natural conditions and other interferences, including theft and vandalism, should be considered. There is no specific recommendation from WMO to secure the stations from theft or vandalism. However, in this section, some recommendations are provided based on experience in different solar energy projects around the world.

The main recommendation to avoid theft and vandalism is to install the stations on private properties with surveillance, rooftops of schools and public entities, and similar facilities where the public access is restricted. Depending upon the location, security may be a significant consideration. Security is both for the protection of the site against vandalism and theft, and for the protection against the harm of would-be intruders.

If the stations are installed in the open field, at a minimum the measurement site should be well fenced against intruders, both human and animal. However, the fence will not avoid its destruction or jumping for someone with the intention to steal in the station. The fencing can prevent entrance to the enclosure of animals (cows, sheep, and similar animals) that could cause damage to the installation ranging from dropping the tower or solar tracker, to biting cables or instruments. Rodents and squirrels are notorious for gnawing on cables or unlocked cabinets. Cabinets should have locks, and cables should be in conduit to prevent this from happening. Bolts or screws securing instruments or equipment should require specialized screwdrivers or tools to dismantle.

The pyranometers and pyrheliometers should be installed in an altitude higher than the fence or any nearby obstacle to avoid losing the field of view of the instrument or reflection. Mirrors, greenhouses, or nearby white walls can be especially troubling.

Further security measures may include alarm systems, security lights (on buildings, but away from the instrumentation), and video camera systems.

Some meteorological stations have a digital signal which can generate an alarm activated by a relay. It can be installed in the door of the fence or another entryway to the station. The alarm can generate an SMS mobile text message of intrusion in the installation. This SMS can be sent to several mobile phone numbers of the people responsible for the station.

Another option to protect the station is with the installation of a security camera. The field of view should be the installation area, and it should have the capability to be activated from the control center whenever an intrusion is detected. The operator of the central control facility will receive immediately the images and take the appropriate actions depending on what is happening at the site. In some locales, special security should be considered against burrowing and gnawing rodents.

Finally, a poster in the cupboard of the station with the message ¡¡¡WARNING HIGH VOLTAGE!!!! may persuade the burglars from robbing.

In summary, the main measures to guarantee the security of the station are the following:

- Fencing the enclosure of the station
- Alarm to avoid intrusions (with detection when the main door of the fence or the door of the cupboard of the station is opened)
- Taking pictures by order from the central control after an alarm is detected.

Figures 2 and 3 are examples of well-fenced monitoring stations.

5 Station Maintenance

Regular station maintenance is essential to achieve the best quality data. This require not only a list of maintenance tasks that are performed, but log sheets that record when the maintenance was performed, what tasks were undertaken, and who

Fig. 2 Example of a meteorological and radiometric station protected with fencing to avoid intrusions. Geonica©

performed the task. When analyzing the data, log sheets help identify unusual events such as dips in readings when the instrument was cleaned, it keeps track of when problems were fixed, what instruments were at the station, and when calibrations were performed.

Maintenance tasks can be broken down into a task that is done during every visit to the station and those that are only done periodically. Maintenance tasks are shown in Table 3 for tasks done with every site visit and Table 4 for tasks that are done periodically. These tables can be used to create log sheets that record activities at the station. Many activities can be accomplished by a comment section, but the more information contained in the log sheet, the more tasks will actually be accomplished.

There are many maintenance tasks that should be performed periodically but do not have to be done on a daily basis. For example, the levels of the instruments should be checked periodically. Structures or platforms can settle or shift so that the level of instruments needs to be tracked. Some tasks like checking the level of a pyranometer or the state of the desiccant may require removing the shield. Unless there is a reason to check the level, such as after a severe store, levels do not change rapidly. Therefore, only periodic checking is necessary. The same period should also serve to check the state of the desiccant. Table 4 lists tasks that should be done periodically.

Fig. 3 Example of a meteorological and radiometric station protected with barbed fencing to avoid intrusions. Geonica©

An example of a maintenance log sheet is shown in Fig. 6. Maintenance log sheets on tablets or laptop are becoming more popular as the results can be filed electronically and sent to a centralized facility.

6 Data Logging

The solar monitoring station requires a versatile data acquisition system to gather the variety of signals from sensors and instruments. The data logger should also be able to operate over diverse climate conditions experienced at the site. Specifically, it should be able to operate over a wide range of temperature and humidity that are expected while maintaining accurate readings. The precision and accuracy of the data logger should be compatible with the instruments used. In addition, the data logger should be able to download data to a central location for analysis, use, and archival.

Considerations for a data logger:

- Operate and maintain accuracy in a temperature range from −40 °C to +50 °C.

  - The local condition may require a different range. Be sure that the data logger meets local conditions.

Table 3 Regular site visit maintenance tasks	**Record date and time of maintenance**: Sign or initial the report
	Physical station condition: Check the general state of the station. Record weather conditions such as rain, snow, or sun. If something is amiss, such as animal intrusions, write that in the comment section. Note the condition of the wind tower, fencing, and supports
	Check instrument alignment: Record the alignment of the pyrheliometer(s) and shadow disks on the shaded instruments (see Fig. 4 as an example)
	Clean sensors: Record any problems such as bird droppings on the dome
	Check ventilators while cleaning: Record and fix any problems observed
	Record activities at the station: Examples are calibrations, painting, mowing the lawn

Table 4 Periodic maintenance checks	**Check level of instruments**: Make sure the leveling bubble is centered (see Fig. 5)
	Check desiccant: Replace if necessary
	Sun tracker: Level, fixing, and orientation of the solar tracker should be checked. The base should be in good conditions
	Data logger: Check data logger and associated communication equipment. Check the panel LED lights to confirm operation condition of equipment
	Solar PV panel: PV panel should be firmly fixed with good orientation and cleaned to remove any moisture or soiling.
	Cables and equipment connectors: All cables should be checked to ensure they are well connected and their exterior plastic jackets are in a good state
	Battery: Battery should be in a good state. Check that connections with cables are OK and properly greased to avoid any bad contact. Check that there is no loss of liquid from the battery

- Can withstand a wide range of relative humidity.

  - A climate-controlled box is an option. However, one must be aware of condensation on the interior of the box and electronic instruments should be away from areas where condensation settles or drips on the equipment.
  - Desiccant packs or ventilation should be considered.

- Ability to handle the number of instruments at the station.

  - The data logger should be versatile and be able to hand a variety of inputs and scales.
  - The ability to handle pulses can be very useful.

Fig. 4 Sunspot on target is slightly low and to the left. This is well within the tolerance of the pyrheliometer, but if the sunspot moves across the black ring of the target, the instrument needs to be realigned. Specification of the tracker indicates when realignment is needed

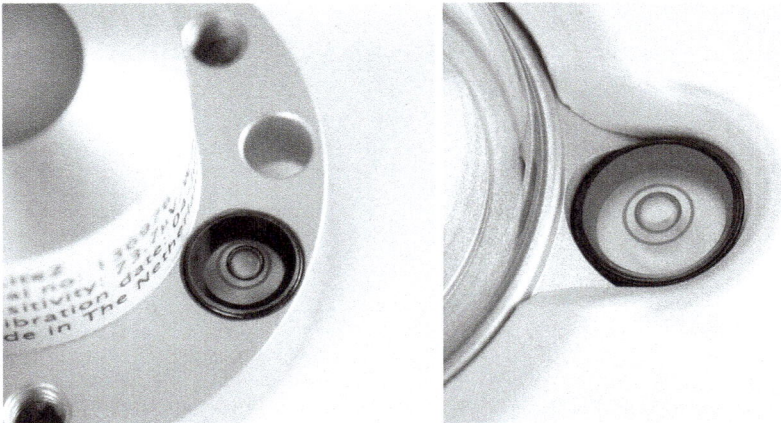

Fig. 5 The instrument on the level shows the bubble in the level to one side. The instrument should be releveled. The instrument on the right has the bubble inside the circle of the level. It is hard to get the bubble more centered than this example. One should test the sensitivity of the bubble level

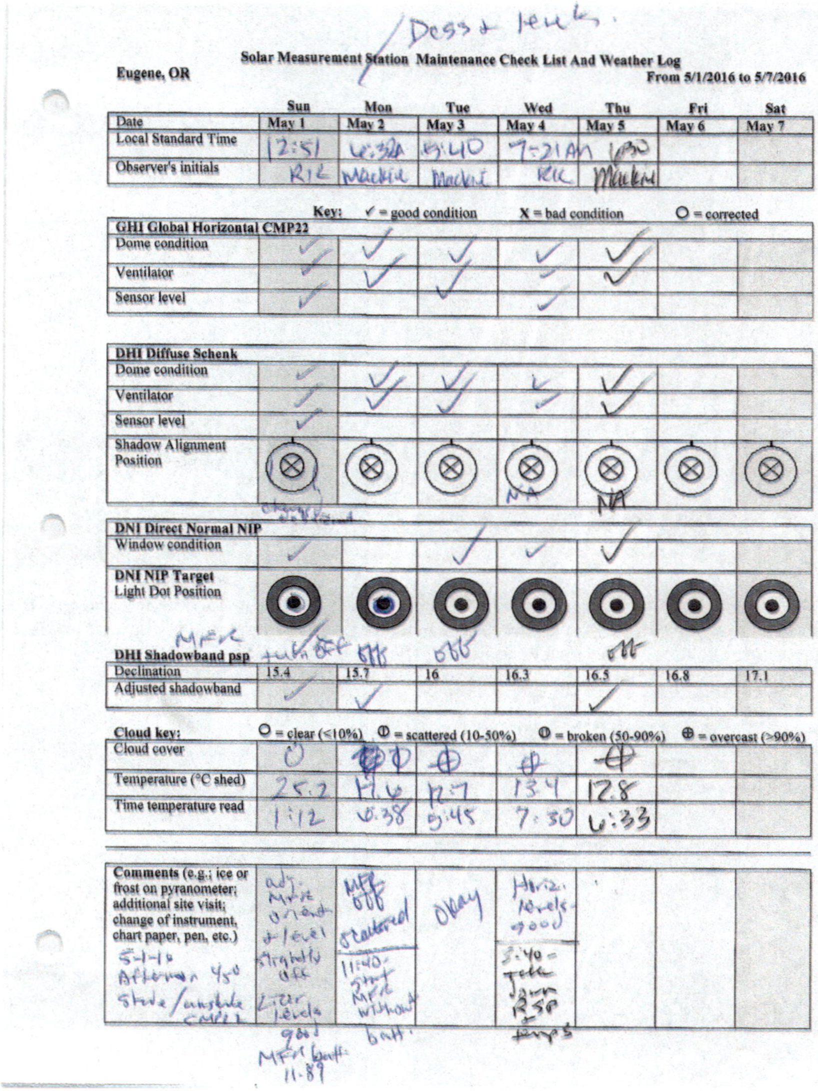

Fig. 6 Image of a log sheet used at the UO SRML station in Eugene. The number of items on the maintenance log sheet can be customized for a specific station. On this log sheet, there are two diffuse sensors: one with a shade disk and one with a shadow band. The position on the shadow is shown in the row labeled shadow alignment position. Several days were overcast, and no shadow could be observed. The MFRSR instrument was having problems at the time, and an added line with the label MFR was used to indicate the problem

- The ability to download data and the ability to modify the embedded program from a central location.

  - Communications with the data logger are important. In addition, to be able to access the information on the data logger locally, communications to remote locations are very important. These communications can be handled over the Internet, via cell phone, or other communication devices.
  - Error-checking code for communications ensures reliable data downloading.
  - Downloading the data at least daily helps ensure a more complete dataset.

- The data logger should have enough memory to hold at least several days' worth of data.
- The ability to handle a variety of time intervals for sampling and registering data is useful. Data loggers now have the ability to collect and store one-minute data from a large number of instruments. The ability to generate sub-minute data is useful during calibrations and may be needed to address future needs.
- Data loggers should sample every second or two to produce integrated values.
- The accuracy of the data logger should be 0.5% or preferably better. Any uncertainty associated with the data logger adds to the uncertainty in the measurements, and for the least total uncertainty, the uncertainty associated with the data logger should be well below the uncertainty associated with the instruments.
- A programmable data logger is most useful and enables one to change configurations if the need arises as it often does.
- The ability to calibrate the data logger is important. While data loggers can be sent back to the factory for calibration, it is useful if one can calibrate the data logger while it is operating.
- The data logger should be able to operate on batteries. This maintains operation when the power goes down.

  - The data logger should be able to automatically restart itself if the power fails.

- Data loggers are liable to remain in the field for long periods. Therefore, a sturdy data logger is recommended.
- Data loggers should have a good clock that can be rest remotely or by a GPS attached to the data logger. Clocks drift with time, and the data logger time should be checked on a regular basis.
- Documentation that comes with the data logger should be comprehensive and easy to read.

Some data loggers have associated software for programming the data logger, downloading data, and/or handling the data files produced. This software may also have features that can check on the communications with the data logger. Data loggers that come with software packages that are easy to learn and use are the most useful. Clear documentation of the software is also very important.

If several stations are in the network, then a backup spare data logger comes in handy for field calibrations and backup if one data logger fails. This reduces downtime and also enables testing of the problem that may arise.

If climate-controlled areas are available, more elaborate data loggers may be an option. High-quality data loggers are useful for sites with many instruments or sites that perform calibrations periodically.

7 Measurement Intervals

Interest has moved to shorter and shorter time intervals as storage capabilities have increased and the ability to measure short-time data has improved. Utilities for years have used 15-minute data as standard. Five-minute and one-minute data have become common for engineering studies and a better understanding of how systems operate during periods of rapidly changing irradiance. Very short time intervals in the millisecond range have been used to study the variation of the output of photovoltaic systems. Some studies of thermal shock use data gathered over several seconds, but such data are not widely generated possibly because the thermopile-based sensors have response times that are several to tens of seconds.

Currently, the one-minute dataset satisfies most needs and three-minute average data can generally produce just as useful results. Fifteen-minute data files do not exhibit enough detail to fully characterize the solar resource. Models do exist that can use fifteen-minute or hourly data files to generate one-minute data files, but measured data are more precise than modeled data even when they exhibit the same statistical distributions.

8 Auxiliary Meteorological Measurements

In addition to irradiance measurements, most solar monitoring stations have an array of other sensors to measure many other meteorological parameters. These other sensors range instruments that measure ambient temperature to anemometers to measure wind speed. A comprehensive discussion of some of these instruments can be found in WMO (2017) and Vignola (2012). For analysis of a site's solar electric potential, the following measurements are useful:

1. Ambient temperature measurement

   a. A radiation shield or aspirator is useful to obtain the most accurate temperature measurements.
   b. The sensor should be 1.2–2 m above the surface, and a grass surface is preferred for consistency. Temperature measurements are often made above a variety of surfaces ranging from desert sands to concrete pads. In many locations, it is difficult to maintain a well-groomed grass surface.

Therefore, one must keep in mind any limitations or biases in temperature measurements. However, have the surface similar to the surface of a solar electric facility will probably yield a more accurate assessment of the temperature the system will face.

2. Relative humidity sensors

   a. Relative humidity sensors should also be housed in a radiation shield or aspirator and are often colocated with the temperature sensor. As with temperature sensors, the relative humidity sensor should be between 1.2 and 2 m above the ground.
   b. Relative humidity sensors require calibration from time to time, and the range of relative humidity values should be checked periodically to make sure the instrument is working well.
   c. If the station is on a roof of a building, the sensor should be located away from vents or other structures that will affect the relative humidity.

3. Wind speed and wind direction

   a. Useful wind measurements require wind towers, and standard wind towers are about 10 m tall. The standard altitudes to measure wind speed are 3 and 10 m. These wind towers should be located away from the path of the sun across the sky (north in the northern latitudes and south in the southern latitudes) (see Figs. 2 and 3).
   b. The wind tower itself should not be highly reflective, and a dull metal surface works OK.
   c. One should be sure that the guy-wires do not block the sun's path from the irradiance sensors.
   d. One should pay attention to maintenance of the anemometer and wind vane. Bearings need replacement, and one should be able to lower the wind tower to do this.
   e. Calibration of wind sensors is also important and account should be taken of how field calibrations are to be conducted for these sensors before final designs for the station are made.
   f. Signal cables connecting the sensors to the data logger should be in conduit, but cables with weather-resistant jackets can work if there are no potential rodent problems. In addition, the cable runs should not present a trip hazard.

4. Pressure sensors

   a. They can be located where their access to the atmosphere is free.
   b. The tube that goes to the sensor should be such that they do not let rainwater to enter and as far away from sources of dust as possible.

5. Rain gauges

   a. They can provide useful information related to the maintenance and cleaning of mirrors and/or PV arrays.

b. Rain gauges can be mounted on wind towers or other structures.
c. They should be mounted high enough so that drifting snow does not affect the gauge, but not so high that the gauge cannot be easily cleaned or maintained.

All sensors need periodic checks on their calibrations. This should be done when irradiance sensors are calibrated. For quality stations, irradiance instruments should be calibrated annually, but calibrations every other year are acceptable. Other annual maintenance should be conducted during site visits. For example, if debris clogs the filter or screens on ventilators, it should be removed and the filter or screens are cleaned.

Regular cleaning helps maintain the quality of the measurements. For the most confidence in the data, irradiance sensors should be cleaned every working day. Of course, records of maintenance are important and notebooks or computer records should be kept of these activities.

9 Other Useful Sensors

9.1 Pyrgeometers

Pyrgeometers that measure sky temperature and downward long-wave radiation are useful additions to a solar monitoring station. Pyrgeometers require two or three data logger channels to monitor all the sensors associated with the instrument. In addition to measuring sky temperature and downward long-wave radiation, data from a pyrgeometer can help identify and quantify thermal offset effects that add uncertainty to thermopile pyranometer measurements. The more recent designs for pyranometers help minimize this effect, but the thermal offset still can present a problem.

It is useful to mount pyrgeometers on an automatic tracker with a shading disk to reduce the effect of direct sunlight on the sensor. Some pyrgeometers are better than others operating in direct sunlight, but keeping the pyrgeometer out of direct sunlight is an optimum solution. Pyrgeometers should also be mounted in ventilators just like pyranometers. The ventilator keeps dust, snow, or ice from building up on the dome and helps maintain the instrument at ambient temperature.

9.2 Net Radiometers

Measuring the up-welling short- and long-wave radiation is also important for studying the earth's water balance. This net radiation (the difference between down-welling and up-welling radiations) can be measured by devices specifically designed for this purpose, but this information can also be measured with downward-facing pyranometers and pyrgeometers. Leveling of the instruments is important, especially near sunrise and sunset. Instruments measuring up-welling radiation can be mounted on the wind tower well below the wind sensors. The instruments should be mounted at least a meter from the tower and the further from the tower the better. One has to be careful that the structure is stable under wind conditions likely to be experienced at the station. It is necessary to take these sensors off the wind tower to calibrate them.

9.3   Spectral Measurements

There are two methods used to obtain spectral measurements. One is to use an instrument like a sun photometer that measures the solar spectrum over a narrow band. These instruments are often used to determine the atmospheric aerosols. One can also use this data with spectral models to obtain an estimate of the solar spectrum.

Another way to measure the solar spectrum is the use of spectroradiometers that are specifically designed to measure the solar spectrum over a wide range of wavelengths. These instruments are similar to pyranometers and pyrheliometers except that they measure the solar spectrum instead of the integrated or broadband irradiance. The exact precision of the wavelength measurements depends on the instruments and the wavelengths involved. Most spectral radiometers use silicon-based sensors and measure the solar spectrum in the 320–1050 nm range. Instrument measuring the spectrum in other ranges are also available. Spectroradiometers are several times more expensive than broadband pyranometers and pyrheliometers, but they can be incorporated into the station design much like other radiometers.

Because of the expense, there is a need to justify the inclusion of these instruments into the station. For research stations and those stations involved in testing the performance of photovoltaic modules, the use of spectroradiometers is easier to justify. One also has to keep in mind that these spectroradiometers must be calibrated periodically. Because these calibrations require comparison with a certified standard lamp that has a limited lifetime, the best calibrations can be expensive. Spectroradiometers should be calibrated about once every other year. A calibrated spectroradiometer can be used to characterize a non-certified lamp. This non-certified lamp can be used to check that the calibration of the spectroradiometer is not changing significantly. However, spectroradiometers do require certified calibrations, and budgets should include funding for calibrations.

Typically, spectroradiometer calibrations are performed with the sensor perpendicular to a standard lamp. This method is well suited for spectroradiometers that measure the direct normal spectrum. However, for spectroradiometers that measure global spectral irradiance, a shade/unshade calibration is needed to determine the deviation from a true cosine response. Therefore, if one has a spectroradiometer to measure the DNI spectrum and one to measure the GHI spectrum, then the DNI spectroradiometer can be calibrated against the standard lamp and the GHI spectroradiometer can be calibrated by the shade/unshade method using the calibrated DNI spectroradiometer as the reference DNI value.

EKO is testing a prototype rotating shadowband spectroradiometer. This instrument is able to measure the GHI, DNI, and DIF spectrums using one spectroradiometer. Tests will determine how well this instrument is able to measure the three spectral components.

The configuration of the spectroradiometers should be much like that for broadband radiometers. These instruments are usually heavier than broadband instruments. Before mounting spectroradiometers on a tracker, the specifications of the tracker should be checked to ensure that it can handle the heavier instruments.

If one has a DNI and a GHI spectroradiometer, the automatic tracker should be strong enough to support the spectroradiometers, a pyranometer, a pyrheliometer, and a pyrgeometer. The DIF spectroradiometer, the pyranometer for DIF, and the pyrgeometer should all be on the tracker under shade disks. With this configuration, the DIF and the spectral DIF can be measured. The GHI spectrum can be obtained by adding the DNI spectral values projected onto the horizontal surface and the DIF spectral values. The same can be said for the broadband instruments. Of course, a broadband pyranometer can also be mounted nearby away from the tracker along with a spectroradiometer that measures the GHI spectral values.

9.4 Visibility Sensors

For concentrating solar power facilities that use heliostats to reflect sunlight on a central tower, visibility is an important parameter. Aerosol and dust in the atmosphere can significantly affect the light incident on the central receiver. While visibility sensors are used to calculate the visual range, visibility sensors first measure the extinction coefficient of light generated by high-intensity xenon strobe directed at a volume of air close to the sensor, typically a photodiode. From the photodiode reading, the extinction coefficient can be obtained. This extinction coefficient is used to calculate the meteorological optical range. The meteorological optical range is the visibility range of a human observer and relates to only the visible portion of the spectrum. For concentrating systems, the extinction over all wavelengths is of interest.

Visibility sensors are fairly large and require AC power. They should be located away from areas where foot or road traffic will stir up dust. They should also be located north of the solar sensors and low enough so that they do not block any significant part of the horizon.

9.5 Sky Imagers

Most irradiance sensors provide an averaged view of the sun or the sky and do not provide any information on the distribution of the clouds across the sky. There are a variety of cloud distributions that can produce the same DIF value. There has been a desire to have more detailed information about the distribution of clouds, especially when forecasting irradiance in the short term. This need has led to the development and improvement of sky imagers. These devices take images of the sky usually with the aid of a fish-eye lens to capture images of the cloud cover. The images are digitized and analyzed to produce a map of the clouds across the sky. This information is fed back to a central location where they can be used to evaluate cloud type, thickness, and movement, three parameters that affect future irradiance.

Like a pyranometer, the sky imager needs a level platform, a clear field of view, and its dome should be cleaned regularly. In addition, the sky camera needs the power that can be supplied by an AC connection or a photovoltaic panel and a battery. The ability to communicate with the central location is essential, and this can be done either through Wi-fi, Ethernet, 4G, or another telecommunication technology. For large facilities that use sky cameras for forecasting, more than one sky imager may help to provide timely estimates of cloud cover, especially if the clouds come from a variety of directions.

10 Ground and Shielding

A solid electrical ground is important for lightning protection and signals noise reduction. A good ground is obtained by driving a copper-coated steel ground rod 1.5–3 m in length and between 10 and 20 mm in diameter into the soil. In most cases, a heavy pipe with an end cap can be used to drive the ground rod into the soil. In areas where lightning strikes are frequent, a more robust grounding strategy might be needed. Morrison (1998) is a good source for detailed information on grounding and shielding.

The ground rod should have at least one lug that makes secure contact with the grounding rod and a place to secure wires from all masts and metal poles. Ground wires should be 10 to 12 gauge (2.0–2.6 mm in diameter) and connected in a manner that will allow a clear path for lightning to follow to the ground. A lightning finial should be considered for areas subject to frequent lightning strikes. Surge protectors such as metal oxide varistors or avalanche diodes can be used for lightning protection. These surge protectors degrade over time, and manufacturers' recommendations should be followed to determine when these surge protectors need to be replaced (Vignola 2012).

All sensitive equipment should have ground wires that connect to the ground rod. To avoid ground loops, sensors should be isolated so that they have only one path to ground. For example, the data logger should be grounded (have a wire connecting the data logger to the ground rod). A pyranometer has a signal cable that goes from the pyranometer to the data logger. This wire should be a shielded twisted pair cable and connected to the data logger as a differential measurement. The shield in the cable should be connected to the data logger ground. The shield should not be connected to the pyranometer ground to avoid a ground loop. By isolating low-voltage signals from the ground, some of the noise is eliminated.

It is best to isolate signal cables from power cables. They should be run through separate conduits. Electromagnetic radiation from power cables can be picked up on signal cables. Using differential measurements helps reduce this interference, but even a little noise can be rectified at any junction. If power and signal cables have to physically cross, they should cross at right angles to minimize the induced interference.

11 Physical Layout of a Solar Monitoring Station

The design of a solar monitoring station has long-term consequences in the quality of the data, the efficiency in maintaining the station, and the ability to adapt to future changes. When designing the layout of a solar monitoring station, several factors need to be taken into consideration.

1. What is the planned lifetime of the station?
2. Type of instrument used for DNI measurements and other solar irradiance measurements.
3. Source and location of power.
4. Security of location.

   a. Is a security fence needed?
   b. If so, how high does it need to be?

5. Will wind measurements be made?
6. What other measurements are contemplated?
7. Ease of maintaining the station.
8. Ease of calibrating the instruments.

   a. Will field calibrations be made or will the sensors be sent to a testing facility for testing?

Photographs of three solar monitoring stations are shown to provide a glimpse of different station types. Figure 7 is of the Solar Radiation Research Laboratory (SRRL) at NREL, Fig. 8 is the reference solar monitoring station on the roof of Pacific Hall at the University of Oregon in Eugene, Oregon, and Fig. 9 is an AgriMet station used for monitoring solar and meteorological parameters for agricultural use.

Fig. 7 Solar radiation monitoring platform of the Solar Radiation Research Laboratory (SRRL) at NREL looking to the west. A variety of irradiance instruments are on the left. Toward the west end of the platform, several automatic trackers make DNI and DIF measurements. A second set of shelves on the right support pyranometers undergoing testing or calibration. The platform and tables are gridded to minimize the accumulation of snow. The platform is on the west side above an office building for the SRRL complex. Photograph courtesy of Tom Stoffel

The SRRL solar monitoring station at Golden, Colorado, USA, is designed to accommodate a wide of variety of new and existing solar instruments and to test, compare, and calibrate a large number of instruments at one time. The SRRL office building and platform to the west and above the office building provide easy installation and maintenance of a large number of instruments. This complex replaced the original SRRL that performed similar activities since the early days of the Solar Energy Research Institute, now the National Renewable Energy Laboratory. This facility was designed by people with years of experience in monitoring solar radiation, and the building has offices for those staffing SRRL. If one is considering building such a facility, one should visit a facility like SRRL to get a feel for the strengths and difficulties associated with such a facility.

Solar Radiation Monitoring Laboratory (SRML), University of Oregon, Fig. 8, is designed for making a wide variety of solar and other meteorological measurements along with testing equipment used in the SRML regional network and calibration of instruments. This facility was put together on the roof that used to house a small observatory that moved to the deserts of Eastern Oregon at Pine Mountain

Fig. 8 Pyranometers, pyrgeometer, and pyrheliometers mounted on an automatic tracker at the UO SRML station in Eugene, Oregon. The pyranometers are in ventilators. The tilted pyranometer on the back shelf had a shade ring to provide a small but uniform ground-reflected component. Data loggers, AC power, and Internet connection are in the shack in the lower background. The very back shelf has room for other instruments to be calibrated. All pyranometer sensors are at the same level. The wind tower with anemometer and pyranometer to measure ground reflection, not shown in this photograph, are to the northwest out of the path of the sun. Photograph taken by Rich Kessler for the UO's Solar Radiation Monitoring Laboratory

for clearer skies. The SRML facility was built piecemeal over time on a relatively scant budget. Unless roofs are part of a grander scheme, they are not ideal locations for solar monitoring stations. Fortunately, access and power to the roof were established when the observatory was located there.

The AgriMet station, Fig. 9, is one of the number of similarly structured stations in the AgriMet network (https://www.usbr.gov/pn/agrimet/). It is designed to be self-sufficient in power and located in areas near agricultural fields. The stand is designed for a limited number of sensors and would enable the instruments to be quickly calibrated once a year in the field. Maintenance is limited, although people near the station can be asked to do the limited cleaning.

The following is a sample layout for a good-quality monitoring station. The station parameters are that it will monitor GHI, DNI, and DIF irradiation on an automatic tracker. In addition, it will have a 10-m wind tower and monitor wind speed, wind direction. Also, ambient temperature, relative humidity, rainfall, and barometric pressure will be monitored. It will have AC power.

Fig. 9 Setup of an AgriMet station looking from the southeast. Instruments are mounted on a frame. Other sensors are in the enclosure. This station is used by an agricultural research center in Kimberly (Twin Falls), Idaho. Photograph taken by Rich Kessler for the UO's Solar Radiation Monitoring Laboratory

A generic layout for a solar monitoring station using an automatic tracker and a wind tower is shown in Fig. 10. This is a sample layout that shows the basic components and relative location of instruments. The automatic tracker in this schematic has a pyranometer for GHI measurements, a pyranometer with a minimal thermal offset for the DIF measurements, and a pyrgeometer for measuring down-welling long-wave radiation and sky temperature. A shade disk blocks direct sunlight from the DIF pyranometer and the pyrgeometer. The pyrheliometer is mounted on the side of the tracker. Objects around the automatic tracker should be located far enough away so that they would not interfere with the motion of the automatic tracker. The data logger and power source would be located in an enclosure behind the automatic tracker. The cables from the instruments on the automatic tracker should come to a fixed location before they go down in a bundle to the ground and then to the enclosure with the data logger. Make sure that the cables from the instruments to the fixed point on the automatic tracker shelf have some slack and a drip loop to inhibit water running down the cable into the wire bundle. The wrapping for the wire bundle should be easily removed in case the signal wires are replaced. Also, ensure that the bundle of signal wires has enough slack so that the

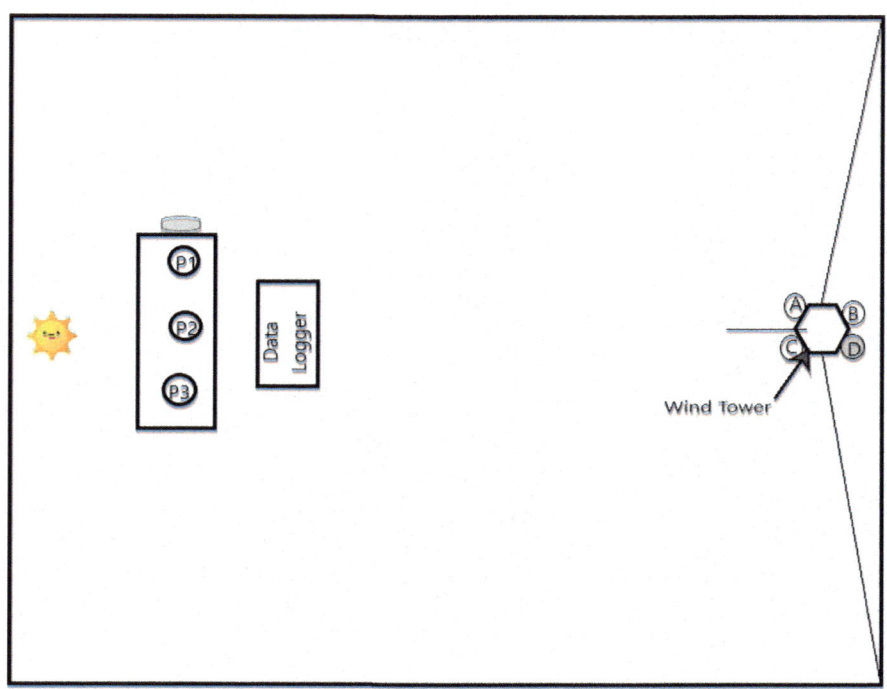

Fig. 10 Sample layout for a solar monitoring station. This assumes an automatic tracker for measuring the solar irradiance components and a wind tower for wind speed and direction. The black outline is the fence on the perimeter of the site. This fence can be as high as necessary for security. All instruments should be mounted above the fence level. The three instruments on the top of the tracker are a pyranometer (P3) for GHI measurement, a pyranometer with minimal or no thermal offset (P2) for DIF measurements, and a pyrgeometer (P1) for measuring sky temperature and downward long-wave radiation. The DIF pyranometer will be shaded by a shade disk as well as the pyrgeometer. The pyrheliometer for DNI measurements is on the side of the tracker below the shelf. Behind the automatic tracker is an enclosure that contains the data logger and connections to the AC power. The wind tower is on the north side of the enclosure with guy wires securing the tower. The wind speed and wind direction sensors are at the top of the tower. Also mounted on the tower are four instrument packages. They are the temperature and relative humidity sensor (A), the barometric pressure sensor (B), a rain gauge (C), and possibly a net radiometer (D). A lightning rod would be located on the top of the tower

automatic tracker can swing freely without putting stress on the signal wires. Any obstruction that might snag the signal wire bundle should be removed. The wires to data logger enclosure should come up from the bottom, and at no point should water be able to run down the wires into the data logger enclosure.

Ventilators are recommended for the instruments on the tracker's shelf. The power cable to these ventilators should be separated from the signal wires. Again, care should be taken to show that water will not get into the enclosure with the power strip. Also, these cables should also be long enough so that they do not interfere with the motion of the tracker.

Inside the enclosure, the data logger should be mounted above the bottom of the enclosure and the data logger inputs should be oriented to minimize the surface area possibly exposed to dripping condensation. If a battery is being charged in the enclosure, vents should be included to prevent the buildup of gas. A thermostatically controlled heater may be added to the enclosure if extreme cold temperatures are experienced.

A shelf should be installed on top of the enclosure on which reference instruments can be mounted for field calibrations. This shelf should be high enough so that the calibration instruments are at the same height as the instruments are calibrated.

The top of the tracker shelf should be such that a person of normal height can look down, see the domes of the instruments, and easily clean the instruments. If the shelf is much above ground level, stairs can be included so that the person maintaining the instruments has easy access to them. Remember that the desiccants need to be checked periodically. Therefore, the instruments need to be accessible from the front and the back.

The wind tower is located on the north side of the monitoring enclosure (Fig. 10). The guy-wires should be run so that they do not block direct access to the solar disk by the solar sensors. The wind direction and wind speed sensors should be at the top of the tower or on an arm near the top of the tower. The tower should be stable under most wind scenarios. A lightning rod should go on the top of the tower, and a lightning finial can be mounted there also.

The wind tower is also a good place to mount other meteorological instruments. A ventilator for the ambient temperature and relative humidity sensor can be mounted on an arm from the tower (labeled A in Fig. 10). The rain gauge and barometric pressure gauge can also be mounted on another lever arm from the tower. Higher on the tower, a net radiometer can be mounted on an arm from the tower. It is important to secure the tower and make sure that the instruments level and do not oscillate. When installing guy wires make sure they are located in places that don't present trip hazards.

To maintain the wind instruments, the tower has to be lowered every year or two. A tower on hinges works well, and one has to ensure that the tower can be lowered without interfering with other instruments. For example, don't place the hinges on the south side of the tower as this would likely cause the top of the tower to hit the data logger enclosure or tracker. An opening in the site fencing may be necessary to adequately lower the tower.

The main lightning protection should be nearer the wind tower because that is the likely area to be struck by lightning. Make sure all the grounding wires connect securely to the ground rod. It may be necessary to get a second lug to securely attach all the wires.

References

Augustyn J, Geer T, Stoffel T et al (2002) Improving the accuracy of low cost measurement of direct normal solar irradiance. In: Proceedings of the solar conference, pp 329–334

Augustyn J, Geer T, Stoffel T et al (2004) Update of the algorithm to correct direct normal Irradiance measurements made with a rotating shadow band pyranometer. In: Campbell-Howe R, Wilkins-Crowder B (eds) Proceedings of the American solar energy society. American Solar Energy Society, Boulder, Colorado, USA

Blanc P, Espinar B, Geuder N et al (2014) Direct normal irradiance related definitions and applications: the circumsolar issue. Sol Energy 110:561–577. https://doi.org/10.1016/j.solener.2014.10.001

Michalsky JJ, Perez R, Stewart R et al (1988) Design and development of a rotating shadowband radiometer solar radiation/daylight network. Sol Energy 41:577–581. https://doi.org/10.1016/0038-092X(88)90060-6

Michalsky JJ, Kutchenreiter M, Long CN (2017) Significant improvements in pyranometer nighttime offsets using high-flow DC ventilation. J Atmos Ocean Technol. https://doi.org/10.1175/jtech-d-16-0224.1

Morrison R (1998). Grounding and shielding techniques, 4th ed. New York: John Wiley & Sons

Pape B, Batlles J, Geuder N et al (2009) Soiling impact and correction formulas in solar measurements for CSP projects. In: 15th SolarPaces international symposium. SolarPaces, Berlin, Germany

Vignola F (2006) Removing systematic errors from rotating shadowband pyranometer data. In: Campbell-Howe R (ed) Proceedings of the American solar energy society. American Solar Energy Society, Boulder, CO

Vignola, F, Michalsky JJ, Stoffel T (2012) Solar and Infrared Radiation Measurements, CRC Press, Taylor & Francis Group, p 394

Vignola F, Derocher Z, Peterson J et al (2016) Effects of changing spectral radiation distribution on the performance of photodiode pyranometers. Sol Energy. https://doi.org/10.1016/j.solener.2016.01.047

Wilbert S, Geuder N, Schwandt M et al (2015) Task 46: best practices for solar irradiance measurements with rotating shadowband irradiometers. IEA SHC Sol Updat Newsl 62:10–11

WMO (2017) Guide to Meteorological Instruments and Methods of Observation 2014 updated 2017, World Meteorological Organization, WMO No. 8.

Chapter 4
Quality Assurance of Solar Radiation Measurements

José P. Silva, José L. Balenzategui, Luis Martín-Pomares, Stefan Wilbert and Jesús Polo

Abstract Solar radiation measurements are necessary for every solar energy project to evaluate solar resource assessment studies. Quality assurance of solar radiation measurements is essential in all the stages of solar resource analysis. Model development and assessment, improvement of models and characterisation of the uncertainty, among others features, depend strongly on the accuracy and quality efforts in designing and operating the solar radiation ground station. This chapter summarises several aspects involved in ensuring the quality of solar radiation measurements, addressing the requirements for instrument selection and the quality methods applied to solar radiation data. This chapter has been written intended to be useful for project or group leaders involved in solar resource assessment and not a rigorous scientist text since the chapter presents a summary of many manuals for quality control.

J. P. Silva (✉) · J. L. Balenzategui · J. Polo
Photovoltaic Solar Energy Unit, Renewable Energy Division (Energy Department)
of CIEMAT, Avda Computense 40, 28040 Madrid, Spain
e-mail: josepedro.silva@ciemat.es

J. L. Balenzategui
e-mail: jl.balenzategui@ciemat.es

J. Polo
e-mail: jesus.polo@ciemat.es

L. Martín-Pomares
Qatar Environment and Energy Research Institute, Hamad Bin Khalifa University,
P.O. Box 5825, Doha, Qatar
e-mail: lpomares@hbku.edu.qa

S. Wilbert
Deutsches Zentrum für Luft- und Raumfahrt (DLR), Institute of Solar Research,
Ctra. de Senés s/n km 4, Apartado 39, 04200 Tabernas, Spain
e-mail: Stefan.Wilbert@dlr.de

© Springer Nature Switzerland AG 2019
J. Polo et al. (eds.), *Solar Resources Mapping*, Green Energy and Technology,
https://doi.org/10.1007/978-3-319-97484-2_4

1   Introduction

Quality of solar radiation measurements comprises multiple points of view. It is connected with many other organizational resources like infrastructure, personnel, facilities, equipment, management, and procedures and training. The concepts concerning control over the quality of measured data and quality control typically have the following parts (NREL 1993):

- Control in preparations for data collection (selection of the best location for the station, the instruments, calibration and installation of the equipment and radiometric instruments).
- Control during the measurement process (inspection, calibration and maintenance of instruments).
- Control during the transmission and recording of numerical values (data acquisition systems, data archival and subsequent management).
- Controlling quality improvements by limited retrospective enhancement of measurements in cases of obvious and rectifiable mistakes.

The implementation of quality systems exceeds the design of technical tasks, isolated from the rest of the measuring system. It also depends on the coordinated work of the meteorological network, the data centre, the testing and calibration laboratory and the organization itself and its policies. To assure the desired level of quality, organisations are encouraged to establish a global quality management system, which implements the appropriate management mechanisms and technical procedures within a coherent framework, in full agreement with international standards of quality. The aim is that solar radiation measurements are accurate, representative and adequate for the use given.

According to this approach, interested organisms can obtain a recognized certification, expressing their commitment to quality standards. In this sense, ISO 9001 (2015) describes the minimum requirements for a quality management system "when an organization needs to demonstrate its ability to consistently provide products and services that meet customer and applicable statutory and regulatory requirements", following ISO's statements.

Also, (ISO 17025 2017) establishes the general requirements for the competence, impartiality and consistent operation of laboratories, which includes both management and technical requisites. It applies to all organizations performing laboratory activities, regardless of the number of personnel. In this way, (ISO 17025 2017) accreditation extends the (ISO 9001 2015) certification scope, describing additional requirements to assure technical competence besides the managerial ones. Obtaining an ISO 9001 certification (i.e. implementing an effective quality system) is a valid path to guarantee the fulfilment of the management requisites to achieve the (ISO 17025 2017) accreditation.

Itemizing the underlying criteria defined within the frame of a quality management system (QMS), these are usually specified regarding requirements applied to the whole organization or subset and can be grouped as follows:

- General requirements, based on establishing a comprehensive set of good practices, ruled by impartiality and confidentiality principles.
- Structural or organizational requirements, based on developing a coherent system, identifying the responsibilities and duties of all personnel involved in the system. All subsequent changes in the QMS must guarantee that its integrity is preserved.
- Human resources. These requirements comprehend the evaluation of competence and experience, and also the design of training plans for all the personnel involved. The success of a quality system regarding efficiency is based on adequate capacitation and training programs for technicians and managers, improving knowledge and commitment. These can be implemented in the form of courses and seminars.
- Material resources. Concerning the infrastructure, facilities and equipment, requirements lead to establishing standardised procedures concerning acceptance terms, handling and use of equipment and installations, transport requirements, labelling criteria, preservation of traceability of reference device calibration, testing or calibration methods, verification and maintenance procedures, declaration of equipment status, etc.
- Processes. System processes must be identified and coordinated. There should be specific procedures concerning requesting forms, estimates, contracts and agreements, validation of measuring methods, manipulation of equipment, records, uncertainty estimation, final reports, data and information management, complaints, nonconformities, etc.
- Requirements of the management system focused on the control of documents and records, design and implementation of preventive, corrective and improvement actions, management reviews, internal and external audits, evaluation of risks and opportunities, etc.
- In addition, the recent approach of ISO (ISO 17025 2017) to evaluate quality systems is focused on formulating purposes and aims, performing process analysis and flux diagrams, risk and opportunities assessment, evaluation of changes and the way to assure the continuous improvement of the quality system, including training and continuous evaluation of the personnel. Another feature of the newest approach is the tendency to evaluate activities and duties rather than structural aspects.

Finally, applied research should be considered as a valuable task situated at the core of the system, enhancing the background knowledge and supporting the technical improvement of methods and procedures. Some crucial tasks like uncertainty estimation, inspection and preventive maintenance and data handling can be substantially improved using research activities. Moreover, concerning the managerial part of the system, research tasks may lead to reconfigure or readapt the processes taking part in the system and to detect new opportunities for development.

As a manner of grouping, the activities performed to assure and assess the quality of measurements, and the subsequent results, the concepts of quality assurance, quality assessment and the evaluation of the system's performance are defined.

First of all, quality assurance activities prescribe the regular activities to implement reliable procedures for measuring and fulfil the quality requirements for a specific application. The aim of quality assurance is precisely to assert, the quality and reliability of the method for obtaining and logging data, maximising the accuracy and minimising errors, drifts and uncertainty sources. According to this, quality assurance prescriptions have an impact on technical procedures, equipment, ranges of validity, etc. Examples of quality assurance tasks are the calibration of sensors or the selection and setting of the equipment according to the requisites of the application. These requirements will be addressed in the next section.

Second, the consistency of acquired data must be assessed; obtained values must be coherent according to their nature, the type of equipment and settings, environmental conditions, etc. They also have to be intrinsically consistent and in logical agreement with other sets of data collected at the same time and location, depending on the specific type of measurements and application. Concerning these requirements, quality assessment activities (see Sect. 3) establish adequate procedures and multi-step checks, to verify the compliance with the quality requirements for the specific application. In this way, the development of quality assessment techniques is focused on characterizing, handling and flagging data, performing corrections when necessary or even removing part of the data set when its quality is under suspicion. The aim is to guarantee the quality of data and results before they are delivered to the client or the public.

Third, a quality system must also exert control over the measuring system itself. This is achieved by designing tasks which are enclosed by the expression system performance control (see Sect. 4). These activities aim to ensure that the operation of the measuring system is correct. Examples of quality performance tasks are the inspection of facilities and infrastructure, verifications of functioning and signals, cleaning and maintenance, etc.

Finally, the main aspects concerning the calibration of solar sensors like pyranometers and pyrheliometers are described in Sect. 5. In particular, it will be addressed the determination of the mean sensitivity of a field device during calibration by comparison to a reference device. The general process of determination of the total uncertainty will also be addressed, according to the guidelines prescribed in the Guide for Uncertainty Measurement (GUM), developed by the Joint Committee for Guides in Metrology (JCGM) (ISO/IEC 2008).

The following terminology definitions help to understand its content in this chapter:

- Quality control: It is the whole integrated and dynamic process which comprehends all the measurement activities beginning with the identification of the most suitable instruments for our application, following the process of data acquisition and maintenance procedures, calibration of the equipment, finishing when the last storing of the values is made. This also involves other processes designed to enhance the quality of the data recorded in the past, being measured and all the plans to improve the future measurements.

- Quality assessment: It is the process of deriving the quality level of each measurement trough flags. In this chapter, we propose a series of internal procedures which compare the data with itself. We will present a set of quality checks selected from the literature which will allow detecting errors in the measurements even in the case we do not have in-depth knowledge about the conditions of the station or the maintenance logs when the values where registered.
- Quality assurance: The maintenance of a desired level of quality in the measurement process, especially using attention to every stage of the process. It is a way of preventing errors in the measurements and avoiding the delivery of wrong data to be used during the resource assessment and in the corrections of the modelled data.
- Quality enhancement: When the data is measured and quality assessed using automatic and semiautomatic procedures, the next phase is the analysis if the data is wrong and attempting to enhance the quality of already recorded data. This is the only phase in which the recorded data is changed. The modification or enhancement of the data should be done only by the scientist of the station or personnel with access to all the maintenance logs and in contact with technicians in charge of the maintenance to avoid erroneous interpretation and modification of the measured data to wrong values which after cannot be retro propagated to original values if needed. The person with access to all the information relevant to a questionable measurement will have the best judgment to provide solutions for the data quality enhancement. If any change is made, it is recommendable to keep the original raw values registered by the data acquisition system without data quality flagging.

Figure 1 summarizes the relationship between the three processes which conform the quality control process.

Fig. 1 Quality control process and the relationship with quality data assurance, assessment and enhancement

When the quality assessment is done in near real time or soon after the measurement process is completed, it can be a useful input for the quality control of measurements in the future, such as the display of the data in real time (Long 1996). Just a few good quality measurements spread widely along the whole day have much less value than a set of measurements done systematically for the same day. Consequently, in essence, quality assurance has a particularly dynamic character, something that scientists analysing radiometric data must be aware.

2 Quality Assurance

The principal requirements of a quality management system, as those described in the previous section concerning the organization's policy, structure, resources, processes and aims, are encompassed in a so-called quality manual. The technical requisites are usually addressed in the so-called technical procedures, accompanied by the subsequent specific technical processes and a variety of related documentation as manuals, technical specifications, records, etc. Technical procedures describe not only the methods used for measuring but also the associated activities and facts to consider as the setup of equipment, preparatory tasks, magnitudes' intervals of validity, etc. Within this framework, the activities enclosed by quality assurance tasks are addressed and introduced.

2.1 Calibration and Measurement Procedures

The quality of solar radiation measurements is solidly based on a reliable calibration of irradiance sensors. To achieve that, reference sensors must be calibrated so that measurements are traceable to the S.I. radiometric scale.

Concerning standardized methods, calibration of field pyranometers can be performed following a technical procedure based on ISO 9847 Standard "Calibration of field pyranometers by comparison to a reference pyranometer" (Technical Committee: ISO/TC 180/SC 1 1992). However, when the calibrated pyranometers are to be used as reference instruments in comparisons, ISO 9846 "Calibration of a pyranometer using a pyrheliometer" (Technical Committee: ISO/TC 180/SC 1 1993) is mandatory. Concerning the calibration of pyrheliometers, it should be done following ISO 9059: "Calibration of field pyrheliometers by comparison to a reference pyrheliometer" (Technical Committee: ISO/TC 180/SC 1 1990a).

When dealing with solar irradiance measurements, the primary scale considered is the World Radiometric Scale (WRR), although its traceability to S.I. units it is still under discussion (it is currently accepted a mean deviation of 0.3%). The WRR is defined by the so-called World Standard Group (WSG), physically constituted by a group of absolute cavity pyrheliometers. Their results while measuring solar irradiance form an average value and a subsequent specific coefficient, relative to the average, which is assigned at each device and applied in subsequent calibrations to maintain the WRR traceability. This traceability to the WRR is spread in comparisons made in the International Pyrheliometer Comparison (IPC), being held in PMOD-WMO (Davos, Switzerland) every five-year period.

In this way, absolute cavity pyrheliometers (ACRs) out of the WSG but participating in an IPC event are considered as primary references. Therefore, a secondary reference pyrheliometer can be calibrated by comparison with a central reference following ISO 9059 standard (Technical Committee: ISO/TC 180/SC 1 1990a). Similarly, a secondary reference pyranometer can also be calibrated by comparison with a primary reference pyrheliometer (there are not primary reference pyranometers), following ISO 9846 (Technical Committee: ISO/TC 180/SC 1 1993).

Finally, it has to be noted that the calibration of devices for measuring is not mandatory only for reference sensors but also for the rest of the associated equipment like data acquisition units. Data loggers need to be calibrated to assure the accuracy of the voltage, current and electrical resistance signals.

2.2 Instrument Selection

Instrument selection is a critical step, which must be done based on the quality requirements for the selected applications. It is necessary to take into account not only the current tasks but also the sort and middle-term plans, projects and activities, which eventually will require future investments. The acquisition of equipment will define the basic technical capabilities of the measuring system. Therefore, some managerial tasks like prospective evaluations, realistic market assessments and inter-annual objectives plans will contribute to the best use of material resources. Summarizing, instrument selection should be consistent with technical applications, quality requirements and the aims of the organization.

In general, factors like accuracy, reliability, robustness, resistance to wearing and long-term degradation of active components must be evaluated for the equipment necessary in every technical procedure.

Concerning irradiance sensors, special care must be taken to select a device while evaluating technical specifications like its principle of functioning, off-set, linearity, spectral range, thermal response and angle response. In this way, ISO 9060 (Technical Committee: ISO/TC 180/SC 1 1990b) standard describes the classification of pyrheliometers and pyranometers. Each class is defined by tolerance or a maximum deviation for the main technical features for each type of sensor, expressed in units of solar irradiance or percentage of reading. Naturally, high-quality equipment means lower uncertainty ranges and higher accuracy, besides longer device durability and reliability of results. When irradiance sensors are used for applications like high-accuracy purposes or accurate calibrations performed at the highest metrological level, uncertainty ranges must be minimized, and the use of top-quality sensors is mandatory. However, the most accurate sensors might not be the best choice for all applications. In particular for solar resource assessment at remote sites, the most precise instruments might not be a good choice as they are less robust than other simple measurement systems. Different kinds of accessories like thermal stabilisers are sometimes required to minimize thermal oscillations during measurements.

On the other hand, data acquisition units are usually defined by physical features like power supply (e.g. allowing batteries for stand-alone applications or DC supply from photovoltaic panels), endurance for outdoor applications, the number of measuring channels (simultaneous signals to measure and register) or communication options. Other technical specifications to be evaluated, affecting directly to the quality of measurements are accuracy, ranges for voltage, sampling rates, current or electrical resistance, errors (offsets, measurement, communication, A/D conversion, etc.) and thermal drift. It also must be noted that signals can be affected by electromagnetic noise, which can be minimized with the election of shielded wires and connectors.

2.3   Data Acquisition and Basic Handling

The design of procedures and settings for data acquisition depends on the specific application, and it is a very sensitive task. Hence, a vast variety of intrinsic and environmental factors that affect the quality of measurements must be taken into account. Among these factors are the requirements of accuracy, the setting of time intervals, scale settings, the amount of data generated (in terms of transmission intervals, time and effort to handle them, etc.), necessary weather variables, basic redundancy of measurements, reference or control signals, threshold alarms, surrounding electromagnetic noise, presence of external thermal sources, vibrations, emergency power supplies, backup of data files, etc. Incorrect settings may even ruin the efforts invested in measuring campaigns. Some basic criteria can contribute to minimize errors:

- The requirement of data accuracy affects integration time and subsequently reduces the maximum possible sampling rate. When higher accuracy is needed, more time is then used for every single reading.
- Sampling rates (temporal rate of reading from the radiometers at which the datalogger is configured) and averaging intervals (final temporal average range at which the measurements are recorded from the sampled data) of readings use to be previously designed, keeping an equilibrium between sufficient but not excessive, according to specific application; as an example, measurements performed for high-accuracy calibration purposes may require shorter sampling rates than those made in order to estimate the yearly total solar irradiation.

On the one hand, when setting great sampling rates, it is obtained a coarse data file, which would be detrimental regarding accuracy if interpolated values have to be calculated. Also, some transitory or anomalous readings would be masked (not detected), which is prejudicial to data flagging or filtering. On the contrary, excessively short sampling rates usually lead to unnecessarily burdensome data files, increasing the complexity of data handling (chunking, resizing, time resampling, etc.). Usually, sub-minute sampling rates are averaged to 1-min values and then reported. For example, BSRN recommends to sample the data at the 1 s interval and record the data at 1-min interval including the average for the period, the maximum value, the minimum and the standard deviation (McArthur 2004). On the other hand, the time interval (the time between two consecutive data records in the output file) is set as a function of time-resolution requirements (e.g. daily, monthly or hourly data), calculated as aggregated results from 1-min data.

- Measurement range settings should be fixed and known for each channel, avoiding auto range settings when possible. Precision errors of data acquisition units are usually expressed regarding the percentage of the scale used. Hence, scale range must be the minimum possible one, while being higher than the maximum expected signal.
- Acquisition of environmental variables, when measuring solar radiation or calibrating radiation sensors, is not a secondary issue but a fundamental one. Mainly, weather conditions use to be logged to establish the intervals in which measurements were made; that is, they define a range of validity for measurements or calibrations. Moreover, they may be necessary to perform some corrections of results based on weather conditions like ambient or device temperature and relative humidity.
- Reference or control signals are fundamental during a subsequent data assessment, as anomalous records can be detected, identified, corrected or removed from valid data sets.
- Raw data files should be saved in (at least) two independent storage devices. Centralized repositories, where backups are made as a routine job, are then highly recommendable for data safety.
- To prevent fatal mistakes, raw data should always be kept unmodified in the original files, and data handling should be made on subsequent copies.

After data acquisition, a primary data handling has to be performed, in the sense of that precise data handling prescribed by the standardized method or technical procedure, usually based on absolute values and maximum admissible deviations. As an emblematic example, ISO 9847 specifies that calibration of field pyranometers in clear sky days has to be performed using readings above 600 W/m^2 (700 W/m^2 in the case of pyrheliometers calibrated following ISO 9059). Data handling does not have to be considered as part of the data quality assessment as no conclusions are drawn on data quality itself.

2.4   Software Validation

Most data acquisition and data handling techniques depend on mathematical algorithms, which are finally implemented using some application or specially designed and developed software. A global approach to quality assurance also comprehends an initial evaluation of these tools. Some basic requirements concerning the quality of software are the following ones:

- Software has to be reliable, that is, has to perform the right tasks based on the exact criteria defined by the method. The user has to be informed of any task or subtask affecting the data (e.g. a data filtering technique). This is known as "logging", which is a means of tracking events that happen when the software runs. Eventual exceptions have to be identified.
- Software has to keep or generate the necessary information to replicate the same result from the same raw data. That is, results have to be traceable. To achieve that, all settings, configuration parameters, environmental or surrounding variables, partial results, exceptional events, etc., have to be released as a separate report or data file.
- Software versions have to be identified, and all changes which are respecting the last release have to be described. Partial and final reports must contain the reference of the version used.

Based on these criteria, the intrinsic logic of the software has to be evaluated using both real and pre-designed input data. On the one hand, real input data contains much of data features that software will have to deal with. Results must be consistent with the expectation, repetitive (it delivers the same results when introducing the same inputs) and consistent with any intentional change of input data or settings. On the other hand, the software has to respond as expected when handling pre-designed inputs, like a cero-signal input, a unity signal, a linear or geometrical progression. When possible, independent calculi should be performed (whether using necessary calculus sheets or even handmade calculus) before the new software is applied in real cases, to verify its reliability (i.e. that both well-known methods and new tools will reproduce the same results).

3   Quality Assessment

Despite the design and implementation of a quality system, technical procedures and operational techniques, the totality of acquired measurements must be checked in a real time or near real-time basis, e.g. week daily. Consistency must be assessed by performing a comprehensive set of checkpoints and introducing validation criteria. Otherwise, partial or total failures can occur and propagate unnoticed, causing the ruin of the project and efforts. These sets of tasks and related activities can be globally named with the expression quality assessment.

3.1 Quality Assessment of Solar Radiation Measurements

The use of automatic tools and visual expert control to evaluate the quality of solar radiation measurements leads to reliable scientific studies and more accurate estimations of energy production.

Evaluation of solar radiation measurements is performed to assess their reliability. Several quality control tests have been proposed and implemented. The sets of tests proposed by the Baseline Surface Research Network (BSRN) project (Long and Dutton 2002), developed under the Climate Research Programme of the World Meteorological Organization, and the SERI quality control program, proposed by the National Renewable Energy Laboratory (NREL) (NREL 1993) are examples of widely used methodologies, which establish basic criteria to be matched by solar radiation data in order to be accepted as valid. Nevertheless, automatic implementation of these criteria is not free of difficulties and errors, so further efforts have been made to improve quality requirements, especially in cases where these methodologies are not applicable (e.g. at low values of solar irradiance and solar elevation angles), and bibliography is abundant.

In general terms, quality control tests implement a list of conditions to be matched by solar radiation data. Many of them are based in the enclosure equation of the direct normal (DNI) and diffuse (DHI) components of the global horizontal irradiance (GHI). When GHI, DNI and DHI are measured independently, there is a redundancy that can be verified. Conditions are also based on physically possible limits and extremely rare limits for the mentioned components and also the upwelling (LWup) and downwelling (LWdn) longwave radiation, measured by a pyrgeometer. Finally, the consistency of some other comparisons (e.g. LWup and LWdn to air temperature) can be evaluated. Total results can be assessed by assigning a value or flag to the records matching a specific criterion and one in case of failure. The flags can then be classified and graphically represented. It has to be considered that the matching some conditions can automatically validate others, and vice versa, so following the prescribed order is an essential point of the methodology.

The methodology for conducting quality control can be divided into the following group of procedures:

- **Control of the data recording time**. Having a correct temporal stamp, especially for hourly and sub-hourly time series, is important since all calculations are a function of the solar geometry. For each of the days with data available, it is important to check the correctness of the timestamp for the corresponding measured data. This can be done visualizing the data in graphics or with automatic procedures recently developed (Moreno-Tejera et al. 2015).
- **Visual inspection of solar radiation components**. Once the temporal reference of the data has been checked, the recorded values should be displayed day by day to assess them by an expert as valid or invalid. The checking allows the detection of possible problems that are not detected using a numerical method,

for example, artificial horizons created with nearby objects which can have a significant artificial influence on the final total values of the solar resource (Table 1).

- *Physically possible* test intended to detect values out of the physical limits of solar radiation reaching one location of the Earth surface. The radiation data falling in the intervals defined in Table 2 are considered "physically possible".
- *Extremely rare test*. The limits in the "extremely rare" procedure are narrower than those of the "physical possible" test. Radiation data which violate these limits may occur over very short time periods under rare conditions. These limits are given in Table 3. Data of "good quality" are assumed to be inside the "extremely rare" limits.
- *Across Quantities* or *Internal consistency checks*: procedures capture smaller errors that have not been detected by the previous quality checks. These tests are based on empirical relationships of the different quantities and solar components measured around the world. The restrictions are defined in Tables 4 and 5.

3.2   Completeness of Data

The first analysis we should do in the measured data is to calculate the percentage of sub-hourly data available to calculate the aggregated values. The next graphic shows an example for one station. Table 6 shows the percentages of one-minute data available for each month analysed from the beginning of the measurement.

3.3   Control of the Data Recording Time and Visual Inspection of the Data

Before proceeding to the quality analysis of the measurements, we have to apply the expression to transform the temporal register from local time to true solar time (TST), which is a temporal reference which strictly depends on the position of the Sun concerning the site where the measurement has been acquired. The change to TST is performed by two corrections; the first one considers the difference in longitude between the meridian of the observer and the meridian of the temporal reference. The second includes Earth orbital effects through the equation of time.

The second procedure is a visual inspection of irradiance time series plots at various time resolutions and in combination with other information. This procedure is probably the most effective test, as no automated check can replace the experienced eye. Figures 2, 3 and 4 show an example for a station showing the validity of the GHI, DNI and DIF data day by day for the entire period analysed by the following colour scale.

Table 1 Table of completeness data for each month

June 2018	July 2018	August 2018	September 2018	October 2018	November 2018	December 2018	January 2018	February 2018	March 2018
98.02%	99.97%	100%	99.95%	98.9%	100%	99.82%	99.99%	99.94%	99.93%

Table 2 Physical limits of the solar radiation component

Parameter	Minimum	Flag for minimum	Maximum	Flag for maximum
Global horizontal irradiance (GHI)	-4 Wm^{-2}	2	$I_{SC}\varepsilon 1.5(\cos Z)^{1.2} + 100$ W/m^2	3
Diffuse horizontal irradiance (DIF)	–	–	700 W/m^2	4
Diffuse horizontal irradiance (DIF)	-4 Wm^{-2}	5	$I_{SC}\varepsilon 0.95(\cos Z)^{1.2} + 50$ W/m^2	6
Direct normal irradiance (DNI)	-4 Wm^{-2}	7	$I_{SC}\varepsilon$	8
Direct normal irradiance (DNI)	–	–	DNI Clear Sky (Bcs)	9

I_{SC} Solar constant (1361 Wm^{-2}); ε Eccentricity of the Earth orbit around the Sun; Z Solar zenith angle

Table 3 Conditions for the cross-component flagging

Parameter	Conditions	Limits	Flags
$\frac{GHI}{DIF + DNI\cos Z}$	$Z < 75°$, $\mathbf{DIF} + \mathbf{DNI}\cos \theta_z > 50$ W/m^2	$1 \pm 8\%$	10
$\frac{GHI}{DIF + DNI\cos Z}$	$75° < Z < 93°$, $\mathbf{DIF} + \mathbf{DNI}\cos Z > 50$ W/m^2	$1 \pm 15\%$	11
$\frac{DIF}{GHI}$	$Z < 75°$, $\mathbf{GHI} > 50$ W/m^2	<1.05	12
$\frac{DIF}{GHI}$	$75° < Z < 93°$, $\mathbf{GHI} > 50$ W/m^2	<1.10	13

Table 4 Conditions for the second group of cross-component filters

Parameter	Lower limit	Upper limit	Flags
DNI cos Z	(GHI–DIF) $-$ 50 W/m^{-2}	(GHI–DIF) $+$ 50 W/m^{-2}	14
GHI-DIF	DNI cos Z $-$ 50 W/m^{-2}	DNI cos Z $+$ 50 W/m^{-2}	15

Table 5 Relation between quality flags and each radiation component

Solar radiation component	Flag
GHI	2, 3, 10, 11, 12, 13, 15, 17, 18, 19, 21
DNI	7, 8, 9, 10, 11, 14, 19, 20
DIF	4, 5, 10, 11, 12, 13, 15, 16, 17, 18, 19

To check if the data is correct in TST or there is any error in the temporal reference, we can visualize in graphics the different solar radiation components measured against the clear sky or extraterrestrial solar radiation. Figure 5 presents an example. The temporal checking can be done with automatic procedures as mentioned before (Moreno-Tejera et al. 2015).

Table 6 Flags of error control

Flag	Conditions
0	The Sun elevation angle is less than 0.01 (night hours)
1	The data is **correct**
2	The GHI data is below the minimum physical limit (-4 Wm^{-2})
3	The GHI data exceeds the maximum physical limit for conservative clear sky conditions
4	DIF is higher than 700 W/m^2
5	The DIF data is below the minimum physical limit (-4 Wm^{-2})
6	The DIF data exceeds the maximum physical limit
7	The DNI data is below the minimum physical limit (-4 Wm^{-2})
8	The DNI data exceeds the maximum physical limit of the solar constant multiplied by the eccentricity
9	The DNI is higher than clear sky DNI
10	The condition relating the three components, GHI, DIF and DNI
11	The condition relating the three components, GHI, DIF and DNI
12	The relationship between DIF and GHI for zenith angles lower than 75° and higher than 50 Wm^{-2}
13	The relationship between DIF and GHI for zenith angles lower than 93° and higher than 75° higher than 50 Wm^{-2}
14	Direct horizontal is within a limit compared with GHI and DIF
15	The difference of GHI and DIF is within a limit of DNI
16	Diffuse fraction limit (K_{d_ext}) calculated from Extraterrestrial solar radiation
17	Clearness index and DIF/GHI limit for cloudy condition
18	Clearness index and DIF/GHI limit for clear sky condition
19	Tracker-off filter
20	Direct normal transmittance upper limit
21	Clearness index normalized by air mass of 1 higher limit

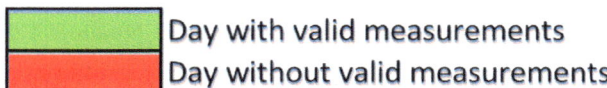

Fig. 2 Results of visual inspection_GHI

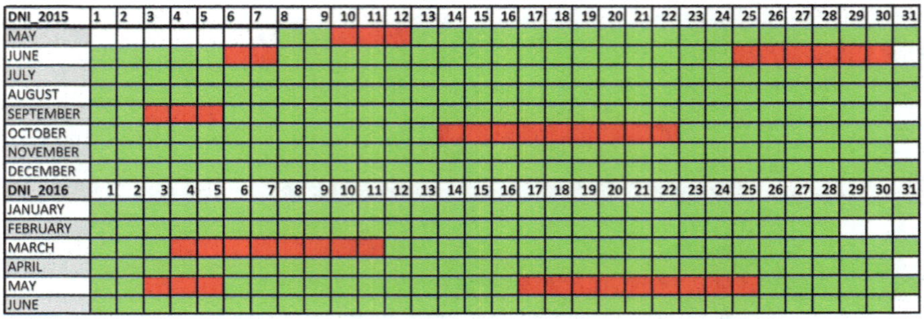

Fig. 3 Results of visual inspection_DNI

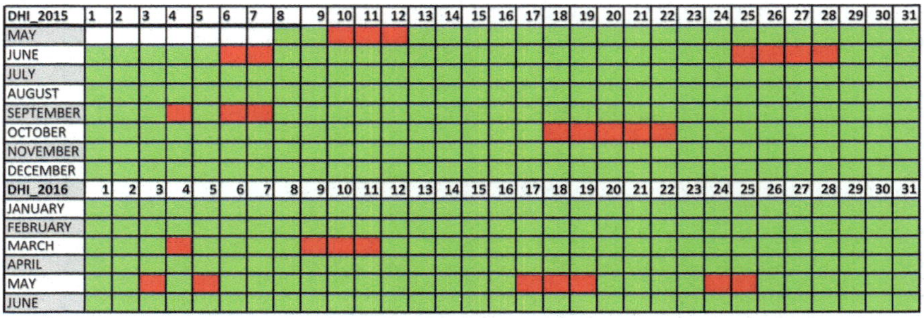

Fig. 4 Results of visual inspection_DHI

3.4 Quality Analysis

The quality analysis with physical filters refers to the verification of the recorded values of the different components of solar radiation, considering its physical sense and not exceeding its value, therefore, limits physically possible. Table 2 presents the physical limits imposed on each component of the solar radiation according to the recommendations of the BSRN (Long and Dutton 2002).

The quality analysis of component cross filters is used to check that the measured data meets the interrelationship between the three components (GHI, DIF and DNI). Failure to pass these filters establishes a supposition that any of the components were poorly measured or that the solar tracker does not point to the Sun correctly. Table 3 shows the conditions imposed on the cross-component analysis.

The next procedures interrelate the three components but using tighter conditions. Table 4 defines the limits for this procedure:

The next procedure relates the diffuse component (DIF) and extraterrestrial irradiance (I_{ext}) using the diffuse index defined as (Long and Shi 2006):

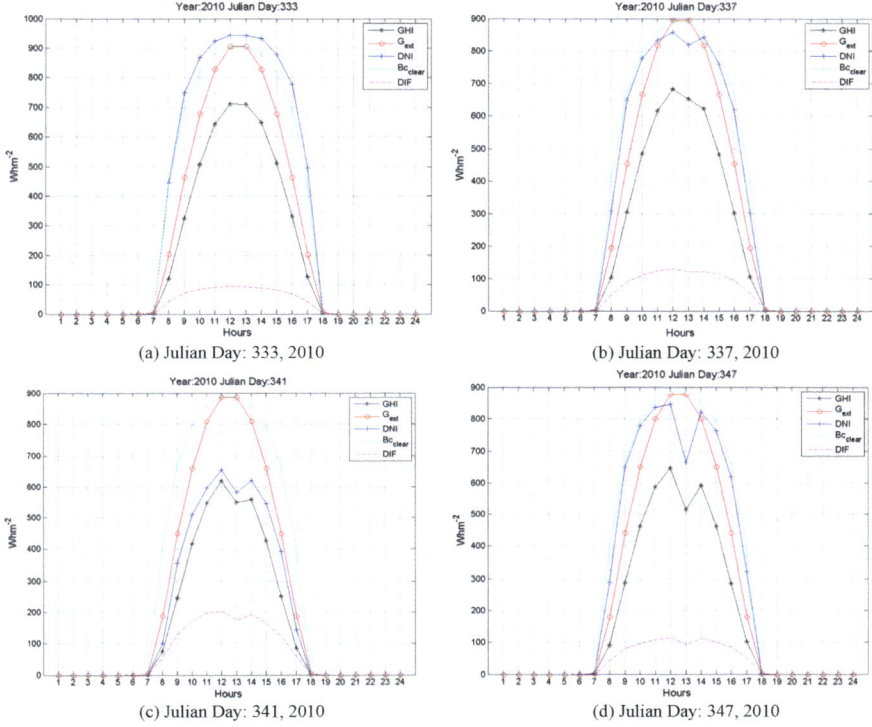

Fig. 5 Measured hourly GHI (black star), DNI (blue), DIF (dashed pink), direct normal clear sky Bc$_{clear}$ (turquoise) and extraterrestrial horizontal irradiance G$_{ext}$ (circle red) plotted day by day to check the temporal reference of the data **a** Julian Day: 333, 2010, **b** Julian Day: 337, 2010, **c** Julian Day: 341, 2010 and **d** Julian Day: 347, 2010

$$K_{d\_ext} = \frac{\text{DIF}}{I_{ext}}$$

A higher limit of 0.6 is given to this filter, and in case it is not fulfilled, the flag number 16 is activated. The next procedure makes use of clearness index (K_t) which is defined as the quotient between ground measured global horizontal irradiance (GHI) and extraterrestrial solar irradiance (I_{ext}). In this procedure, we establish the next condition for the activation of flag number 17:

If K_t is lower than 0.2 and DIF/GHI is lower than 0.9, then flag 17 is activated.

The flag number 18 uses the same variables as the last filter but with the following conditions:

If K_t is higher than 0.5 and DIF/GHI is higher than 0.8, then flag 18 is activated.

The filter with flag number 19 is named as the *tracker-off* filter, and it is used to detect when the solar tracker is not working correctly. First, the global horizontal irradiance (Sum_SW) is estimated from measured diffuse solar irradiance and measured direct normal irradiance using the expression which relates the three solar radiation components. Then, the following condition is established using clear sky global horizontal irradiance (GHI$_{clear}$) estimated with a clear sky model.

For DIF > 50 W/m^2,

If (Sum_SW)/GHI$_{clear}$> 0.85 and if DIF/(Sum_SW), the tracker is not properly following the Sun.

This filter only works under clear sky conditions.

The quantities to compare are based in clearness indices. The clear sky index (K_t) defined as:

$$K_t = \frac{\text{GHI}}{I_{\text{ext}}}$$

where GHI is the horizontal global irradiance and I_{ext} is extraterrestrial solar radiation in the upper border of the atmosphere for a horizontal plane.

Clear sky index (K_c) is defined as:

$$K_c = \frac{\text{GHI}}{\text{GHI}_{\text{cs}}}$$

where GHI$_{\text{cs}}$ is the global horizontal irradiance for clear sky conditions.

A similar test can be done with the beam clearness index K_b defined as (flag number 20):

$$K_b = \frac{\text{DNI}}{I_{\text{sc}}}$$

To use the clearness index as a reliable sky condition descriptor, Perez et al. (1990) modified this parameter to make it independent of the solar elevation angle. The formulation is the following:

$$K_t^* = \frac{K_t}{(1.031 \ \exp(-1.4/(0.9 + 9.4\text{AM})) + 0.1)}$$

where AM is the optical air mass as defined by (Kasten 1980). Values of K_t^* higher than 1 are flagged with the value flag number 21. We could have used the K_t directly as a filter, but as we will see in the next figures for low solar elevations the extinction of solar radiation due to the high air mass has an exponential form and these values are limited and never reach the value 1. With the K_t^*, we can normalize this dependency with AM removing it and have a range of values between 0 and 1 for all solar elevations.

Fig. 6 Relationship between K_t and Sun elevation (1-min data)

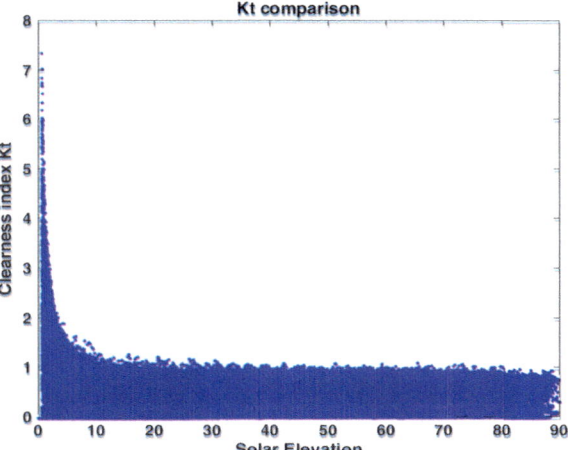

Table 5 shows a summary of the flags that indicate to which component of solar radiation (GHI, DNI or DHI) they apply.

Table 6 presents a description of the different filters described previously.

The next graphics show the relationship between K_t and K_t^* and Sun elevation for 1-min data where the value of K_t is higher than 1 for low solar elevations. These erroneous measurements could be due to several reasons, among them, problems with a temporal reference, so the solar radiation divided by the extraterrestrial component is not well synchronised, issues in the level of the pyranometers, artificial lights from the surroundings or reflection due to walls or white objects in the surroundings (Fig. 6).

Figure 7 shows the same information as the previous, but instead of showing K_t it shows K_t^*. As we can see, the magnitude of the wrong values (values higher than 1) is amplified.

The next graphics show the relationship between K_t and sun elevation for valid data (green colour) and not accurate data for 1-min temporal resolution once the data is filtered (Fig. 8).

The next graphics show the same information but for hourly raw values and filtered ones (Fig. 9).

3.5   Other Quality Control Procedures

In the literature, there are other quality control procedures. One of them is the SERI QC procedures from NREL (NREL 1993). The three solar radiation data components (GHI, DNI and DIF) are quality checked using SERI QC, a procedural and software package developed by NREL. SERI QC[1] defines ranges of acceptable

---

[1]https://www.nrel.gov/grid/solar-resource/seri-qc.html

Fig. 7 Relationship between K_t^* (normalized for 1 AM) and Sun elevation (1-min data)

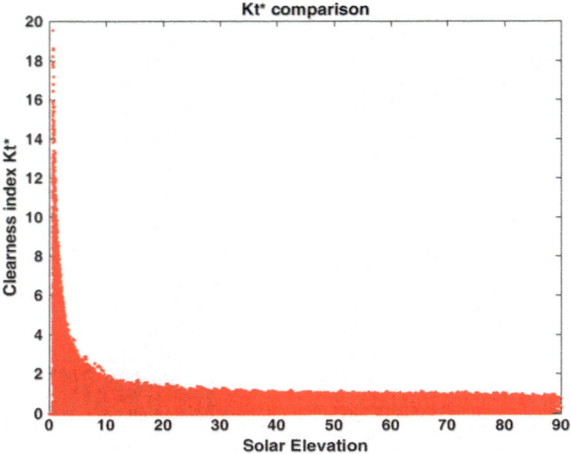

Fig. 8 Relationship between K_t and Sun elevation for valid data (green colour) and raw data (red colour) (1-min data)

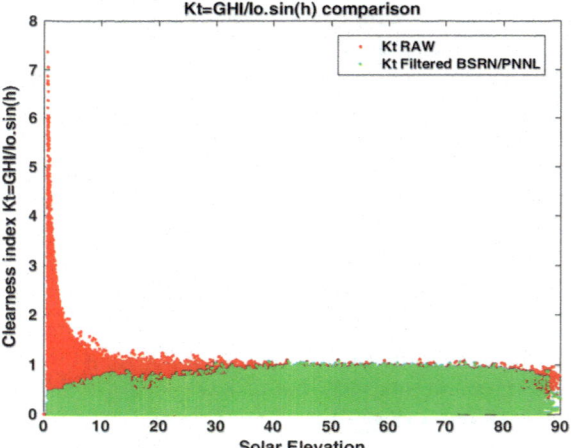

data, depending on whether one, two or all three hourly data elements are present. Ranges are defined based on dimensionless parameters normalized concerning extraterrestrial radiation. Figure 10 presents an example of the filters proposed by SERI QC where two-element test by specifying a range of acceptable values within boundaries is established based on K_t, K_d or K_n.

Several additional frameworks for quality check ground measured data have been proposed (Younes et al. 2005). A web service for controlling the quality of measurements of global solar irradiation was developed (Geiger et al. 2002). Management and Exploitation of Solar Resource Knowledge (MESoR) project proposed a specific framework which improved the basic BSRN filters (Beyer et al. 2009). Furthermore, quality control of global solar radiation was done using

Fig. 9 Relationship between K_t and Sun elevation for valid data (green colour) and raw data (red colour) (hourly)

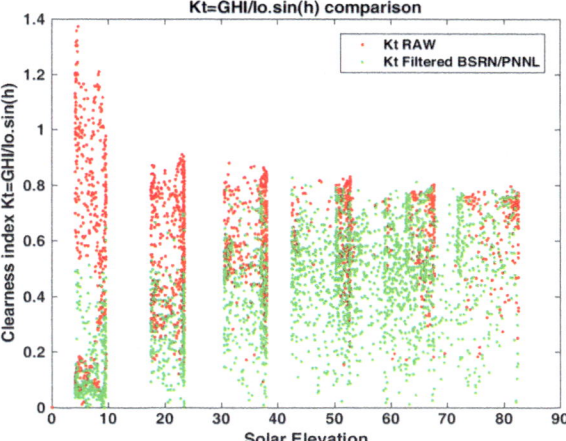

Fig. 10 SERI QC data boundaries for two-element quality assessment (NREL 1993)

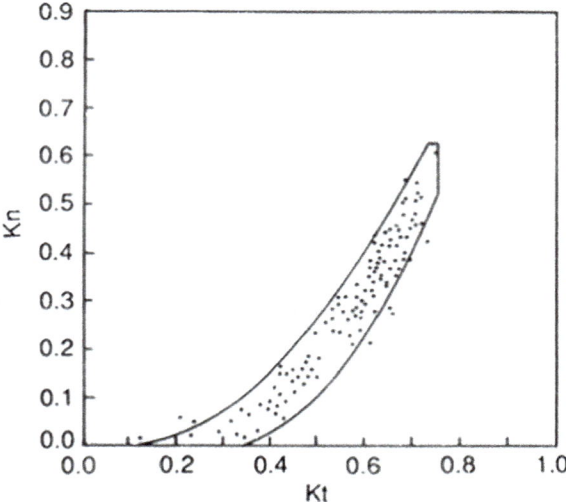

sunshine duration hours (Moradi 2009). A major effort has been undertaken at the Royal Meteorological Institute of Belgium (RMIB) to develop procedures and software for performing post-measurement quality control of solar data from the radiometric measurements with additional procedures to fill missing values (data initially lacking or removed via quality checks) (Journée and Bertrand 2011). The Harmonization and Qualification of Meteorological Data Project (ENDORSE) made a review of the quality checks proposed in the literature (Espinar et al. 2011; Dumortier 2012). Quality control and estimation of hourly solar irradiation were presented for China (Tang et al. 2010) and South Korea (Lee et al. 2013). The measured data of global solar irradiation on a horizontal surface, the number of bright sunshine hours, and the amount of cloud cover for major cities of South Korea, during the period (1986–2005) were analysed (Lee et al. 2013). Following best practices for quality assessment tests, such routines are implemented at the Solar Radiation Resource Assessment (SRRA) project of the Ministry of New and Renewable Energy, India, in a network of 51 automatic solar radiation monitoring stations across India (Kumar et al. 2013; Schwandt et al. 2014). A methodology for quick assessment of timestamp and quality control results of solar radiation data was proposed (Moreno-Tejera et al. 2015). The topics of visual expert inspection and automatic screening and flagging of solar irradiation, other meteorological parameters and auxiliary data (battery voltage, etc.) are also discussed (Geuder et al. 2015). Satellite estimations were used to quality check ground measurements (Urraca et al. 2017). Finally, improved quality checks combining BSRN and PNNL were also proposed utilizing an approach similar to the one presented in this chapter (Perez-Astudillo et al. 2018a). Effect of Solar Position Calculations on Filtering was also analysed (Perez-Astudillo et al. 2018b).

3.6 Analysis of the Results from the Filtering

The results of the quality flags can be plotted in monthly graphics for each hour and each day or percentage of data flagged erroneously for each day. As an example, Figs. 11 and 12 show for one station the flag values detected individually for each hour and each month. The flag number 0 indicates that the Sun elevation is lower than 0, night hours. In background green colour, it is shown that the registration of the three components of solar correctly is correctly measured. With the red colour in the background, it is shown that some or all of the solar irradiance components (GHI, DIF or DNI) are suspicious to be wrongly measured.

The results of the quality checks can also be plotted as the percentage of data which does not pass the tests with a colour scale. Figures 13, 14 and 15 shows an example applied to different components of solar radiation showing the percentages of data flagged for each day and different months.

Another way to show the results of the test is with a graphic of contour. The following figures present the results for each hour and day for a whole year with the percentage of values which have active the flag. For two quality flags, Fig. 16 presents the percentage of values which pass the filters. Most of the good values are in the central hours for the whole day. We can see that the rate of values not detected as correct or with flag 1, appear in flag 5 in Fig. 17 which shows that there are problems in that station during sunrise and sunset.

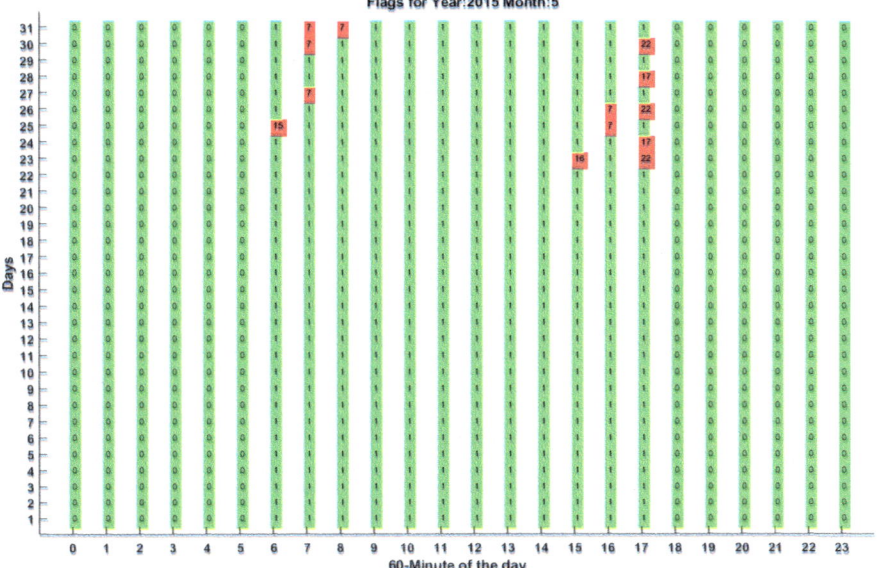

Fig. 11 Flags for each hour and day. Month: June. Year: 2015

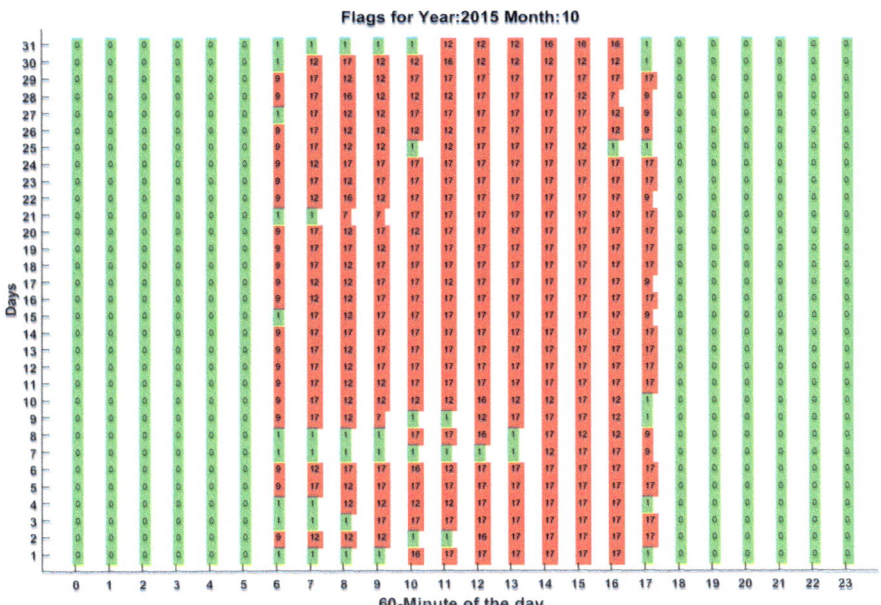

Fig. 12 Flags for each hour and day. Month: November. Year: 2015

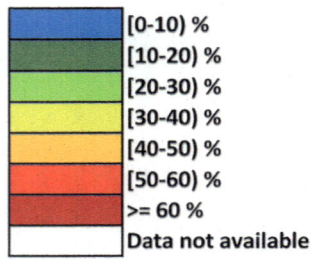

Physically possible ranges	Test 1
Extremely rare ranges	Test 2
Cross-component relationships	Test 3

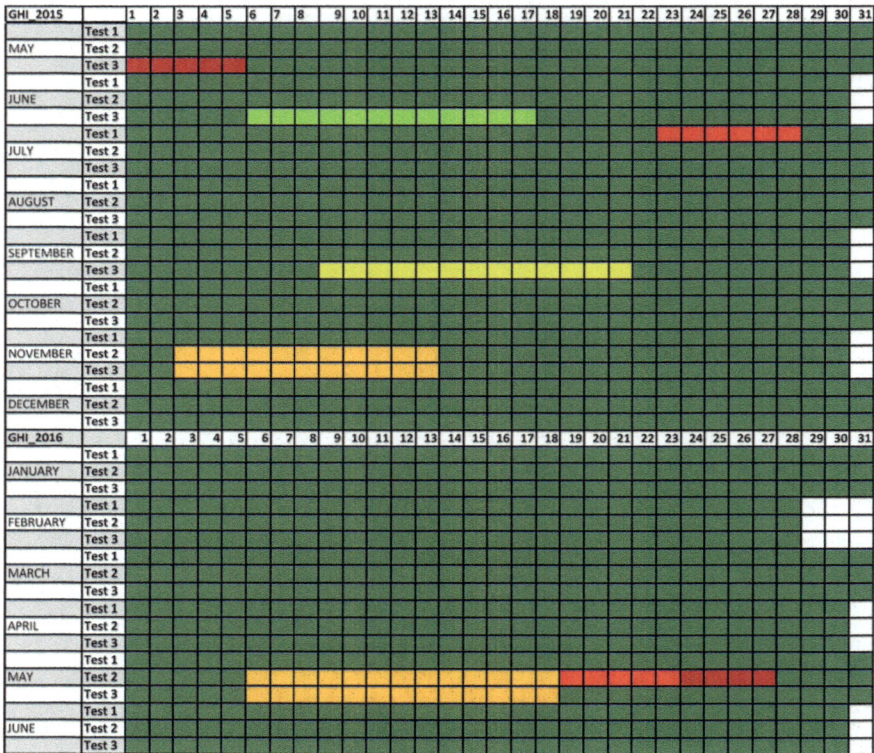

Fig. 13 Results of automatic tests_GHI

3.7 Daily Graphics from the Quality Checks

Another way to present the results of the quality checks to analyse the reasons why the data was detected as suspicious is plotting the values of the flags in graphics day by day. In this section, we present some examples. The variables which are included in the next samples are the following: global horizontal irradiance (GHI) in black colour (—*— GHI), global extraterrestrial irradiance (G_{ext}) in red colour

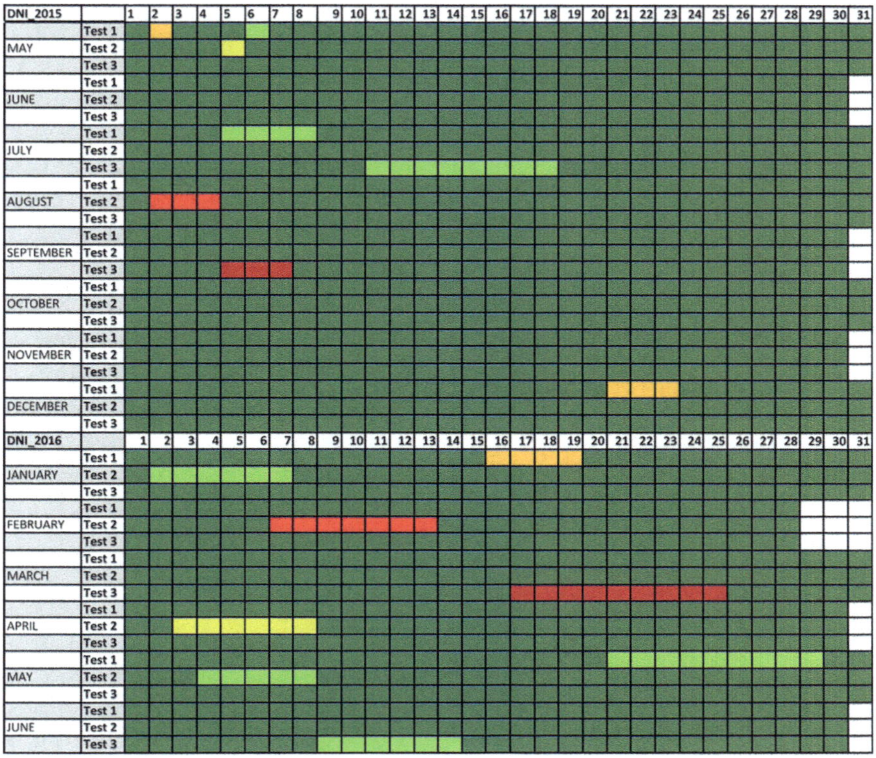

Fig. 14 Results of automatic tests_DNI

($-\!\!\circ\!\!-$ G_{ext}), direct normal irradiance (DNI) ($-\!+\!-$ DNI), clear sky direct normal irradiance (Bc_{clear}) ($-\!\!\!-$ Bc_{clear}), diffuse irradiance (DIF) ($----$ DIF), clear sky diffuse irradiance (Dc_{clear}) ($----$ Dc_{clear}), estimated global horizontal irradiance (Sum SW) obtained from measured diffuse irradiance and direct normal irradiance using the following equation DIF + DIR cos θ ($-\!+\!-$ Sum SW), estimated direct normal irradiance (Ib_{est}) obtained from measured global horizontal and diffuse irradiance (Ib_est) and clear sky global horizontal irradiance (Gc_{clear}) ($-\!\!\!-$ Gc_{clear}). Besides, the flag code indicates which is the quality control flag value detected for the solar irradiance measured. The green colour in this number means that the measurement is right, and the red colour shows that the measurements are *suspicious* to be wrong or are wrong. Downside another row is included to indicate if the solar tracker is working properly. The code 0 with green colour in the background shows that the solar tracker is working ok, and in red colour, with code 1 it suggests that it is not functioning properly. The malfunctioning of the solar tracker is only identified when there is a clear sky day condition as explained before (Fig. 18).

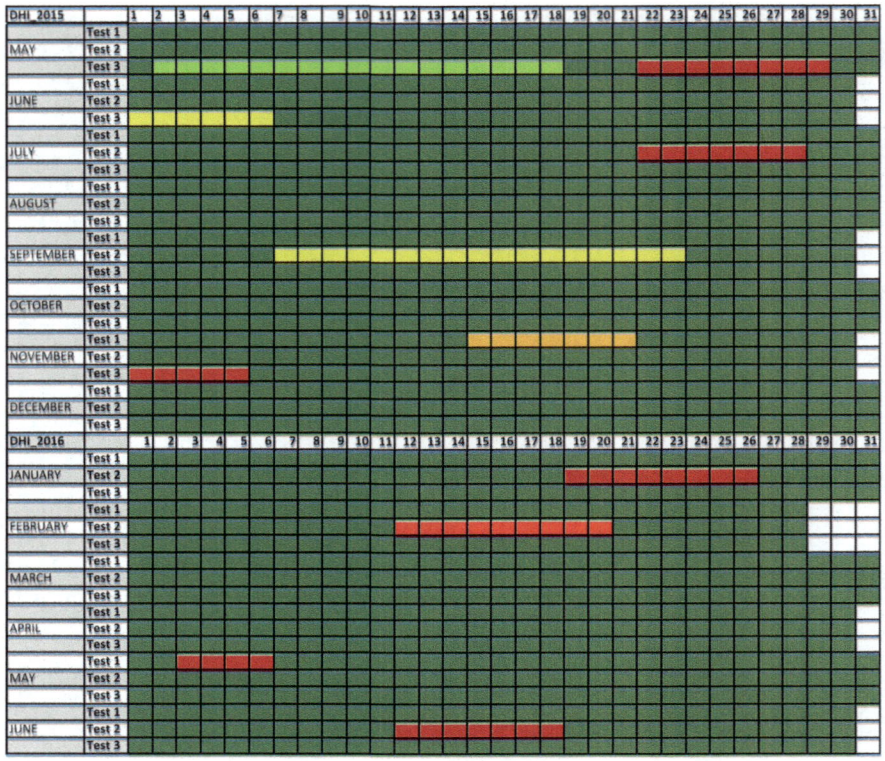

Fig. 15 Results of automatic tests_DHI

Figure 19 shows another example where solar radiation has been measured correctly during all hours of the day.

3.8   Quality Check of Meteorological Ground Measured Data

The procedures to check the quality (quality assurance) of the meteorological data are divided into the following groups based on physical magnitudes on a global scale. For locations with specific climatological conditions, the ranges of maximum and minimum values can be different.

- **Temperature**. The physical limits are defined as 60 °C as higher limit and −20 °C as the lower limit. Seasonal and climatological limits will depend on the specific conditions of the site.

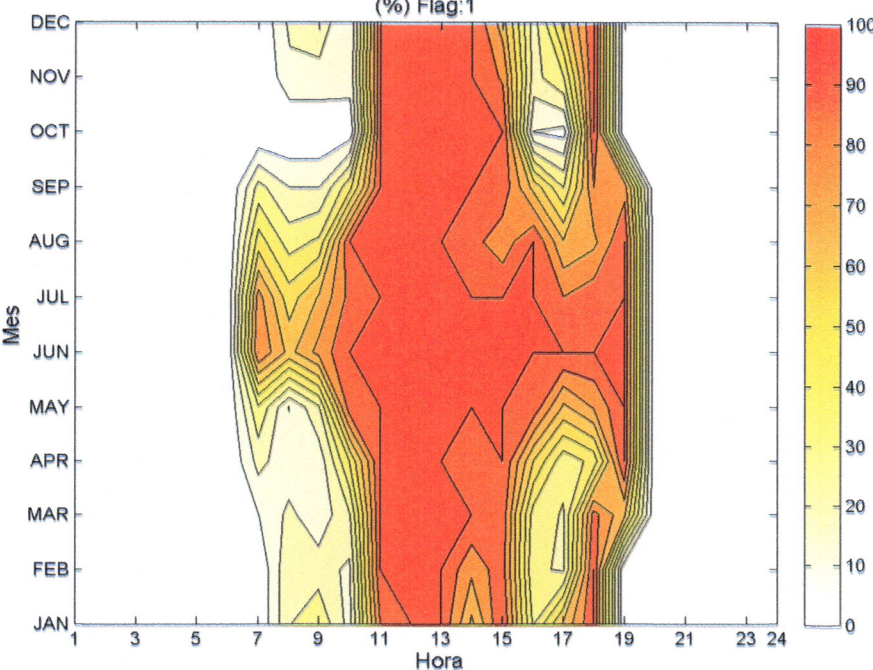

Fig. 16 Flag contour diagram 1 (%)

- **Relative humidity**. The physical, seasonal and climatological limits are the same for this variable. The lower limit is 0, and the higher limit is 100.
- **Pressure**. The limits for this variable will depend on the altitude of the location above sea level. Air pressure is not uniform across the Earth. However, the normal range of the Earth's air pressure is from 980 millibars (mb) to 1050 mb. These differences are the result of low and high air pressure systems which are caused by unequal heating across the Earth's surface and the pressure gradient force.
- **Wind Speed**. The physical limits are defined as 60 m/s as higher limit and 0 m/s as the lower limit. Seasonal and climatological limits will depend on the specific conditions of the site.
- **Wind direction**. The physical, seasonal and climatological limits are the same for this variable. The lower limit is 0°, and the higher limit is 360°.

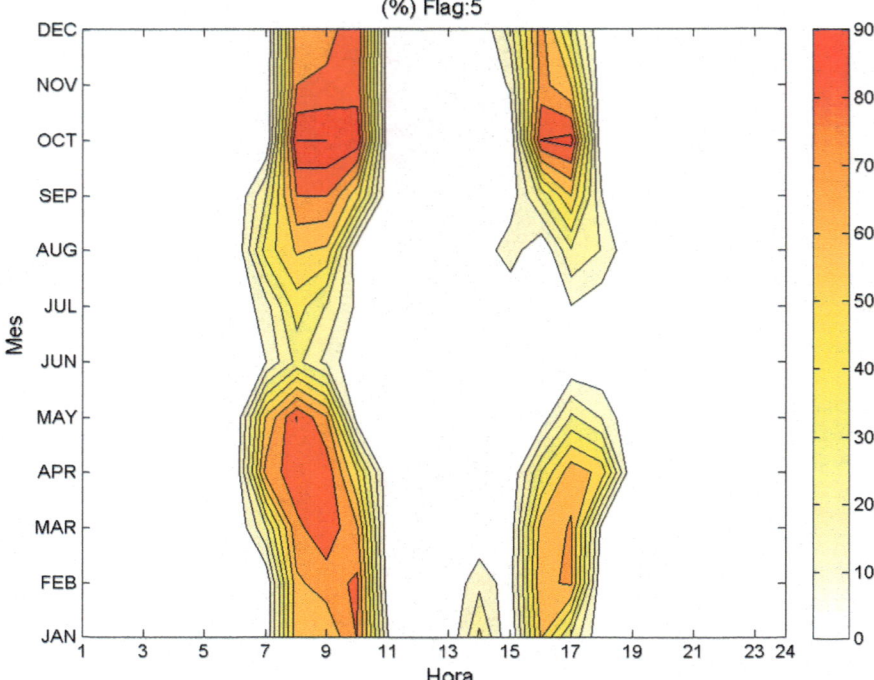

Fig. 17 Flag contour diagram 5 (%)

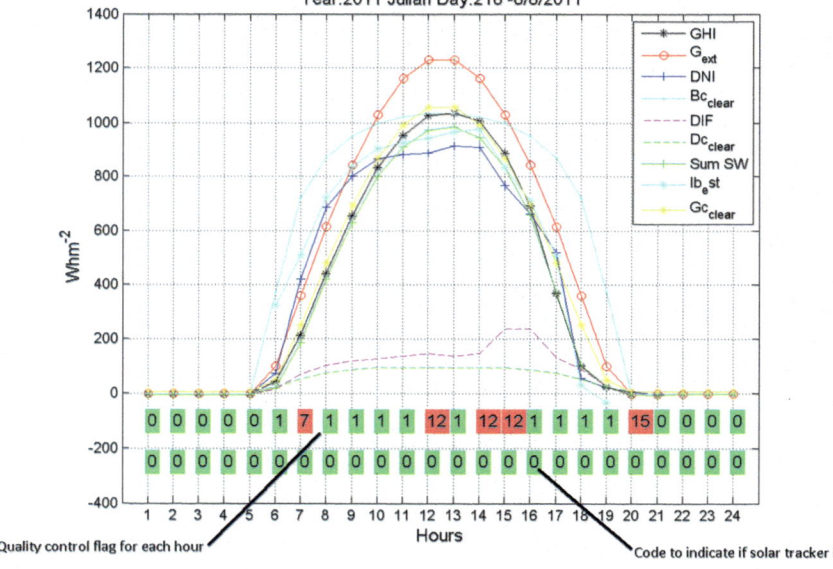

Fig. 18 Example to explain the meaning of the codes representing the quality flags

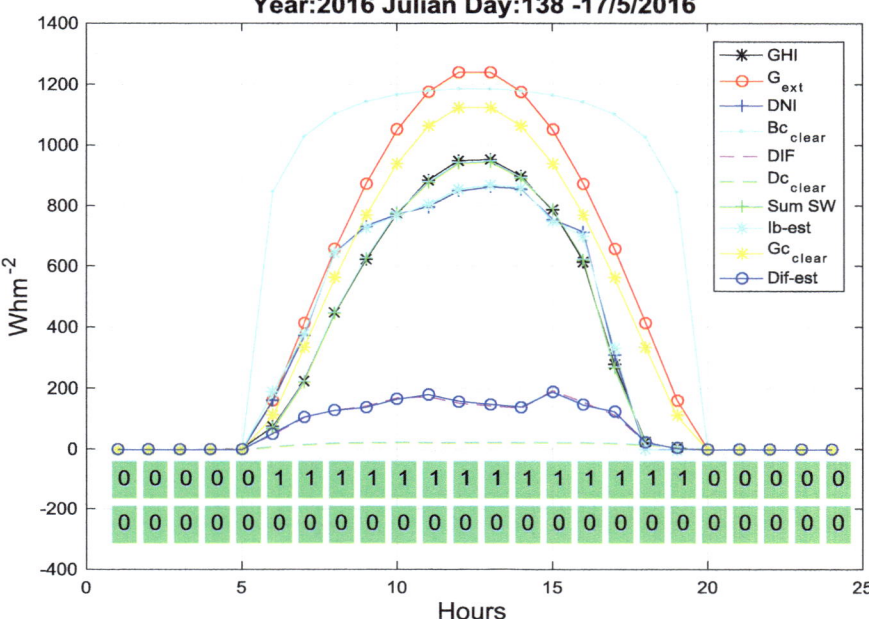

Fig. 19   Example to explain the meaning of the codes representing the quality flags. Julian Day: 138, Year: 2016

3.9   Consistency Checks of Meteorological and Radiometric Data

The internal consistency checks enforce reasonable, meteorological relationships among observations measured at a single station. For example, a dew point temperature observation must not exceed the temperature observation made at the same station. If it does, both the dew point and temperature observation are flagged as failing their internal consistency check. Pressure internal consistency checks include a comparison of pressure change observations at each station with the difference of the current station pressure and the station pressure three hours before and a comparison of the reported sea-level pressure with a sea-level pressure estimated from the station pressure and the 12 h mean surface temperature. In the former check, if the reported 3 h pressure change observation does not match the calculated observation, then only the reported observation is flagged as bad. In the latter test, however, if the reported sea-level pressure does not match the computed observation, then both the sea-level and station pressure observations are flagged as failing.

Another type of consistency check is that related to the maximum allowed variability of an instantaneous value; i.e., if the current value differs from the prior one by more than a specific limit, then the current instantaneous value fails the check. Possible limits of a maximum variability can be air temperature: 3 °C;

relative humidity: 10%; atmospheric pressure: 0.5 hPa; wind speed: 20 ms^{-1}; solar radiation: 1000 Wm^{-2}.

A spatial consistency check involves comparing the data from a set of stations and determining the median of the values. The checked value should not differ from the median value by a certain percentage. The median is used instead of the mean so as not to allow a defective station to affect the validation checks of other stations. When performing this form of spatial checking, it is essential that the set of stations chosen possess comparable characteristics. This is not just dependent on the location of the station and elevation but can be influenced by other physical factors (e.g. proximity to a mountain range or the seaside) that may affect the prevailing weather conditions at the station. The suitability of station to form spatial check set with other stations needs to be determined on an element by element basis. For example, while a stations pressure reading may be comparable to another station, they may have different wind characteristics, and therefore should not be spatially checked for similarity in experience, by careful analysis of historical data.

The statistical spatial consistency check uses weekly quality checks (QC) statistics to mark observations as failed if they failed any QC check 75% of the time during the previous 7 days. These observations will continue to be labelled as failed by this check until the failure rate falls below 25% in the weekly statistics.

The consistency check of balance meteorological data is performed by ensuring that all the atmospheric, seasonal and illumination conditions are represented in the set of estimated and measured data.

All these consistency checks are inspired by recommendations from the World Meteorological Organization (WMO).

3.10 System Performance

By mentioning the expression system performance, we are usually referring to those activities focused on evaluating the measuring system itself. In the frame of a quality management system, this kind of work is designed using verification and maintenance programs, where individual tasks are conveniently scheduled during the current and successive months or even years.

The general aim of system performance is preventing systematic errors and drifts caused by undesired factors like malfunctioning, misalignment, ageing, excessive wearing, dust, unnoticed shadings, thermal stress, condensation, electromagnetic noise, vibrations or any other potential trouble that possibly affects the quality of measurements. By implementing system performance activities, the uncertainty ranges, deviations and drifts due to the functioning of equipment are minimized to nominal values. Otherwise, the factors mentioned above can diminish the accuracy of data and results, sometimes dramatically, even finally causing the ruin of the project.

Activities like inspection of infrastructure and facilities, verification of equipment and signal monitoring are examples of system performance. In most cases, these activities lead to perform another type of tasks like re-locating of installations, re-positioning or alignment of sensors, reparation, cleaning, maintenance, thermal or electromagnetic shielding, grounding, lubrication of mechanisms, and removal of dust and humidity. When possible, implementing these tasks simultaneously with other necessary works to improve installations and renew equipment may lead to saving time, resources and efforts.

4 Quality Enhancement

4.1 Calculation of Solar Radiation Data from Other Components

Gaps and incorrect data are usually in the measurements. It is not a conventional process the identification and replacement of missing or incorrect data. The typical action with this data is to discard it in the analysis. After quality tests have been performed on the measured ground data, the gaps can be filled. The gap procedure can be done for each instant (1 min value) using the components which are flagged as correct values:

1. If DIF and GHI are correct (flag 1), they are used to calculate DNI.
2. If DIF is not correct (flag higher than 1) and DNI and GHI are OK (flag 1), DIF is calculated from measured DNI and GHI.
3. If DIF and GHI are not OK (flag higher than 1), then DNI is OK (flag 1). The GHI is filtered with the value of a clear sky model considering a Rayleigh scattering. Afterwards, DIF is calculated from filtered GHI and measured DNI which is OK (flag 1).

When direct normal irradiance is not measured, it can be estimated (Ib_{est}) from measured GHI and DIF using the following expression:

$$Ib_{est} = \frac{GHI(1 - k_{do})}{\sin(\gamma)}$$

where k_{do} is defined as:

$$k_{do} = \frac{DIF}{GHI}$$

And γ is the angle of solar elevation with respect to the horizon.

4.2 Data Aggregations

The 1-min values which do not pass the filters or are not gap filled can be marked as NULL or not valid values depending on the programming or scripting language used.

For each hour, the mean hourly value can be calculated from 1-min values with flag 1 for each solar radiation component. To calculate the hourly values, it is recommendable to have almost 50% of the 1-min values with flag 1 for each hour. If there is not enough percentage of 1-min values which are OK (flag 1), a correct hourly value can not be calculated. Usually, the quality filters are applied with less uncertainty to hourly values than 1-min values. This is due to the different time response of the radiometers thermopiles which can affect the consistency in the relationship between the three solar radiation components.

The daily values can be calculated if there are almost 75% of correct hourly values for the sunlight hours. An interpolation of the hourly values can be done regarding clearness index normalized for air mass 1 (K_t^*) for the days which fulfil the last condition to have values for all sunlight hours for each day.

The monthly average values could be calculated if there are more than 20% of correct daily values for each month.

4.3   Enhancement of Measured Data from Interpolated Calibration Constant

The application of fresh instrument calibration constants when calculating irradiance values is one way of adding quality to data by applying retrospective methods (Esterhuyse 2004). The black surface of a radiometer thermopile becomes more reflective with prolonged solar exposure; hence, less radiation is absorbed by the sensing element, and the instrument's calibration factor gradually becomes smaller (Esterhuyse 2004). Regular calibration (the recommended frequency is once every six months) and subsequent updating of the radiometer's calibration constant is therefore imperative if the recorded irradiance values are expected to be an accurate reflection of the irradiance values (Esterhuyse 2004). If the irradiance values are calculated while thermopile measurements are sampled, care must be exercised that the latest (freshest) calibration constant is applied featured in the system for real-time quality control on values as close as possible to the actual values (Esterhuyse 2004). If this is not done, necessary adjustments must be made on recorded data, so that archived irradiance values reflect the latest thermopile sensitivity (Esterhuyse 2004). Simple interpolation of the calibration factors between calibration episodes helps to keep the values as close as possible to the actual sensitivity during a specific month (Esterhuyse 2004).

In case calibration is not done for an extended period, new calibration constant can be obtained and an interpolation of the previous last calibration constant, and the recent one can be obtained to be applied retrospectively to the signal of the thermopile pyranometer. The next graphics show examples of stations which have been recalibrated after several years and the intermediate calibration constant obtained from the interpolation (Figs. 20, 21 and 22).

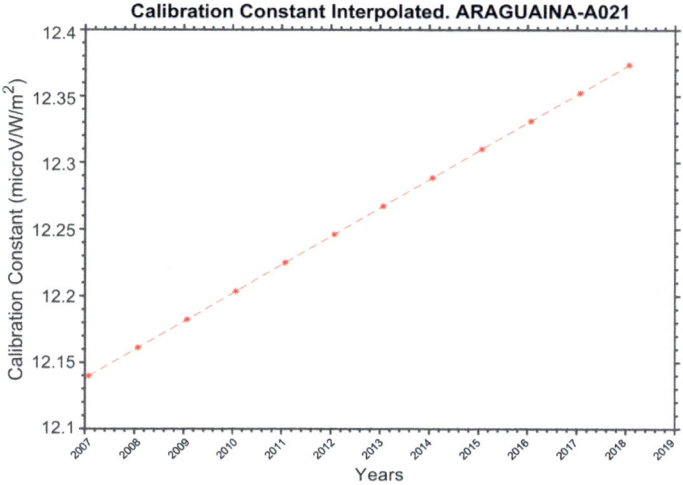

Fig. 20 Temporal interpolation of the calibration constant of the pyranometer. Station: A021. Araguaina

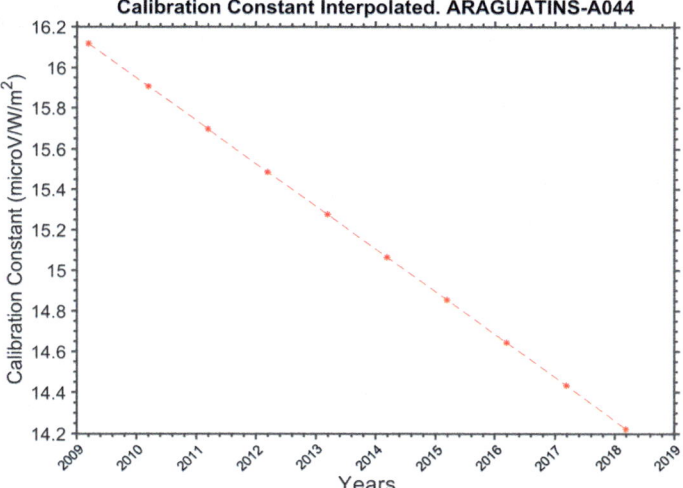

Fig. 21 Temporal interpolation of the calibration constant of the pyranometer. Station: A044. Araguatins

The recalibration of the values registered can reduce the dispersion between satellites estimated and ground measured data. As an example, Fig. 23 shows a comparison of the evolution of dispersion parameter rRMSD for monthly values in

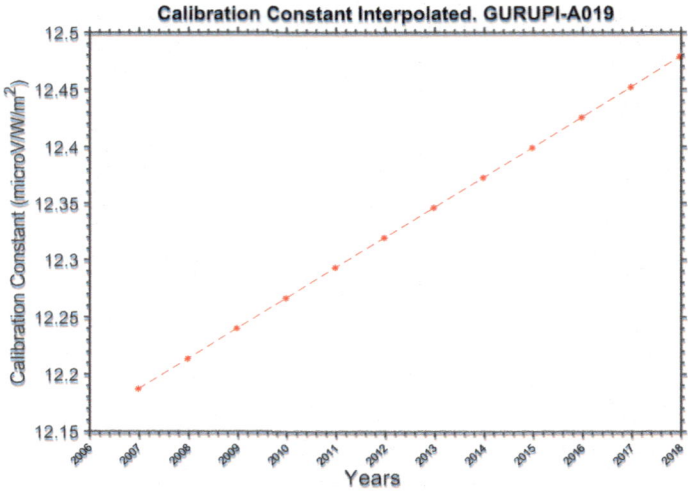

Fig. 22 Temporal interpolation of the calibration constant of the pyranometer. Station: A019. Gurupi

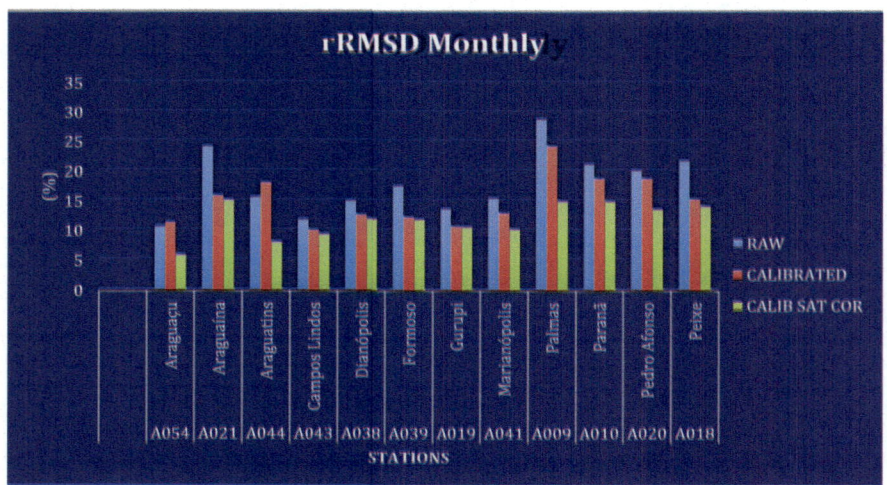

Fig. 23 Evolution of values of dispersion parameter rRMSD for INMET stations in Brazil compared with SARAHv2 satellite estimations for RAW uncalibrated measurements (blue), recalibrated retrospectively measurements from interpolated calibration constant (red) and recalibrated retrospectively measurements from interpolated calibration constant and site-adapted satellite estimations to produce the final solar maps (green) (Polo et al. 2016)

three different circumstances to produce the final solar maps for a group of stations in Brazil.

5 Conclusions

Quality control is defined as the overall technical activities that are used to fulfil requirements for quality. Quality assurance is defined as the management activities involving planning, implementation, assessment, reporting, and quality improvement to ensure that measurements are of the type and quality needed and expected by our solar mapping application. Unfortunately, we do not know the correct answers, and we only see a fraction of the (obviously) incorrect values. This means that even after all the checking, we cannot be 100% that the data is correct. Quality assurance procedures provide us guides to flag data we believe to be suspect. In some cases, aided by operator notes, we can assure data is bad. In other cases, the best we can do is educated guesses based on the experience of the scientist.

Measurements which do not fall into the physically possible limits should be taken as highly suspect. Measures beyond the extremely rare cases should be visually inspected. If no physical reasons—such as cloud multi-reflection, extreme weather conditions, etc.—can be found, these data should be excluded. When comparing measurements, especially solar radiation components (GHI, DNI or DIF), with each other, it is unclear which of the two (or more) values caused the inconsistency. We should compare estimated and measured solar radiation components (GHI, DNI and DIF) based on the equation which relates them and cosines of zenithal angle. Further investigations should be done based on our experience.

An automatic replacement of the data exceeding the test limits could result in a loss of realistic measurements, and it is thus not recommended. This way, wrong values, which have a flag value higher than 1, should be excluded from the dataset used to make the site adaptation and improvement of the long-term satellite and numerical modelled data to create the final solar maps.

Acknowledgements This work has been partially supported by the Spanish National Funding Program for Scientific and Technical Research of Excellence, Generation of Knowledge Subprogram, 2017 call, DEPRISACR project (reference CGL2017-87299-P).

References

Beyer HG, Polo MJ, Suri M et al (2009) Report on benchmarking of radiation products. Management and exploitation of solar resource knowledge project report D 1.1.3. CA – Contract No. 038665

Dumortier D (2012) Project ENDORSE, 1–9. doi:Grant agreement no. 262892

Espinar B, Wald L, Blanc P et al (2011) Report on the harmonization and qualification of meteorological data: Project ENDORSE—Energy downstream service: providing energy components for GMES—Grant agreement no. 262892, Paris, France

Esterhuyse DJ (2004) Establishment of the South African baseline surface radiation network station at De Aar

Geiger M, Diabaté L, Ménard L, Wald L (2002) A web service for controlling the quality of measurements of global solar irradiation. Sol Energy. https://doi.org/10.1016/s0038-092x(02) 00121-4

Geuder N, Wolfertstetter F, Wilbert S et al (2015) Screening and flagging of solar irradiation and ancillary meteorological data. Energy Procedia 69:1989–1998

ISO/IEC (2008) Uncertainty of measurement—Guide to the expression of uncertainty in measurement (GUM:1995). ISO/IEC Guide 98-32008. https://doi.org/10.1373/clinchem.2003. 030528

ISO 17025 (2017) ISO/IEC 17025:2017—Technical committee: ISO/CASCO & Committee on conformity assessment—General requirements for the competence of testing and calibration laboratories

ISO 9001 (2015) ISO 9001: 2015—Systems, technical committee ISO/TC176/SC2 Quality— Quality management systems-requirements

Journée M, Bertrand C (2011) Quality control of solar radiation data within the RMIB solar measurements network. Sol Energy. https://doi.org/10.1016/j.solener.2010.10.021

Kasten F (1980) A simple parameterisation of the pyrheliometric formula for determining the Linke turbidity factor. Meteorol Rundschau 33:124–127

Kumar A, Gomathinayagam S, Giridhar G et al (2013) Field experiences with the operation of solar radiation resource assessment stations in India. Energy Procedia 49:2351–2361

Lee K, Yoo H, Levermore GJ (2013) Quality control and estimation hourly solar irradiation on inclined surfaces in South Korea. Renew Energy 57:190–199. https://doi.org/10.1016/j.renene. 2013.01.028

Long CN (1996) Report on broadband solar radiometer inconsistencies at the atmospheric radiation measurement (ARM) Southern Great Plains (SGP) Central facility during the arm enhanced shortwave experiment (ARESE). ARM-TR 003. Richland, Washington, USA

Long CN, Dutton EG (2002) BSRN global network recommended QC tests, V2.0. BSRN Technical Report

Long CN, Shi Y (2006) The QCRad value added product: surface radiation measurement quality control testing, including climatology configurable limits. Office of science, office of biological and environmental research

McArthur LBJ (2004) Baseline surface radiation network (BSRN): operations manual (Version 2.1)

Moradi I (2009) Quality control of global solar radiation using sunshine duration hours. Energy. https://doi.org/10.1016/j.energy.2008.09.006

Moreno-Tejera S, Ramírez-Santigosa L, Silva-Pérez MA (2015) A proposed methodology for quick assessment of timestamp and quality control results of solar radiation data. Renew Energy 78:531–537. https://doi.org/10.1016/j.renene.2015.01.031

NREL (1993) User's manual for SERI_QC software—Assessing the quality of solar radiation data. NREL/TP-463-5608. Golden, CO

Perez-Astudillo D, Bachour D, Martin-Pomares L (2018a) Improved quality control protocols on solar radiation measurements. Sol Energy 169:425–433. https://doi.org/10.1016/j.solener.2018. 05.028

Perez-Astudillo D, Bachour D, Martin-Pomares L, Perez-Astudillo D (2018b) Effect of solar position calculations on filtering, pp 2–3

Perez R, Ineichen P, Seals R, Zelenka A (1990) Making full use of the clearness index for parameterizing hourly insolation conditions. Sol Energy 45:111–114

Polo J, Wilbert S, Ruiz-Arias JA et al (2016) Preliminary survey on site-adaptation techniques for satellite-derived and reanalysis solar radiation datasets. Sol Energy 132:25–37. https://doi.org/ 10.1016/j.solener.2016.03.001

Schwandt M, Chhatbar K, Meyer R et al (2014) Quality check procedures and statistics for the Indian SRRA solar radiation measurement network. Energy Procedia 57:1227–1236

Tang W, Yang K, He J, Qin J (2010) Quality control and estimation of global solar radiation in China. Sol Energy. https://doi.org/10.1016/j.solener.2010.01.006

Technical Committee: ISO/TC 180/SC 1 & C-M and data (1992) ISO 9847: 1992—Solar energy —Calibration of field pyranometers by comparison to a reference pyranometer

Technical Committee: ISO/TC 180/SC 1 & C-M and data (1993) ISO 9846: 1993—Solar energy —Calibration of a pyranometer using a pyrheliometer

Technical Committee: ISO/TC 180/SC 1 & C-M and data (1990a) ISO 9059: 1990—Solar energy —Calibration of field pyrheliometers by comparison to a reference pyrheliometer

Technical Committee: ISO/TC 180/SC 1 & C-M and data (1990b) ISO 9060: 1990—Solar energy —Specification and classification of instruments for measuring hemispherical solar and direct solar radiation

Urraca R, Gracia-Amillo AM, Huld T, et al (2017) Quality control of global solar radiation data with satellite-based products. Sol Energy. https://doi.org/10.1016/j.solener.2017.09.032

Younes S, Claywell R, Muneer T (2005) Quality control of solar radiation data: Present status and proposed new approaches. Energy 30:1533–1549

Chapter 5
Clear-Sky Radiation Models and Aerosol Effects

Christian A. Gueymard

Abstract This chapter offers a description of the main factors that affect the transmission of solar radiation through the cloudless atmosphere, and the corresponding modeling approaches. The limitations of broadband modeling are discussed, and methodological improvements are described. A detailed discussion of the various inputs required by different clear-sky radiation models, and how to obtain such data, is provided so that the reader can operate these models with appropriate inputs, depending on the application and geographical coverage. In particular, the benefits of using atmospheric data provided by recent reanalyses are described. The impact of aerosol attenuation on the different irradiance components is discussed, with a focus on the aerosol optical depth. Its methods of measurement, properties, reduction methods, accuracy, and spatiotemporal variability are described. The error propagation between aerosol data and the predicted irradiance is quantified, and examples are provided. Seven models of the literature are selected for further discussion and validation. This validation is performed using high-quality radiometric data from Tamanrasset, Algeria, and is done in two different ways: an ideal validation based on the best possible (locally measured) aerosol information and a practical method (generalizable anywhere) based on reanalysis data. A sensible degradation of performance is obvious when using the second approach. Finally, some likely or desirable future developments in the field are described.

1 Introduction

Modeling solar radiation incident at the surface is a complex task, owing to the variety of atmospheric processes involved and their spatiotemporal variability. This is particularly the case of the impact of clouds, which can considerably alter the transmitted irradiance over short or long periods. The general (all-sky) irradiance

C. A. Gueymard (✉)
Solar Consulting Services, Colebrook, NH 03576, USA
e-mail: Chris@SolarConsultingServices.com

© Springer Nature Switzerland AG 2019   137
J. Polo et al. (eds.), *Solar Resources Mapping*, Green Energy and Technology,
https://doi.org/10.1007/978-3-319-97484-2_5

modeling challenge is thus normally simplified by considering two independent parts: (i) modeling under assumed clear-sky (cloudless) conditions and (ii) modeling of the superimposed effect of clouds, if present. Hence, the all-sky irradiance is assumed a direct function of its clear-sky counterpart, even when the sky is completely covered by clouds. Nearly all radiation models of the literature are constructed around this basic superimposition principle, whose validity has been demonstrated recently (Oumbe et al. 2014; Xie et al. 2016). Notable exceptions do exist, including in a few types of extremely simplistic and empirical modeling approaches, such as those based on sunshine to directly estimate the all-sky global horizontal irradiance (GHI) on a daily or monthly basis. Such low-performance models are now considered obsolete and will not be considered further here.

The evaluation of clear-sky irradiance is necessary for a variety of solar-related applications, such as (i) preparing time series of satellite-derived irradiance data; (ii) providing the basis for solar radiation forecasts; (iii) developing specialized solar resource databases for the design of concentrating solar systems; (iv) testing the validity of irradiance observations in quality-control algorithms; or (v) calculating the cooling loads of buildings. All these applications require a calculation of solar radiation at frequent intervals, typically from 1 min to 1 h.

The present chapter focuses on the modeling of the clear-sky irradiance for such applications, and most particularly from the perspective of solar resource quantification and assessment. The modeling principles and main clear-sky radiation models of the literature are reviewed, and the availability of their necessary inputs is discussed. In particular, the existing data about atmospheric aerosols and the latter's impact on solar irradiance are described in detail. Finally, the accuracy of some prominent clear-sky radiation models is evaluated using a well-established validation method.

2 Recent Evolution in Solar Data

In recent years, rapid progress has occurred in the field of solar resource assessment and can be attributed in large part to the convergence of interests between the solar, remote sensing, and meteorological communities. The latter, in particular, is now seeing the former as an important stakeholder, owing to the rapid development of renewable energy sources and accompanying need for both retrospective and forecasted meteorological data. Consequently, the solar community has rapidly delved into the available meteorological resources and has asked for more relevant data. In parallel, improvements in atmospheric reanalysis have provided new sources of data that are heavily used by the solar community to derive solar radiation products.

Reanalysis is a systematic approach that assimilates actual weather observations into numerical weather prediction (NWP) or forecast models to produce consistent datasets for climate monitoring and research (More information and comparative lists of current reanalysis projects are available online.[1]). These datasets are typically developed by major meteorological institutions to retrospectively assess weather and climate. Actual observations from the surface and spaceborne sensors are combined using NWP models into a process called data assimilation, ultimately resulting in the reanalysis dataset. Typically, reanalyses span back over decades throughout the entire planet. Some of them are even being expanded continuously with only a few weeks or months of delay to include the most recent periods. Among the existing reanalyses, two recent ones are particularly worth mentioning here: NASA's MERRA-2[2] and ECMWF's ERA5.[3]

- NASA's Modern-Era Retrospective Analysis for Research and Applications, version 2 (MERRA-2 for short) is a reanalysis based on the Global Earth Observing System Version 5 (GEOS-5) research atmospheric model. MERRA-2 provides a large number of weather-related variables since 1980, at a spatial resolution of $0.5 \times 0.625°$ over the whole world and at two distinct temporal resolutions (one hour and one month). Predictions of both clear-sky and all-sky GHI are offered, per internal calculations performed with the Rapid Radiative Transfer Model for GCMs (RRTMG), which is openly available. RRTMG is a physical multiband model that resolves the radiative transfer equation with a two-stream solver and appropriate improvements to make irradiance predictions highly accurate. At each time step, calculations are done at various atmospheric levels, both for assumed ideal cloud-free situations and for the realistic cloudiness predicted by the model. RRTMG's accuracy is well validated (Iacono et al. 2004). Being computationally efficient, it is now used for radiative calculations in many NWP and global circulation models. It has also started to be recognized as a benchmarking reference for the validation of simpler models (Gueymard and Ruiz-Arias 2015; Ruiz-Arias and Gueymard 2015, 2018a). In its MERRA-2 implementation, however, all RRTMG-based GHI calculations are affected by a systematic error in the time stamp. In Eq. 2 of Chap. 1, the ET term is ignored in MERRA-2, which leads to a time error that varies between −14 and +16 min each hour, depending on the period. This translates into a significant sun position error, which seriously affects GHI. For that reason, MERRA-2's GHI predictions (both for clear-sky and all-sky situations) are not recommended. Additional details on MERRA-2's atmospheric outputs are provided in Sect. 6.
- The European Centre for Medium-Range Weather Forecasts (ECMWF) has developed successive reanalyses as part of its long-term European Reanalysis (ERA) project. The most recent version, ERA5, consists of a significant update

---

[1]https://reanalyses.org/atmosphere/comparison-table

[2]https://gmao.gsfc.nasa.gov/reanalysis/MERRA-2/.

[3]https://www.ecmwf.int/en/forecasts/datasets/reanalysis-datasets/era5.

of the previous version, ERA-Interim, and has just started to become available.[4] ERA5 currently covers the period since 2008, but will eventually be reprocessed to start in 1950.[5] The spatial and temporal resolutions are $0.25 \times 0.25°$ and 1 h, respectively. The global and direct horizontal irradiations (GHI and DHI) are calculated with RRTMG and are provided for both the ideal clear-sky case and realistically modeled cloudiness. The ERA5 outputs are currently accessible only by following an elaborate procedure,[6] which might evolve into a more user-friendly service as time progresses.

3 General Concepts

The evaluation of surface solar irradiance is necessary for a wide range of applications, each having specific requirements, such as spectral range and resolution, accuracy, component separation, or temporal resolution. Similarly, the type of radiation model that is most appropriate for each application may also depend on external factors, such as the availability of appropriate input data to cope with the intended application's requirements. In the general practice, however, what is meant by "clear-sky radiation model," hereafter CSRM, is very specific. It is a calculation method to obtain the components of solar irradiance from a number of inputs that characterize the state of an assumed cloudless atmosphere at a specific location and time. The two main components that are sought after are the direct normal irradiance (DNI) and the global horizontal irradiance (GHI). However, most CSRMs separately evaluate only two components: DNI and the diffuse horizontal irradiance (DIF). In such cases, GHI is derived from the first two through the fundamental closure equation:

$$GHI = DIF + DNI \cos Z \tag{1}$$

where Z is the sun's zenith angle, which is calculated with methods described in Chap. 1.

Some specialized CSRMs are devoted to the evaluation of the clear-sky global tilted irradiance (GTI), particularly to evaluate the cooling loads of buildings. In the general case, however, only the all-sky GTI is of interest in solar energy applications. It can be evaluated using one of the many "transposition" models of the literature (Gueymard 2009; Yang 2016), which rely on preliminary determinations of the all-sky GHI and DNI as inputs.

---

[4]http://rtweb.aer.com/rrtm_frame.html.

[5]https://software.ecmwf.int/wiki/display/CKB/ERA5+data+documentation.

[6]https://software.ecmwf.int/wiki/display/CKB/How+to+download+ERA5+data+via+the +ECMWF+Web+API.

3.1 Modeling Approaches

The focus of this book is on the determination of solar irradiance at the earth's surface. Considering the geographical expansion of solar installations over the world, and the constant development of new solar technologies, it must be stressed that the concept of "surface irradiance" needs qualification. First, high-elevation solar applications increase rapidly. Regions like the Tibetan Plateau or the Atacama Desert have a high solar resource, which attracts many solar projects. (The latter region has actually the highest solar resource in the world.) Hence, solar engineers need to be sure that the CSRM they use is accurate under such extreme surface conditions. Moreover, various types of stratospheric airships, also referred to as "high-altitude platforms," are being designed to efficiently collect solar energy at altitudes typically varying from 6 km to more than 20 km, i.e., above essentially all cloud layers (d'Oliveira et al. 2016; Li et al. 2016; Zhang 2016; Zhu et al. 2018). Such applications require a CSRM that can handle the specific conditions of the atmosphere at such altitudes, despite the lack of precise in situ input data.

An essential distinction between the CSRMs used in solar applications and those used in atmospheric sciences (including meteorology and climatology) is that the latter must provide the irradiance at all vertical levels of the atmosphere to evaluate the vertical heating rate profile in addition to the surface irradiance, which is what the solar-centric CSRMs just provide. This distinction explains why these CSRMs are simpler than the detailed radiative transfer models (RTMs) used in atmospheric sciences. To account for the many possible differences between the CSRMs available in the literature, a typology involving five different model classes has been proposed (Gueymard and Ruiz-Arias 2015). Models of Class A are those that are meant to be used primarily in atmospheric sciences, and will not be discussed extensively in this chapter, except to note that they can also be used for benchmarking solar CSRMs (Gueymard and Ruiz-Arias 2015; Ruiz-Arias and Gueymard 2018a). At the other extreme, models of Class E are those that are completely empirical and highly simplified. Owing to their poor performance (Gueymard and Ruiz-Arias 2015), these are not recommended. Common models in intermediate classes are discussed in Sect. 5.

3.2 Spectral Reference

Solar applications primarily depend on the solar resource, which is characterized by the total solar irradiance received at the surface, irrespective of its wavelength. More and more applications, however, also rely on spectral information, either because a specific wave band (such as the UV) is of special interest, or because the process under scrutiny has a variable spectral response. This is particularly the case of photovoltaic (PV) cells. Note also that the RTMs used in atmospheric sciences are usually of a spectral nature, which adds to their complexity.

If spectral details are necessary, a spectral CSRM is required. If not, a broadband CSRM is normally sufficient. The two types of model are discussed further in Sect. 5, with emphasis on broadband models, since they are used more customarily than spectral models in solar applications.

4 Modeling Principles

In the current solar practice, a CSRM is often needed for computer-intensive tasks, such as the calculation of clear-sky irradiance over continents at high spatiotemporal resolution. In such a case, the candidate CSRM is expected to achieve two different goals: accuracy and speed of execution. These are actually conflicting goals because speed calls for highly simplified or empirical algorithms, which can result in incorrect results under a number of circumstances. In contrast, the RTMs used in atmospheric sciences are much slower, but of a physical nature that guarantees their accuracy under any circumstance.

An "exact" solution of the radiative transfer equation typically results in an execution that is many orders of magnitude longer than that of common CSRMs. This is one reason why physical RTMs are normally not used in the solar practice. Instead, parameterizations of the different radiative transfer processes are developed. Historically, the simplest parameterizations were based on the use of the Linke turbidity coefficient, T_L, as the main atmospheric input. Examples of this type of model are those of Class D (Gueymard and Ruiz-Arias 2015), such as the ESRA model (Rigollier et al. 2000). The performance of such models appears significantly inferior to that of various models in Classes A–C in general. A recent benchmarking study (Ruiz-Arias and Gueymard 2018b) concluded that simplified Linke-based CSRMs prevented accurate simultaneous predictions of GHI and DNI. Hence, such models cannot be recommended anymore, especially for applications demanding good accuracy. A potential exception would be for the evaluation of the baseline clear-sky irradiance in the context of short-term DNI forecasting because T_L (which varies only slowly over time) can be extracted from immediately recent on-site pyrheliometer measurements under clear line-of-sight conditions (Chauvin et al. 2018; Inman et al. 2015). The methodology used by CSRMs of Classes B and C is discussed in the next subsections.

4.1 Atmospheric Processes and Optical Masses

Solar radiation is attenuated from the top to the bottom of the atmosphere. At the top of atmosphere (TOA), only direct irradiance exists. Its magnitude is controlled by two different processes: the total solar irradiance (TSI) and the sun–earth distance factor. TSI varies slightly over time due to solar activity, as shown in Fig. 1. Its long-term average value, called the solar constant (SC), is now determined as

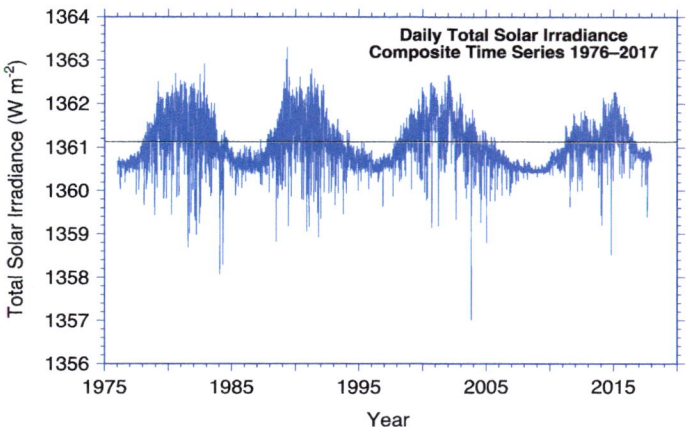

Fig. 1 Daily total solar irradiance time series from a 42-year reconstruction (Gueymard 2018). The horizontal line indicates the long-term mean, or "solar constant" (1361.1 W/m^2)

1361.1 W/m^2 (Gueymard 2018). In solar applications, it is customary to ignore the daily variations of TSI and to consider only a fixed SC value at TOA. The maximum error thus made is about $\pm 0.3\%$, which happens to be the magnitude of the lowest possible uncertainty in irradiance measurements based on the World Radiometric Reference. The sun–earth distance factor adds a deterministic correction, which is normally provided by the sun position algorithm also needed to obtain the sun's zenith angle (Chap. 1). Denoting this correction as S and the solar constant as E_{sc}, the TOA irradiance at normal incidence is simply obtained as $E_{0n} = S\, E_{sc}$, if the daily TSI variability is neglected.

The attenuation processes in the clear atmosphere are of two different kinds: scattering and absorption. The first process is what creates diffuse radiation, which is partly available at the surface. The second process is what provides energy to the atmosphere, in the form of heat. That part is lost to the surface. The processes of interest, whose individual transmittance must be modeled to derive the irradiance incident at the surface, are:

1. Molecular (or Rayleigh) scattering
2. Aerosol scattering
3. Aerosol absorption
4. Water vapor absorption
5. Ozone absorption
6. Mixed gas absorption
7. Nitrogen dioxide absorption.

The Bouguer–Lambert–Beer law can be applied in each case (except for water vapor absorption, which is more complex), so that the spectral transmittance of process i, $T_{i\lambda}$, can be evaluated as $\exp(-m_i \tau_{i\lambda})$, where m_i is the optical mass of process i and $\tau_{i\lambda}$ is its spectral optical depth. The optical air mass (or simply "air mass," often denoted as AM or m) is most familiar to solar analysts. It represents the path-length of the direct beam through the atmosphere relative to the reference case of an overhead sun when considering molecular extinction. Hence, an air mass of 1 (usually noted AM1) denotes an overhead sun ($Z = 0°$). If, for simplification, the atmosphere could be assumed plane-parallel, the optical mass would simply be 1/cos Z. In reality, the spherical nature of the atmosphere makes this only a rough approximation. Moreover, the optical mass depends on the extinction process because each process is mostly active over a specific altitude range, in general. For processes whose majority of constituents is concentrated at high altitude, like ozone, the optical mass tends to be low. Vice versa, the optical mass is high when the constituents are concentrated near the surface, like aerosols or water vapor. The cases of air molecules and mixed gases are intermediate, since they are dispersed over the whole atmosphere. Figure 2 illustrates the difference between various optical masses for $Z > 80°$. A general equation can be used to obtain the optical mass of each process:

$$m_i = 1 / \left[\cos\ Z + a_{i1} Z^{ai2} / (a_{i3} - Z)^{ai4} \right] \tag{2}$$

where the a_{ij} coefficients are provided in Table 1. In most simplified CSRMs, only one optical mass is considered for all extinction processes—the air mass. This accelerates calculations, but tends to lower accuracy under low-sun conditions.

An important issue is that of the effect of surface elevation, or site pressure, on AM. In the early days of solar radiation modeling, particularly when using the Linke approach mentioned above, it was customary to consider an absolute or pressure-corrected air mass, AMp, such that AMp = AM p/p_0, where AM is the relative air mass defined above, p is the site pressure, and p_0 is the standard pressure. The absolute air mass concept is still used by some authors, which is

Fig. 2 Optical masses for various atmospheric processes at high zenith angles ($Z > 80°$). Below 80°, all optical masses are close to 1/cos Z

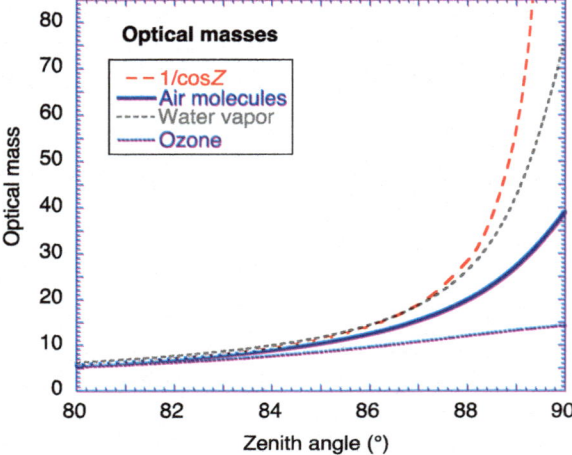

Table 1 Coefficients for Eq. 2, considering each atmospheric extinction process

Extinction process	i	a_{i1}	a_{i2}	a_{i3}	a_{i4}
Molecular scattering	1	0.48353	0.09585	96.741	1.7540
Aerosol scattering	2	0.16851	0.18198	95.318	1.9542
Aerosol absorption	3	0.16851	0.18198	95.318	1.9542
Water vapor absorption	4	0.10648	0.11423	93.781	1.9203
Ozone absorption	5	1.06510	0.63790	101.800	2.2694
Mixed gas absorption	6	0.48353	0.09585	96.741	1.7540
Nitrogen dioxide absorption	7	1.12120	1.61320	111.550	3.2629

confusing and may lead to incorrect modeling assumptions and irradiance predictions because not all atmospheric extinction processes decrease proportionally to pressure. In what follows, only the relative air mass, m, is considered.

4.2 Optical Depths and Transmittances

All atmospheric extinction processes mentioned above are spectrally dependent. For instance, the molecular optical depth, $\tau_{m\lambda}$—which is at the core of the sky's blue color according to Rayleigh's theoretical developments—decreases proportionally to $\lambda^{-\gamma}$, where λ is wavelength (in μm) and $\gamma \approx 4$. Moreover, $\tau_{m\lambda}$ is proportional to the site's pressure. Similarly, the aerosol optical depth (AOD) at wavelength λ, $\tau_{a\lambda}$, is conveniently expressed as

$$\tau_{a\lambda} = \beta\lambda^{-\alpha} \tag{3}$$

where β is the Ångström turbidity coefficient and α is the Ångström exponent. AOD is an optical characteristic of the total columnar amount of aerosols between surface and TOA, but is not directly dependent on pressure (contrarily to $\tau_{m\lambda}$). Equation 3 is known as Ångström's law (Ångström 1929), while aerosol scattering in general is a particular case of the Mie scattering theory. (Note that a small fraction of $\tau_{a\lambda}$ characterizes the aerosol absorption properties.) More details on how aerosols interfere with irradiance are provided in Sect. 7.

Whereas the Rayleigh and Mie scattering processes vary smoothly with wavelength, things are completely different regarding gaseous absorption, which is highly spectrally selective. For instance, ozone absorbs heavily below 0.35 μm, and much less between 0.45 and 0.75 μm, whereas oxygen has a sharp absorption band at 0.76 μm, and water vapor has many absorption bands in the near IR, including large bands around 0.94, 1.15, 1.4, and 1.85 μm. Additional details may be found in (Gueymard and Kambezidis 2004; Ruiz-Arias and Gueymard 2015).

4.3   Interdependence of Broadband Transmittances

At any wavelength in the spectral space, all the extinction processes introduced above can be assumed independent from each other. Hence, for a specific wavelength, the total atmospheric transmittance for direct spectral irradiance can be obtained as the product of each individual spectral transmittance. This is the way that a spectral radiation model like SMARTS (Gueymard 1995, 2001) evaluates the spectral direct irradiance at the surface. In the broadband space covering all wavelengths of the "shortwave" spectrum (0.3–4.0 μm), however, things become more complex because the constituents of each extinction process are concentrated in different layers of the atmosphere. Most authors of transmittance-based CSRMs simply evaluate the broadband transmittance of extinction process i, T_i, as

$$T_i = \int T_{i\lambda} E_{0n\lambda} \mathrm{d}\lambda / \int E_{0n\lambda} \mathrm{d}\lambda \qquad (4)$$

where $T_{i\lambda}$ is the spectral transmittance of process i, $E_{0n\lambda}$ is the spectral irradiance at TOA, and the integration is performed over the whole shortwave spectrum. In reality, Eq. 4 would have general validity only if each extinction process occurred in a single layer just below TOA and over a reasonably short spectral interval. Since this is actually not the case, the broadband transmittances of different processes become somewhat interdependent. Detailed discussions of the problem, and mathematical demonstrations, can be found in (Gueymard 1996, 1998, 2003a; Molineaux and Ineichen 1996). In particular, it is stressed that the correct formulation to obtain the broadband transmittance of process i is actually also a function of all other processes, according to:

$$T_i = \int T_{i\lambda} \prod_{j=1}^{j=i-1} T_{j\lambda} E_{0n\lambda} \mathrm{d}\lambda / \int \prod_{j=1}^{j=i-1} T_{j\lambda} E_{0n\lambda} \mathrm{d}\lambda \qquad (5)$$

Even though the paradigm of independent broadband transmittances (Eq. 4) does simplify calculations, which justifies its assumption, it typically leads to prediction errors for various combinations of constituent concentrations and sun positions.

To obtain accurate broadband irradiance results, four possible alternatives can be considered:

1. Make all calculations with a high-resolution spectral model to derive spectral irradiances at the surface, and then integrate spectrally to obtain the desired broadband irradiances. Although in principle this is the most accurate method, it has not been frequently used in the solar resource practice so far because of the logistics and substantially increased execution time. Nevertheless, there is growing interest for the development of spectral information on a worldwide basis in the context of improving the simulation of PV systems with respect to,

e.g., their actual spectral gains or losses compared to reference conditions (Polo et al. 2017; Xie et al. 2017).

2. Prepare a large quantity of preliminary simulations with a highly accurate model, using a vast number of data points covering the whole range of observable values for all atmospheric inputs, then arrange the results in the form of a large lookup table (LUT) of broadband irradiances, which is stored in a convenient repository for future usage. At execution time, the LUT is queried for the results corresponding to all inputs closest to the actual conditions, and some interpolation is performed. This is the avenue followed by the Zhang model (Zhang et al. 2018) and the McClear model (Lefèvre et al. 2013), for instance. The latter model is particular because it actually presents itself as a centralized service (with online access), rather than as a stand-alone model that anyone can reproduce. The time required to develop the LUT can be extremely long (up to many months), but once this is done, all executions can be very fast.

3. Follow the multilayer-weighted-transmittance modeling scheme (Gueymard 1996, 1998, 2003a), whereby broadband transmittances are developed in a way that the broadband optical depths involved take the extinction layers' interdependence into account. For example, the broadband AOD becomes a function of not only aerosol inputs but also of water vapor inputs. This formalism leads to difficult parameterizations, at least using conventional means. (Advanced fitting methods using, e.g., artificial neural networks could improve this modeling approach.) Models of this type are MLWT1 (Gueymard 1998) and its successor, MLWT2 (Gueymard 2003a), but these provided the broadband surface DNI only.

4. Perform the solar irradiance calculations over a limited number of spectral bands that are wisely selected to minimize interdependence errors. For instance, this avenue was followed for the development of CPCR2 (Gueymard 1989) and its successor, REST2 (Gueymard 2008b), using two bands separated at 700 nm, on the ground that atmospheric extinction is dominated by scattering below that limit, and by absorption above it. Other examples include RRTM and RRTMG, which use a larger number of bands.

It is informative to quantify the interdependence effect on broadband transmittances. For simplification, the broadband transmittance of an atmospheric process is alternatively defined here as the ratio between the actual broadband DNI and the broadband DNI corresponding to an ideal total absence of the constituent driving that process, where all other constituents remain fixed. For demonstration purposes, this exercise is done separately below for the ozone and water vapor transmittances, assuming that the aerosol abundance is the only other constituent that can vary in each case. To calculate the transmittance of ozone, its vertical columnar amount, u_o, is needed. It is typically expressed in atm-cm or in Dobson unit (DU), with 1 atm-cm = 1000 DU, and varies in general between 0.2 and 0.5 atm-cm. (Ozone hole situations arise when $u_o < 0.1$ atm-cm.) The water vapor abundance is measured by a vertical amount of precipitable water (PW or w), which can be expressed in units of height (cm or mm) or specific mass (kg/m^2 or g/cm^2). Note that 1 cm of PW is equivalent to 10 kg/m^2.

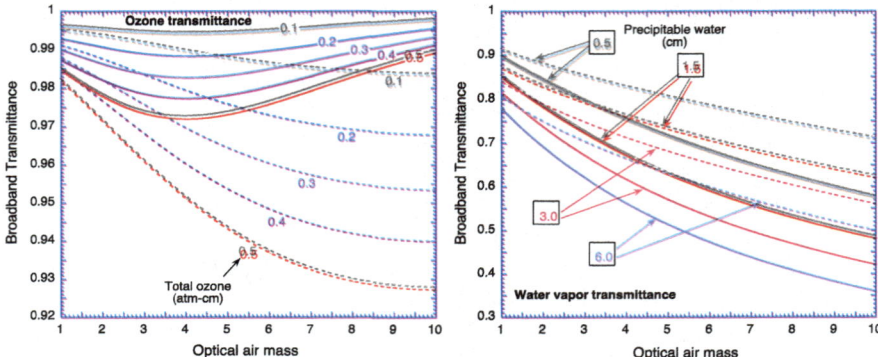

Fig. 3 Broadband ozone transmittance (left) and water vapor transmittance (right) as a function of the optical air mass and various gas concentrations, as calculated using the detailed formalism of the SMARTS spectral radiation model. Dashed lines indicate results pertaining to low-AOD conditions (0.084 at 0.5 μm), whereas solid lines indicate results pertaining to high-AOD conditions (0.54 at 0.5 μm). For simplification, all other atmospheric conditions are fixed

For each absorption process under scrutiny, the corresponding spectral irradiance is first calculated with SMARTS and then integrated to provide the desired broadband irradiance. The ozone transmittance is calculated here for ozone amounts between 0.1 and 0.5 atm-cm, assuming a sea-level site and the US Standard Atmosphere ($w = 1.416$ cm). A rural aerosol model is also assumed, with an AOD of either 0.084 or 0.54 at 0.5 μm. The former value is typical of clean conditions (Gueymard et al. 2002) and was adopted to promulgate reference solar spectra, as standardized in ASTM G173, ASTM G197, and IEC 60904-3. The larger AOD value is representative of hazy conditions and is being used to obtain subordinate reference spectra (Jessen et al. 2018). In parallel, the water vapor transmittance is calculated for w varying from 0.5 (dry conditions) to 6 cm (very humid conditions). A typical ozone amount of 0.35 atm-cm is fixed for that calculation, and the same two AOD values as above are used to characterize widely different aerosol situations. All other atmospheric conditions remain fixed for simplification. The results for the two transmittances are illustrated in Fig. 3 and are plotted as a function of the optical air mass (rather than the ozone mass or water vapor mass) for convenience. The results clearly demonstrate that the broadband transmittance of a specific extinction process is actually not a pure function of that process's constituent.

5 Selecting Clear-Sky Radiation Models

Many CSRMs have been proposed in the literature, and new ones still continue to be developed. Badescu et al. (2012a, b, 2013) have described and tested 54 such models, most of them of a simplified empirical nature. Many other studies have validated various CSRMs, as analyzed or reviewed extensively in the literature, including recent contributions (Engerer and Mills 2015; Gueymard 2012a; Ineichen 2016; Ruiz-Arias and Gueymard 2018b). Considering that the latter studies underline improvements in performance from recent CSRMs, it is desirable to mention here only a small selection of those CSRMs that have good potential in solar applications and that have been recommended for their accuracy and/or universality in at least a few validation studies, including the limited one developed below in Sect. 8. The selected models are succinctly described below, in alphabetical order. All of them are designed to predict the broadband irradiance components on a horizontal surface only. A notable exception is SMARTS, which is designed to predict the spectral irradiance on either horizontal or tilted surfaces.

Since models typically require different sets of inputs, the reader is referred to Table 2 for a comprehensive description of which inputs are required by each model. Note that some of the models described here are still being improved, so that their inputs and performance may both depend on which version is used. Furthermore, it is stressed that the performance of a model strongly depends on the quality of its inputs, as will be demonstrated in Sect. 8. In general, the user is left with the delicate—and potentially error-prone—task of selecting the best possible set of inputs to operate a CSRM. To help the reader in that task, Sect. 6 offers a comprehensive review of the most important sources of input data that are available as of this writing. Since more, and possibly better, data sources can be expected in the future, the reader should remain informed as much as possible.

Operating any given CSRM for the location and period under scrutiny is not an easy task because of the quantity and required quality of the inputs. These must usually be obtained from different sources—an operation that actually involves many preliminary steps because of the frequent requirement of interpolating, extrapolating, correcting, or synchronizing them. To remedy this time-consuming difficulty, an alternate option is to avoid this process altogether and rather obtain precalculated irradiance data. In that avenue, the underlying modeled clear-sky GHI and DNI irradiance predictions are produced by "black boxes," which are discussed

Table 2 Requested atmospheric inputs (besides date, zenith angle, and solar constant) for all models considered here

Model	ρ	p	u_o	u_n	w	τ_{a550}	τ_{a700}	α	β	ϖ	Aerosol type
Bird	•	•	•		•			•	•		
Ineichen (2008)		•			•		•				
Ineichen (2018)		•			•	•					•
REST2v5	•	•	•	•	•			•	•	•	
REST2v9.1	•	•	•		•	•		•		•	
SMARTS	•	•	•	•	•	•		•		•	

Possible inputs are: ρ surface albedo; p site pressure; u_o total ozone abundance; u_n total nitrogen dioxide abundance; w precipitable water; τ_{a550} aerosol optical depth at 550 nm; τ_{a700} aerosol optical depth at 700 nm; α Ångström's wavelength exponent; β Ångström's turbidity coefficient; ϖ aerosol single-scattering albedo; and aerosol type in Sect. 5. In this context, a black box is simply an "inaccessible" radiation model, whose inputs are directly wired to specific atmospheric databases, resulting in a closed calculation system that *cannot* be internally accessed or modified by end users. In such a case, the model's performance cannot be separated from the quality of its inputs, which may not always be a desirable feature.

5.1 Bird Model

The model developed by Bird and collaborators is one of the earliest detailed solar transmittance models of the literature (Bird and Hulstrom 1980, 1981a, b). Its transmittance functions were derived from a more sophisticated spectral model. Its performance has been validated in a number of studies, e.g., Badescu et al. (2012b); Gueymard(2003b, 2012a); Gueymard and Myers (2008); Gueymard and Ruiz-Arias (2015); Ineichen (2006). The aerosol inputs of the original model depend on antiquated spectral data. This prompted a simple modification to accommodate more modern sources of data (Gueymard 2012a).

5.2 Ineichen Models

Similarly to Bird, Ineichen developed a broadband scheme from an elaborate spectral model called SOLIS (Mueller et al. 2004) and referred to its broadband version as "simplified SOLIS" (Ineichen 2008). It is rather referred to here as "Ineichen 2008" to avoid confusion. That model has been validated in various studies (Antonanzas-Torres et al. 2016; Badescu 2013; Badescu et al. 2012b, 2013; Engerer and Mills 2015; Gueymard 2012a) and was generally found to perform very well, except under circumstances that include very high turbidity and/or very low or very high humidity (Gueymard and Ruiz-Arias 2015; Ruiz-Arias and Gueymard 2018a, b; Zhang et al. 2014). Two ways of dealing with the high-AOD situations were explored in (Ruiz-Arias and Gueymard 2018a), but none of them led to satisfactory results simultaneously for DNI and GHI.

To remedy this situation, Ineichen recently proposed an updated version of the model (Ineichen 2018), which is now constituted of parameterizations of results based on the libRadtran spectral radiative transfer package.[7] The new version is now referred to as "High Turbidity Solis Clear Sky Model" by its author (but here as "Ineichen 2018" for simplicity). Although this update is supposed to constitute an improvement over the previous version, this author found two important issues:

- The model requires the *type* of aerosol as an input, to be selected manually by the user from four possible types: rural, maritime, urban, or tropospheric. (Note

---

[7]http://libradtran.org.

that two other important types—dust and smoke—are not represented.) This is cumbersome because, in the real world, aerosols are actually mixtures of different types, which change dynamically over space and time. Hence, Ineichen suggested that the rural aerosol type could be used as a default in all situations, but this remains to be independently verified.

- Many equations are incorrectly typed in the paper, which makes all predictions incorrect. While waiting for a Corrigendum to be published, the spreadsheet format of the model (linked to the paper) should be used.

5.3 REST2 Model

The REST2 model uses the band separation principle described in Sect. 4.3 to reduce the prediction errors caused by interdependent transmittance effects. The shortwave spectrum is split into two bands, with a separation at 0.7 µm. Transmittance parameterizations are based on predictions from the SMARTS model (described in the next subsection) for each of the two bands. REST2 has been in constant development since 2003. Version 5 is described in (Gueymard 2008b) and is compared to the most recent version 9.1 in Sect. 8. The latter version is used to predict clear-sky fluxes as part of more sophisticated solar radiation models (Sengupta et al. 2018; Xie et al. 2016). REST2 has been validated in a number of studies (Antonanzas-Torres et al. 2016; Badescu et al. 2012a, b; Eissa et al. 2018; Engerer and Mills 2015; Gueymard 2012a; Gueymard and Myers 2008; Gueymard and Ruiz-Arias 2015; Ineichen 2016; Zhong and Kleissl 2015).

5.4 SMARTS Model

The simple model of the Atmospheric Radiative Transfer of Sunshine (SMARTS) has been in development since the early 1990s (Gueymard 1995, 2001). More recent versions of the code can be freely obtained from its normal repository[8] for research and education purposes, upon acceptance of its license agreement. Version 2.9.2, released in 2003, was used to derive reference spectra (Gueymard et al. 2002), which then became the basis for three ASTM standards (G173, G177, and G197) and an IEC standard (60904-3). Similar standardization action is also underway at ISO (Jessen et al. 2018). Various improvements were introduced in the more recent versions 2.9.5 (2005) and 2.9.8 (2018). A special and faster version (3.2) has also been specially developed for intensive calculations, such as those described in Sect. 8.

---

[8]https://www.nrel.gov/rredc/smarts/.

SMARTS evaluates the clear-sky spectral distribution of the direct, diffuse, and global components for radiation incident on horizontal or tilted surfaces. The spectral resolution varies between 0.5 and 5 nm in the range 280–4000 nm, for a total of 2002 wavelengths. The model has been extensively validated for the cases of both spectral and broadband irradiance predictions (Gueymard 2005, 2008a; Ruiz-Arias and Gueymard 2018a) and has been used in many studies related to the determination of spectral effects in various solar energy processes, particularly those using planar or concentrating photovoltaic generators (Baig et al. 2016; Fernández et al. 2013; Liu et al. 2016; Polo et al. 2017; Theristis and O'Donovan 2015).

5.5 Special Models

The McClear model has been developed using the LUT approach discussed above (Lefèvre et al. 2013). The model's large LUT (more than 1.7×10^9 points) is populated by irradiance results from libRadtran. The atmospheric inputs are provided by the gridded CAMS reanalysis database of ECMWF (at ≈100-km resolution) on a 3-hourly basis, and the surface albedo values are derived from 16-day mean MODIS estimates. The irradiance results are accessible to registered users of the CAMS Web service,[9] where the user can only select the location's coordinates, start time and end time, time step (between 1 min and 1 month), time reference (UTC or solar time), and file format (CSV or NetCDF). As of this writing, the simulated time series are produced with version 3.1 of the model and are available only since January 01, 2004. McClear has been validated in various subsequent studies (Eissa et al. 2015; Lefèvre and Wald 2016; Zhong and Kleissl 2015).

Another source of precalculated irradiance data comes from atmospheric reanalyses, such as MERRA-2 or ERA5, as discussed in Sect. 2, but their spatiotemporal resolution is currently limited.

6 Atmospheric Input Data

Radiation models need inputs that describe the (variable) state of the atmosphere at any instant. As a general rule, the more sophisticated a model is, the larger number of specific inputs it will require. This explains why CSRMs with no (or just one) atmospheric input cannot perform as well as more detailed models. The number of inputs is not a guarantee of prediction accuracy, however. The latter actually depends on three conditions:

---

[9]http://www.soda-pro.com/web-services/radiation/cams-mcclear.

1. Detailed modeling correctly describing the underlying physics of the extinction processes, for any sun position
2. Specific and appropriate atmospheric input(s) for each significant extinction process
3. Sufficient accuracy of the atmospheric inputs at any time or location.

6.1   Input Requirements

In practice, the last requirement above is usually the most limiting factor. If solar irradiance estimates are necessary at a specific site where a solar project is just at the prospection stage, there are typically no on-site observations of the atmospheric inputs that are necessary to run a CSRM. Since there is no perfect solution to this common problem, any local clear-sky irradiance prediction from a CSRM will suffer from input-induced error propagation. Such modeled irradiance results can thus be expected to have higher uncertainty than a measurement that would follow the best practices (Sengupta et al. 2017). To replace missing on-site atmospheric observations, various imperfect methods can be used, in isolation or combination:

1. Interpolate data between two or more stations at some distance.
2. Spatially extrapolate data from a single station.
3. Use gridded data for the specific cell that contains the site under scrutiny.
4. Interpolate in time if data exists at a lower temporal resolution.
5. Estimate the input data based on some available proxy.

The first four solutions are self-explanatory. An example for the last situation would be if both temperature and relative humidity were measured on-site or nearby. It is then possible, in general, to estimate PW from such measurements with sufficient accuracy (Gueymard 2014a). Another common proxy is site's elevation, h, from which the site pressure, p, can be easily estimated using the scale-height approximation:

$$p = p_0 \exp\left(-h/H_p\right) \tag{6}$$

where $p_0 = 1013.25$ hPa is the standard sea-level pressure, $H_p = 8.4345$ km is the atmospheric scale height, and elevation h is expressed in km above sea level. This approximation makes p a fixed value over time, but this is sufficient because its direct impact on irradiance is known to be small (Gueymard 2003b).

Table 3 describes the approximate magnitude of the impact of different extinction processes on the three irradiance components and associates the typical atmospheric inputs that are necessary to evaluate each effect. Depending on model or author, the AOD ($\tau_{a\lambda}$) can be associated with different wavelengths, most frequently 0.5, 0.55, 0.7, or 1 μm. All corresponding values can be converted into each other through Eq. 3, but α needs to be known at all times. DNI and DIF are

Table 3 Approximate magnitude of different extinction processes on clear-sky irradiance components expressed on a scale of 4

Extinction process	Impact on DNI	Impact on DIF	Impact on GHI	Relevant input
Mixed gas absorption	+	+	+	p
Nitrogen dioxide absorption	+	+	+	u_n
Ozone absorption	+	+	+	u_o
Water vapor absorption	++	++	++	w
Aerosol absorption	+	+	+	ϖ
Aerosol scattering	+++	+++	++	$\tau_{a\lambda}$, α
Molecular scattering	++	++	++	p
Backscattering	—	+	+	ρ

none (—), small (+), moderate (++), and high (+++)
The relevant atmospheric quantity related to each extinction process appears in the last column
Key p site pressure; u_n total nitrogen dioxide columnar amount; u_o total ozone columnar amount; w precipitable water; ϖ aerosol single-scattering albedo; $\tau_{a\lambda}$ aerosol optical depth at wavelength λ; α Ångström exponent; ρ broadband surface albedo

most sensitive to AOD, which is why the latter's specific impact on clear-sky irradiances is discussed more thoroughly in Sect. 7.

6.2   Sources of Data

Based on the discussion above, it is obvious that, with the goal of producing high-quality irradiance outputs, the availability of accurate input data is as important (if not more) as selecting a high-performance CSRM. Solar analysts are thus frequently confronted to the problem of obtaining accurate input data for virtually any location in the world. Various issues make this task difficult and lead to some critical questions:

1. New sources of data appear regularly; which one to choose, and where to look for new data sources?
2. The accuracy of atmospheric data is not always well established or documented; how should that accuracy be characterized in the case of solar applications?
3. Errors tend to propagate from input to output; how does input inaccuracies translate into irradiance prediction uncertainty?
4. Some data might be accurate over some areas or some periods, and less accurate elsewhere or for other periods; how is it possible to obtain more information?
5. Observational time series might not be complete; what to do if they are not?
6. Available ground observations might be from a distant site; what maximum distance can be considered "safe" without interpolation or extrapolation?

7. Gridded data might be available only with a coarse spatial or temporal resolution; what would be the effect of interpolating to match the desired spatiotemporal resolution?
8. The data format might be unusual, without the necessary metadata, or requiring specialized software to decode; how to easily convert gridded data stored in a myriad of individual daily files into a time series for a single site, for instance?

Resolving all these questions is extremely time-consuming and can lead to wrong or suboptimal choices being made just "to save time." The literature dedicated to the validation of CSRMs also shows that different authors will rarely select the same source of input data to operate a specific model. This makes the direct comparison of model performance results of the literature difficult since a significant part of the prediction uncertainty is caused by inaccuracies in the input data. All these inadequacies contribute to the growing interest for special "black-box" models, such as those discussed in Sect. 5.5, and particularly those that are linked to reanalysis datasets, for reasons introduced in Sect. 2.

An important advantage of reanalysis datasets is that they provide long time series (sometimes of 30 years or more) at fine temporal resolution (hourly or 3-hourly) of many atmospheric variables over the whole world. Reanalysis products are provided on a gridded scale with moderate spatial resolution. Newer reanalysis products tend to have a finer resolution (e.g., $0.25°$ or $0.5°$) compared to older products (e.g., $1°$ to $3°$). Such resolutions are still coarser than those of satellite-derived irradiance products (typically $0.03°$ to $0.10°$). Nevertheless, reanalysis products are invaluable to provide the necessary atmospheric inputs to large-scale irradiance modeling. One notable inconvenience, however, is that reanalysis products are stored in a way that makes their access difficult to a majority of solar analysts. The gridded results are compressed into individual daily files (typically in netCDF, HDF, or GRIB format) for the whole world. To derive, e.g., a 15-year time series of atmospheric data for a single site in a more usual CSV format, knowledge of scripting tools (using, e.g., python) is essential. As of this writing, no public-domain set of scripting tools that would be tailored for such a task, and would be adaptable to various major reanalysis sources, seems to exist. The Giovanni Web site[10] constitutes a notable exception: The output of any of its data collections (such as many variables from the MERRA-2 reanalysis) can be requested as a time series in CSV format, at least for single-site queries.

In addition to Giovanni, some large online data repositories exist, with the goal of offering a large variety of data sources for many atmospheric variables, such as from reanalyses or satellite platforms. It is always good to explore these Web sites first to know what they can offer, and to do that regularly because they are often updated as new sources of data become publicly available. Examples of such data repositories include:

---

[10]https://giovanni.gsfc.nasa.gov/giovanni/.

1. GES DISC: https://disc.gsfc.nasa.gov/
2. CAMS: http://apps.ecmwf.int/datasets/data/cams-nrealtime/levtype=sfc/
3. CM SAF: http://www.cmsaf.eu/EN/Home/home_node.html
4. RDA: https://rda.ucar.edu/
5. SoDa: http://soda-pro.com/
6. Webservice-Energy: http://geocatalog.webservice-energy.org
7. AERONET: https://aeronet.gsfc.nasa.gov/.

The next subsections provide a discussion of the main sources of data for each extinction process.

6.3 Site Pressure

As mentioned in Sect. 6.1, pressure is not an essential input. A nearby meteorological station at same elevation can provide such data. World pressure observations are compiled at https://rda.ucar.edu/datasets/ds132.1/. Over flat terrain in a radius of at least 50 km, reanalysis data can provide such data too. For a rapid estimate, Eq. 6 is sufficient.

6.4 Surface Albedo

Under clear-sky conditions, the impact of the backscattering effect (which increases the surface DIF and GHI through inter-reflections between the surface and the scattering particles of the atmosphere) is relatively small, except over very bright surfaces (e.g., desert sand or snow). Many simplified CSRMs actually neglect this effect completely, which constitutes a source of error over bright areas. The MERRA-2, ERA5, and CAMS reanalyses provide detailed albedo information, as part of their normal output. If a finer spatial resolution is required, the MODIS MCD43GF albedo database[11] can be helpful. Observations from the two MODIS instruments (onboard the Terra and Aqua satellites) are merged, gap-filled, and manipulated to remove the effects of clouds and snow (Sun et al. 2017). The spatial resolution is much finer than that of reanalysis (0.0083°), but the temporal resolution is coarser (8 days). The database includes spectral and angular information of the reflectance process. Since this database only provides a snow-free albedo, additional information is needed to derive the albedo of snow-covered areas. Complementary snow and ice analyses are available,[12] but have various limitations.

---

[11]https://www.umb.edu/spectralmass/terra_aqua_modis/modis_brdf_albedo_cmg_gap-filled_snow-free_product_mcd43gf_v005.

[12]http://nsidc.org/data/G02156; http://www.globsnow.info/.

6.5 Ozone

The amount of stratospheric ozone is variable only slowly, so that data at a daily frequency can be considered sufficient. Tropospheric ozone (mostly produced in urban/industrialized areas) is more variable but is only $\approx 10\%$ of the stratospheric amount (Ziemke et al. 2011). The reanalyses mentioned above (MERRA-2, CAMS and ERA5) all provide the total columnar ozone amount (also referred to as "vertical column"). Single-point ozone data can be retrieved since 1978 from the *Ozone Over Your Head* tool.[13] Other sources of data are available from the Giovanni Web site (Sect. 6.2). Alternatively, databases of satellite observations can be explored.[14] Finally, a network of ground-based observing stations also exists.[15]

6.6 Nitrogen Dioxide

The impact of nitrogen dioxide (NO_2) absorption is small, except over polluted areas. In such cases, the amount of tropospheric NO_2 is far superior to that of stratospheric NO_2, contrarily to ozone. Various satellites monitor NO_2, and their data can be retrieved from the Temis and Sciamachy Web sites.[16]

6.7 Aerosols

Table 3 indicates that CSRMs might require various aerosol-related input variables. In terms of the magnitude of the effect they might have on surface irradiance, AOD is the most important one, followed by α. The Giovanni Web site provides spaceborne observations of daily or monthly AOD and α from various platforms. It also provides hourly estimates of AOD at 550 nm (AOD550 or τ_{a550}) and α from MERRA-2 since 1980. In parallel, the MACC and CAMS reanalyses from ECMWF provide AOD at various wavelengths (from which α can be calculated, see Sect. 7.1). An example of how a reanalysis can be used to provide all the necessary inputs to a CSRM is shown in Fig. 4. In this case, the mean annual clear-sky DNI and GHI are evaluated with SMARTS v3.2 using MACC inputs at their native resolution ($1.125° \times 1.125°$). Whereas latitude (which determines the zenith angle)

---

[13]https://ozoneaq.gsfc.nasa.gov/tools/ozonemap/.

[14]http://www.sciamachy.org/products/index.php?species=O3; https://disc.gsfc.nasa.gov/datasets/OMO 3PR_003/summary.

[15]https://woudc.org/data/explore.php?lang=en.

[16]http://www.temis.nl/airpollution/no2.html; http://www.sciamachy.org/products/index.php?species= NO2.

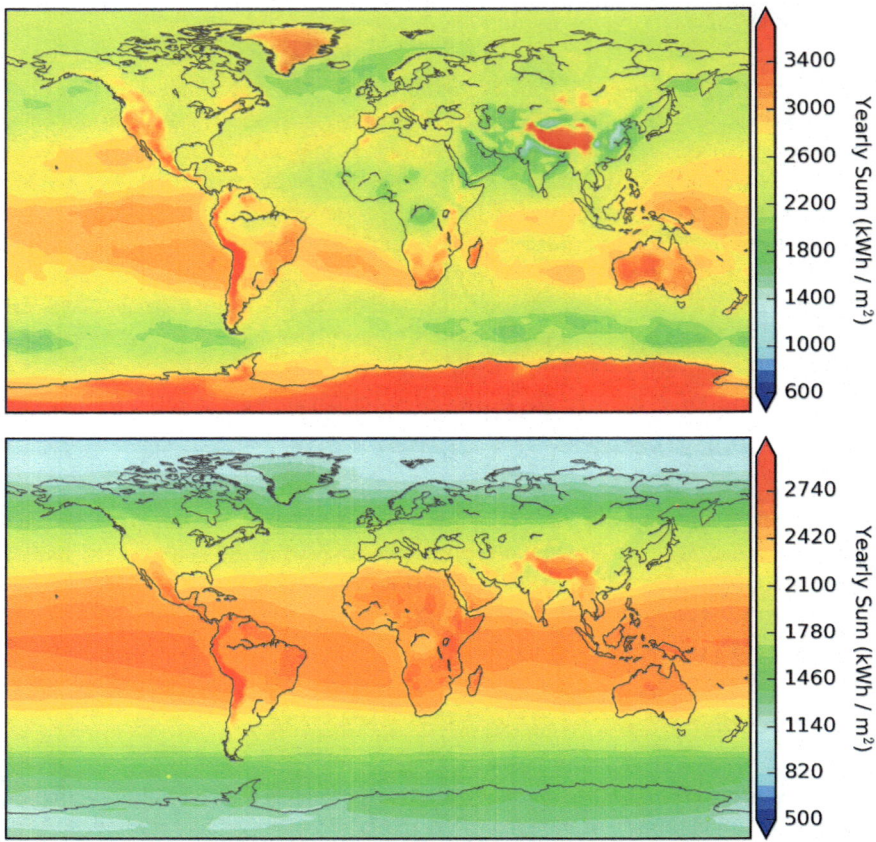

Fig. 4 Worldwide mean annual clear-sky DNI (top) and GHI (bottom) using SMARTS v3.2 and inputs from the MACC reanalysis during 2012. Images courtesy of Dr. J.A. Ruiz-Arias

is the major driver for GHI, DNI is clearly maximized over elevated, dry, and very clear areas, such as the Atacama Plateau, the Tibetan Plateau, or Antarctica.

Note: Under the impulsion of Ångström, AOD was usually reported at 1 μm in the past. This variable is referred to as the Ångström turbidity coefficient, β. In recent decades, however, the more prevalent reporting wavelength has become 550 nm. In contrast, ground-based sunphotometers observe AOD at various wavelengths, but usually *not* at 550 nm. As mentioned earlier, Eq. 3 can be used to convert AOD from a reference wavelength to another. Hence, any source of data can be used to obtain β from τ_{a550}, or vice versa, inasmuch as α is also available.

Besides the modeled data just mentioned, more accurate data can be obtained from ground-based sunphotometric observations. The main world aerosol network is NASA's AERONET (Holben et al. 1998), a federated worldwide network that now counts hundreds of stations, but their records are not continuous.

Aerosol observations are done every $\approx$15 min whenever the sun is high enough ($\geq 8°$). Post-processing of the raw Level-1 data consists in screening observations affected by clouds (Level 1.5), and then recalibrating and quality controlling the data (Level 2). Smaller sunphotometric networks also exist in the world, such as SKYNET,[17] ESR,[18] or GAW-PFR.[19] Additionally, national networks (CARSNET, CSHNET, and SONET) exist over China (Che et al. 2015; Li et al. 2018; Zhang et al. 2012), but without public access to the data.

The aerosol single-scattering albedo can be obtained at AERONET stations. It can also be derived from MERRA-2 as the ratio between the scattering optical depth (variable "TOTSCATAU") and the total aerosol optical depth at 550 nm discussed above ('TOTEXTTAU'). The aerosol type corresponds to ideal modeling assumptions, which arbitrarily separate aerosol mixtures into rural, urban, maritime, smoke, or desert types, for instance. In the real world, many combinations of such types exist and change continuously over space and time. Hence, the specific aerosol type is either only used for CSRM development (as with the McClear model), or must be specified a priori (as with Ineichen's 2018 model).

6.8 Water Vapor

For many decades, the vertical profile of water vapor has been derived from radiosonde records of pressure, temperature, and humidity at various atmospheric levels. Such measurements are typically done once or twice a day at meteorological stations. One major drawback of this type of data is that the total vertical column of water vapor (expressed in terms of precipitable water) needs to be integrated numerically along all pressure levels. This requires intermediate calculation, based on observations that can be affected by bias (Gueymard 2014a), and has only a coarse spatiotemporal resolution. Hence, recourse to radiosonde data is only advisable for the evaluation of clear-sky irradiance for periods of the distant past, when no other direct observation or reliable estimate existed. During the last few decades, meteorological services around the world have been installing networks of "Global Positioning System Meteorology" (GPS-Met) stations that first extract the zenith path delay from the GPS signal, and then back-calculate PW. This PW retrieval is computationally intensive, but is done in near-real time at supercomputer centers. One data portal for GPS-Met data is the Suominet Web site.[20] It covers a large part of North and Central America and provides 30-min PW data for hundreds of stations. In the case of Europe, many countries collaborate to the E-GVAP

---

[17]http://atmos3.cr.chiba-u.jp/skynet/data.html.

[18]http://www.euroskyrad.net/.

[19]http://ebas.nilu.no/default.aspx.

[20]http://www.suominet.ucar.edu/index.html.

network.[21] In addition to observing aerosol variables, the AERONET ground-based sunphotometer network also provides PW. Alternatively, the Giovanni Web site offers gridded hourly PW data from MERRA-2 since 1980. Similarly, the CAMS and ERA5 reanalyses also provide gridded PW data.

7 Aerosol Properties and Their Impacts on Irradiance

The surface solar irradiance cannot be predicted accurately without good knowledge of the prevailing atmospheric conditions. Under clear skies, aerosols usually constitute the main source of atmospheric attenuation. Nevertheless, many simple models that have been, or are still, in use in the solar resource community do not have a satisfactory way of handling the aerosol attenuation (Gueymard 2003a). This typically induces significant errors (Gueymard 2012a).

7.1 Aerosol Optical Depth and Its Measurement

The aerosol optical depth (AOD), also referred to as aerosol optical thickness (AOT), is a columnar quantity that represents the total optical attenuation effect of aerosols on the incident irradiance. This column extends from the top of the atmosphere to the surface (i.e., at the local site's elevation). Ground-based measurements are typically made with a sunphotometer pointing to the sun. Knowing the sun position and the spectral optical depth of each other radiatively active atmospheric constituent, the slant aerosol optical depth is converted to a vertical AOD, which is the desired quantity. (The ratio of the two quantities is the optical aerosol mass, discussed in Sect. 4.1.) Spaceborne observations are made with spectrometers that sense the radiance reflected by the atmosphere and surface. The AOD is inferred using a retrieval algorithm that takes into account the spectral reflectance of all other atmospheric constituents and of the surface. Because the reflected aerosol signal may be very small relative to the reflected signal from the surface (particularly for low-AOD conditions over highly reflective ground), retrievals from spaceborne data are necessarily much more uncertain than ground-based sunphotometric observations.

   Modern ground-based sunphotometers include a number of "channels," typically from 4 to 10, each equipped with a filter that isolates the direct irradiance within a few nanometers around a specified central wavelength. Most of these channels are dedicated to the measurement of AOD, but some channels can also be added to evaluate the total columnar amounts of ozone and/or water vapor. The aerosol signal is sensed in atmospheric windows where no strong gaseous absorption is

---

[21]http://egvap.dmi.dk/.

present. The aerosol-specific wavelengths depend somewhat on the manufacturer. A typical list of central wavelengths for aerosol observation is: 340, 380, 440, 500, 675, 870, and 1020 nm. AERONET sunphotometers also include a 940-nm water vapor channel, and sometimes an additional 1640-nm aerosol channel.

According to Ångström's law (Ångström 1929), the spectral AOD at wavelength λ (μm), $\tau_{a\lambda}$, can be expressed as a function of the AOD at 1 μm, β and the Ångström exponent, α, through Eq. 3. In practice, β and α can be derived from multiwavelength sunphotometric measurements by least-squares fitting of the spectral data to a linearized version of Eq. 3:

$$\ln \tau_{a\lambda} = \ln \beta - \alpha \ln \lambda \qquad (7)$$

The spectral range of validity of Eqs. (3) and (7) is debatable. It is at least 0.4 to 1 μm (Ångström 1929), but it also seems to hold in the UV (below 0.4 μm). As with any least-squares fitting process, the fit normally improves with the number of points, which calls for the largest possible number of sunphotometer channels. On the other hand, the uncertainty in spectral AOD increases somewhat below 0.4 μm and above 0.9 μm. As a result, AERONET considers that the best determination of α is obtained by limiting the data to the interval 0.44–0.87 μm. Whether or not such results are truly better than if more points from an extended interval were used, e.g., 0.34 to 1.02 μm, is still an open question.

7.2 Data Quality

An alternate way of obtaining β and α consists in resolving Eq. 7 using the $\tau_{a\lambda}$ data from only two wavelengths, λ_1 and λ_2. This simpler method generally introduces too much uncertainty and instability, depending on the respective values of λ_1 and λ_2 (Cachorro et al. 1987a, b; Martinez-Lozano et al. 1998), and is therefore not recommended.

Although Ångström's law is empirical, it is verified experimentally in the vast majority of cases, barring any exceptional circumstance in the aerosol size distribution or large experimental uncertainty. An example is shown in Fig. 5 using actual measurements obtained at the AERONET station of Tamanrasset, Algeria. The top curve (for July 28, 2008) is almost flat ($\alpha = 0.005$) and corresponds to a dust episode (very high β), during which most particles have a large size. The observations for November 18, 2006, are for background conditions, during which the particles size is moderate ($\alpha = 1.228$), and the particle concentration is low because of the station's high elevation. Values of α in the range 1.0–1.5 are most frequent and typical of "rural" conditions, which are normally dominated by vegetation-related particles. In contrast, large values of α (1.5–2.5) are encountered in polluted environments because of the preponderance of small particles. (The inverse relationship between α and the mean diameter of particles was already noted by Ångström 1929.) In Fig. 5, the results for February 08, 2010, exemplify the

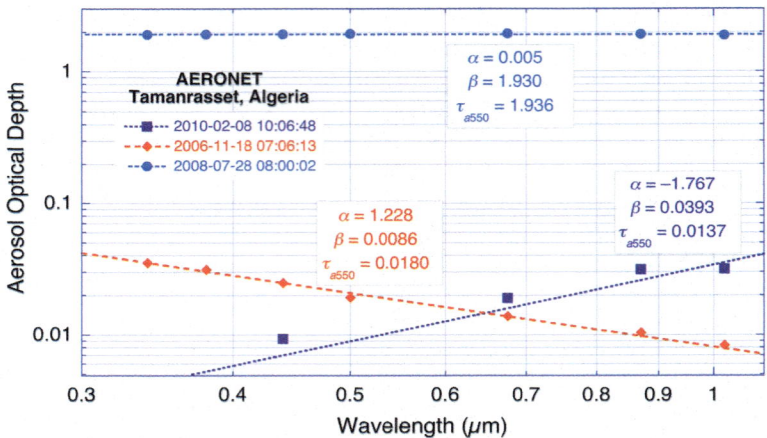

Fig. 5 Measured spectral AOD at Tamanrasset under widely different atmospheric conditions

(relatively rare) case of anomalous observations: α is then strongly negative (−1.767), which should not happen. (Values of α lower than about −0.1 or larger than $\approx$2.5 can be considered highly suspect; moreover, negative values of AOD at any wavelength are unphysical and should be eliminated.) Of the four channels that reported data at that moment, it can be assumed that at least two of them were faulty, for reasons unknown. Although AERONET "Level-2" observations undergo various quality tests conducive to a low uncertainty of $\approx$0.01 (Holben et al. 2001) and are typically considered "perfect" in all validation studies that use them as ground truth, they might actually be highly biased during occasional periods of malfunction. Such periods can be isolated by visual inspection of the β and α time series. An example of this appears in Fig. 6 for two AERONET stations that each experienced four periods of obvious malfunction over the years.

7.3   Impact on Solar Irradiance

The aerosol attenuation follows the Bouguer–Lambert–Beer law throughout the atmosphere, so that DNI is a decreasing exponential function of the corresponding slant attenuation coefficient, $m\tau_a$ (i.e., AM•AOD). Because of the compensation caused by the simultaneously increasing DIF, GHI only decreases moderately with $m\tau_a$ and almost linearly, as shown in Fig. 7. A rural aerosol model is considered here along with a fixed relative air mass of 1.5, two site elevations (sea level and 1500 m), and three PW amounts (0.47, 1.42, and 4.25 cm), while τ_{a550} is allowed to vary between 0 and 1.1. The sea-level case with $\tau_{a550} = 0.0764$ and the intermediate PW value correspond to the atmospheric conditions that were used to define the reference spectra and PV reporting conditions in important standards

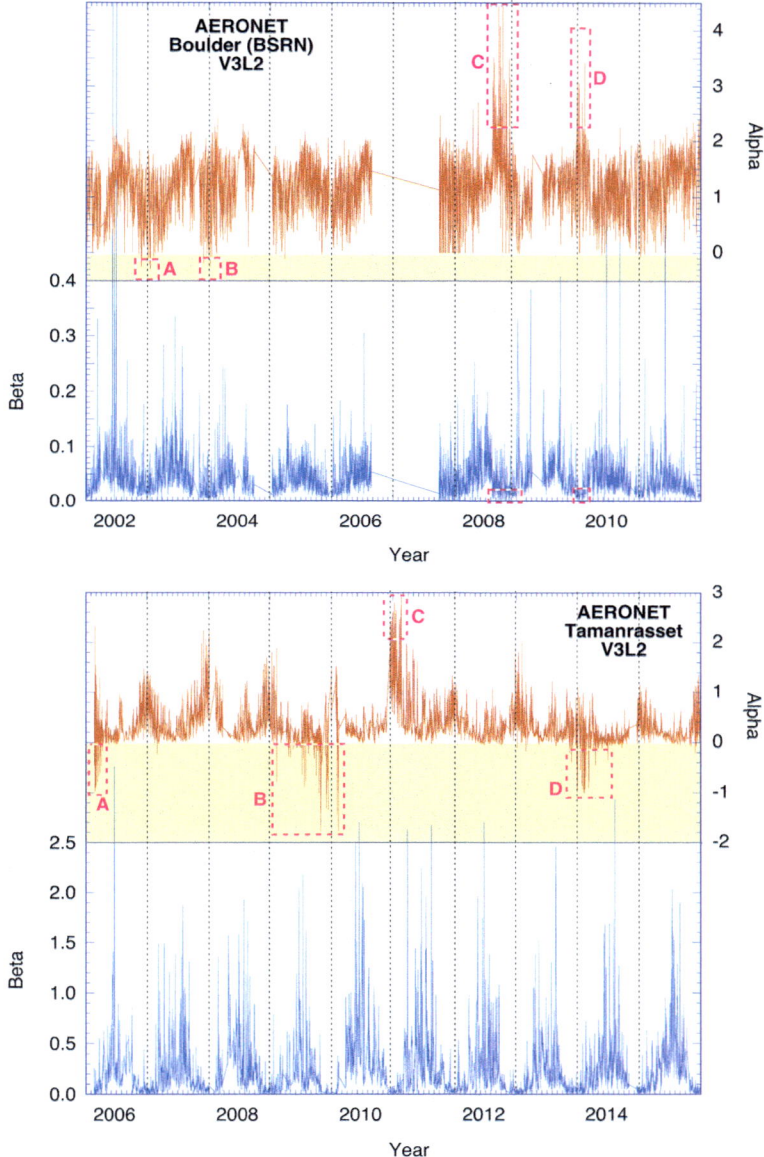

Fig. 6 Time series of the instantaneous Level-2 α and β at Boulder (top) and Tamanrasset (bottom) during 10 years, as obtained with multiwavelength sunphotometers and Version 3 of AERONET's retrieval algorithm. The red rectangles indicate periods of suspicious or anomalous data that were not detected during the quality-control process

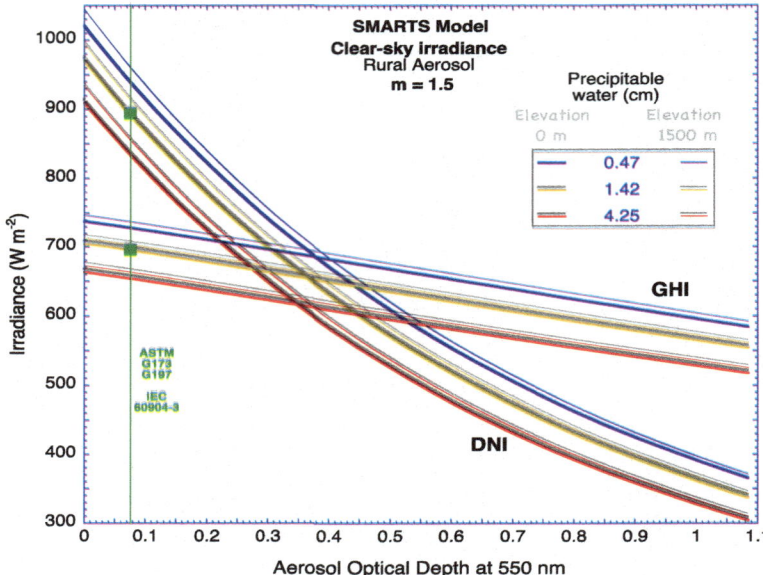

Fig. 7 Impact of AOD at 550 nm (τ_{a550}) on SMARTS predictions of DNI and GHI for a rural aerosol model, an air mass of 1.5, two site elevations, and three precipitable water amounts. The irradiance values corresponding to the atmospheric conditions (including τ_{a550} = 0.0764) of standards ASTM G173, ASTM G197, and IEC 60904-3 are indicated by a green square

(ASTM G173, ASTM G197, and IEC 60904-3). The impact of PW is clearly not as important as that of AOD, particularly in the case of DNI. The impact of site elevation appears even smaller, although it is stressed that the decrease of AOD and PW with elevation is not considered here because it can be site-specific. In general, AOD and PW decrease with elevation following the exponential function described in Eq. 6, but with a scale height of only ≈2 km (Gueymard 1994; Gueymard and Thevenard 2009). Important exceptions to this general rule do exist, however, such as in the case of polluted urban areas located on a high plateau. To apply the general sensitivity results assembled in Fig. 7 to the case of a specific site, the statistical distributions of the important underlying variables at that site must also be accounted for. For instance, it is conceivable that a remote site could experience both a relatively stable AOD and large variations in PW, which would counter-balance the larger AOD sensitivity.

Since DNI and DIF vary in opposite directions when AOD fluctuates, the diffuse–beam ratio, DIF/DNI, is observed to be a strong function of AOD, and only a second-order function of PW or AM, as shown in Fig. 8. Hence, AOD can be derived from that ratio with high confidence (Gueymard and Vignola 1998) even when PW is not known precisely, at least *if* truly cloudless conditions can be ascertained.

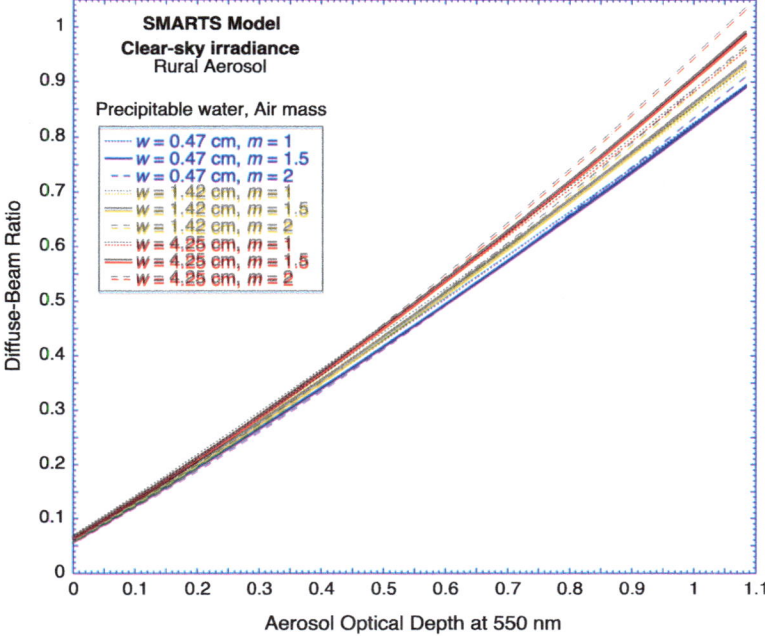

Fig. 8 Variation of the diffuse–beam ratio (DIF/DNI) as a function of aerosol optical depth at 550 nm (τ_{a550}) for three air masses and three precipitable water amounts

7.4   Spatiotemporal Variability and Its Impacts on Irradiance

Aerosol optical properties change rapidly over space and time, albeit most generally not as rapidly as cloud properties. Spatial variability is particularly obvious at the continental scale, as demonstrated in Fig. 9 for the case of South America. The August monthly mean AOD at 550 nm over the whole continent, as obtained from two different gridded sources, is compared to spot measurements from AERONET stations. The gridded sources are the (i) retrievals from MODIS–Terra Collection 6 (C6) at $1° \times 1°$ resolution; and (ii) modeled data from the MERRA-2 reanalysis at $0.5° \times 0.625°$ resolution. During August, high-AOD situations occur north of the equator because of dust transport from northern Africa, and in western Brazil, because that period is the onset of the biomass-burning season (which reaches its peak in September, during which the monthly mean τ_{a550} may locally reach values as high as ≈ 2). The MODIS retrievals display two significant issues: (i) incomplete geographical coverage (which would be even much worse on a daily basis); and (ii) large overestimation of AOD over arid areas known for their atmospheric clarity, such as the Atacama Plateau. In contrast, MERRA-2 is unaffected by both issues. The aerosol information in MERRA-2 is derived from a fully coupled

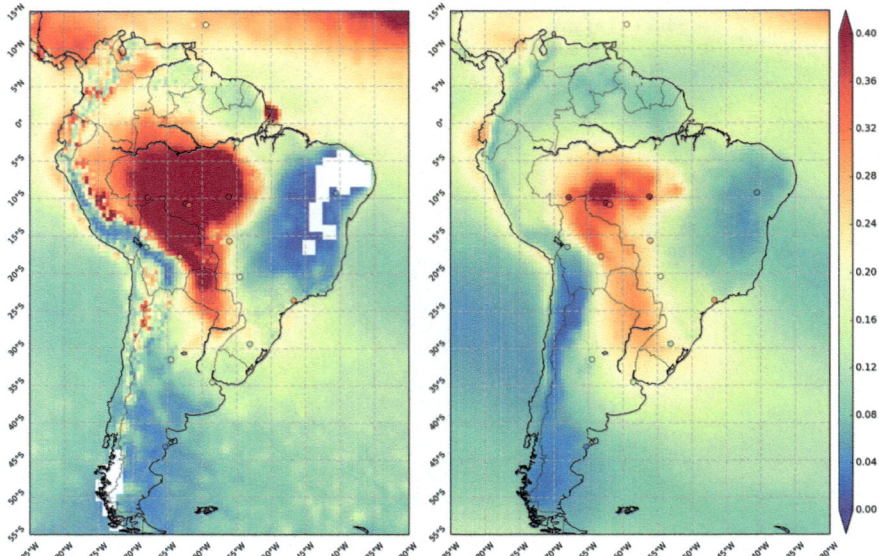

Fig. 9 Long-term mean monthly aerosol optical depth at 550 nm (τ_{a550}) over South America for August as either retrieved from MODIS–Terra C6 (left) or modeled with MERRA-2 (right). Color circles represent matching observations from AERONET stations. White areas in the left image indicate missing values

atmosphere-aerosol modeling combination of the GEOS-5 research forecast model and GOCART aerosol transport model,[22] with daily assimilation of spaceborne and ground-based AOD observations.

The AOD temporal variability is also significant on a daily, monthly, or inter-annual basis, as shown by Gueymard (2012b). Based on worldwide AERONET data, that study used the Aerosol Variability Index (AVI) to compare the temporal variability in AOD at various time scales. Interestingly, AVI was found roughly proportional to the mean annual AOD. This can be related to the observation that large sources of aerosols, such as smoke plumes, dust storms, or pollution out-bursts, are usually highly variable over time, with additional interference from changing weather patterns that can rapidly modify the spatial distribution of the aerosol concentration. In practice, it is observed that, at any site, the AOD temporal variability translates into a characteristic lognormal distribution (Ruiz-Arias et al. 2016), at least when considering daily or subdaily values. Two important direct consequences of this are that: (i) the range of observable AOD values increases with the mean AOD; and (ii) the mean, median, and mode of the distribution differ.

An example of daily variability over a period of 100 days is shown in Fig. 10 for Tamanrasset and for a site of lower turbidity (El Arenosillo, southern Spain).

---

[22]https://gmao.gsfc.nasa.gov/research/aerosol/modeling/nr1_movie/

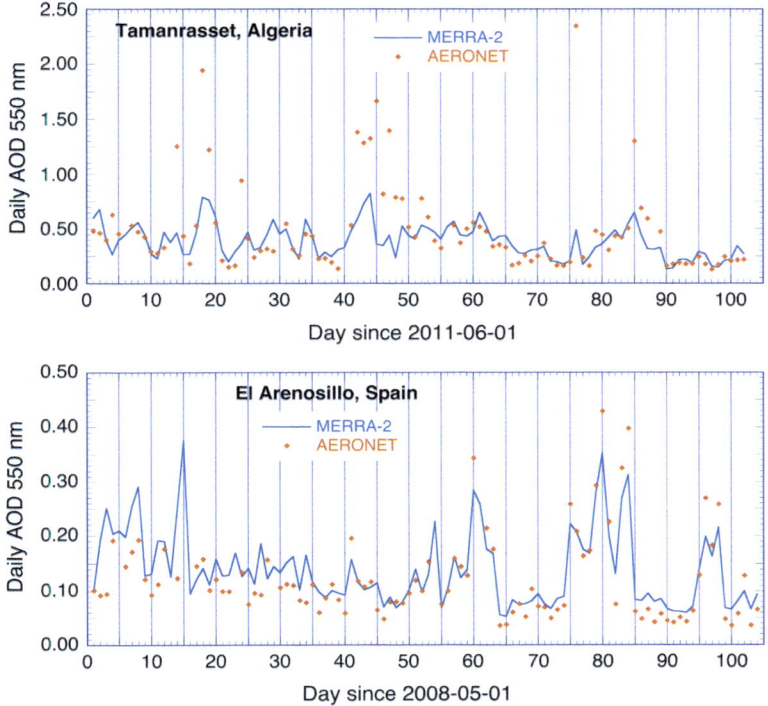

Fig. 10 100-day time series of daily mean AOD at Tamanrasset (top) and El Arenosillo (bottom) as obtained from the MERRA-2 reanalysis and AERONET ground observations

The two sites are occasionally impacted by dust events, with much more strength at the former site since it is closer to the aerosol source. Note the difference in scale, and how MERRA-2 often misses the strong dust events at Tamanrasset.

The impact on DNI of AOD's variability, or of its estimation error, can be described by the Aerosol Sensitivity Index, ASI (Gueymard 2012b), defined as:

$$ASI = -\frac{\Delta E_{bn}/E_{bn}}{\Delta \beta} \tag{8}$$

where E_{bn} stands for DNI. Because of the exponential relationship between DNI and AOD, Eq. 8 stresses that an *absolute* variation (or error) in β translates into a *relative* variation (or error) in DNI of the opposite sign. Based on limited tabulations (Gueymard 2012b), a simple rule of thumb can be derived: ASI ≈ 0.85 m. Hence, the impact on DNI of an error of 0.1 in β would be 12.75% for an air mass of 1.5, which is substantial. Since only estimated AOD values with significant uncertainty are available in the vast majority of cases when the clear-sky irradiance needs to be modeled, it is important to evaluate the magnitude of the predicted irradiance uncertainty caused by the inherent error propagation process.

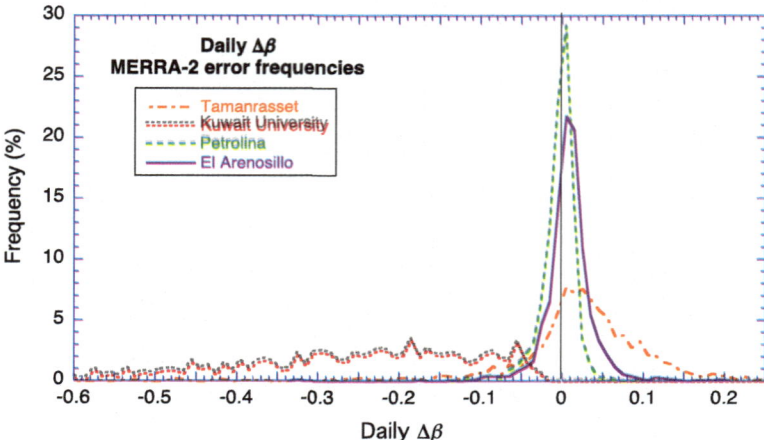

Fig. 11 Frequency distribution of the daily absolute difference between MERRA's β and AERONET's β at four sites

It is found that $\Delta\beta$ varies widely according to climate area and AOD data source. The frequency distribution of the daily $\Delta\beta$ for four AERONET stations is displayed in Fig. 11, where $\Delta\beta$ is evaluated here as the difference between the daily mean β derived from MERRA-2 for sun-up hours and its AERONET counterpart. Sites such as El Arenosillo and Petrolina, Brazil, experience only moderate turbidity, and their $\Delta\beta$ frequency distribution is well centered around 0, with only a small fraction of differences larger than ±0.05. At hazier sites, such as Tamanrasset or Kuwait University, Kuwait, the distribution is much flatter due to the high frequency of high-AOD situations (e.g., dust events) and the difficulty of predicting the correct strength and precise occurrence time of such events by models such as MERRA-2. Significant errors can then be expected in modeled irradiance datasets when using such sources of aerosol data, which confirms earlier results (Gueymard 2011). Such modeling errors in turn have a detrimental effect on the simulation of the energy produced by concentrating solar power plants, for instance (Polo and Estalayo 2015).

7.5 Circumsolar Irradiance

Another aerosol-related source of error in modeled irradiance is caused by potentially inadequate consideration for the impact of circumsolar radiation. This is particularly important when modeled irradiance components (DIF or DNI) are compared to measured data, or when the true DNI (from the sun's disk only) must be derived from experimental pyrheliometer data. This issue is of particular importance in the case of concentrating solar applications because the acceptance angle of concentrators is usually much smaller than the aperture of pyrheliometers used to measure DNI. The reader is referred to the definitions and discussion contained in an extensive study on this topic (Blanc et al. 2014).

Under cloudless conditions, the circumsolar irradiance is a direct function of AOD and of the type of aerosols. Most importantly, dust aerosols tend to generate a larger circumsolar contribution than other aerosol species. To evaluate the magnitude of this effect, Fig. 12 compares the measured DNI at Tamanrasset to the corresponding circumsolar fraction calculated with SMARTS for a typical pyrheliometer aperture, both as a function of the slant aerosol attenuation coefficient, $m\beta$. As this key coefficient increases, DNI decreases exponentially while the circumsolar fraction increases linearly, up to a maximum of $\approx24\%$ for $m\beta \approx 5$, at which point DNI is essentially negligible. In concentrated solar thermal applications, only high-DNI situations (above ≈400 W/m^2) are usually of interest. From Fig. 12, this implies that $m\beta$ is between 0.4 and 1.15 (most typically ≈0.8) for that threshold irradiance, which in turn is conducive to a circumsolar fraction of 2–6% (most typically $\approx4\%$) under this kind of dust aerosol environment. In comparison, the circumsolar fraction would be only $\approx1\%$ for a design DNI value of 800 W/m^2. Depending on circumstance, and in comparison with other sources of error, these estimates may be considered significant or not.

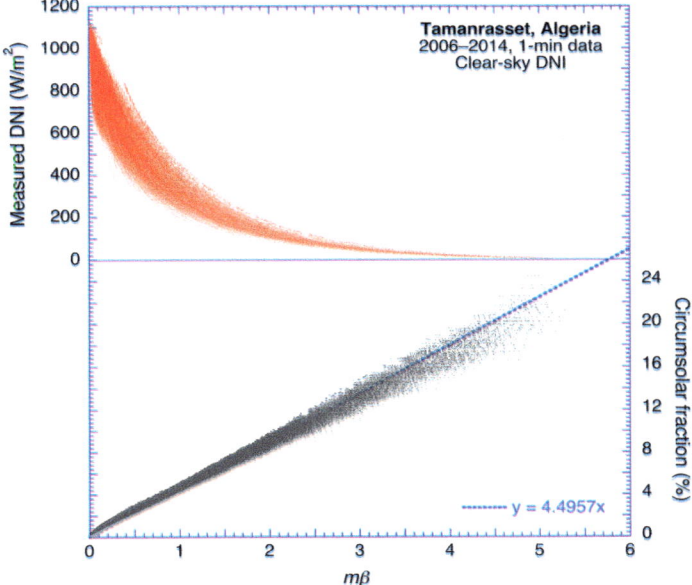

Fig. 12 Experimental DNI at Tamanrasset and its modeled fraction of circumsolar radiation as a function of $m\beta$

8 Validation of Clear-Sky Radiation Models

Most generally, solar analysts have more confidence in measured data—at least if obtained through the best practices (Sengupta et al. 2017)—than in modeled estimates, because the uncertainty in the former is normally lower than in the latter. This uncertainty issue negatively impacts the *bankability* of solar projects if their financial projections are based only on modeled data (Vignola et al. 2012). Solar radiation models, as well as other meteorological or climatological modeled databases (including irradiance), are typically subjected to a thorough validation process to increase the confidence one can have in them, and more specifically (i) quantify the uncertainty of the estimates; (ii) detect when or where anomalous results might be found; and (iii) provide comparative results and the necessary basis for ulterior model improvements.

In the case of CSRMs, many validation studies have appeared in the literature, as recently reviewed (Ruiz-Arias and Gueymard 2018b). Three major difficulties are inherent to this type of exercise: (i) the lack of commonly accepted methodology; (ii) the selection of different sets of models and sources of inputs by each author; and (iii) the imperfections of the method used to screen cloudy periods in observed irradiance time series. All this makes the results from different studies rarely comparable on an equal basis, at least to quantify and compare uncertainties. The critical factors that must be dealt with when attempting to validate CSRMs against measured data are described in the next two subsections. Since this kind of exercise is often an integral part of solar resource assessments, an example is developed in Sect. 8.3.

8.1 Time Stamp, Time Reference, Time Integration, and Spatial Homogeneity

In principle, any CSRM is able to evaluate the clear-sky radiation components on an instantaneous basis. Knowing the sun position, and particularly the sun's zenith angle at any instant, a CSRM can return the GHI and DNI predictions (in W/m^2) for that instant. When comparing such model predictions to radiometric observations, it is important to remember that the latter are not instantaneous, but represent an average over some period, like 1 min, 10 min, or 1 h, which are the most typical time steps currently in use. These observations are thus actually *irradiations* (as defined in Chap. 1), or aggregated irradiances, in Wh/m^2, sometimes divided by the elapsed time. Hence, a measured irradiation of 1000 Wh/m^2 that has been accumulated during 1 h can be considered equivalent to a constant irradiance of 1000 W/m^2 during the entire hour. It is customary to compare such aggregated measurements to an instantaneous irradiance modeled for the central time of the accumulation period. Hence, an irradiation of 1000 Wh/m^2, or a mean hourly irradiance of 1000 W/m^2, observed between 11:00 and 12:00 would be compared to the predicted instantaneous irradiance at 11:30. Because the apparent sun position moves fast around sunrise and sunset, this simple approach may not be always accurate, which is one reason why most studies eliminate low-sun data points. Another solution is to calculate the zenith angle continuously during the whole period, e.g., each minute, and use its hourly average to evaluate the irradiance for that period. A specific discussion on the best way of defining the sun geometry in such cases is proposed elsewhere (Blanc and Wald 2016). In any case, the calculations implied above assume that both the *time stamp* and *time reference* are well defined. This actually requires attention, as examined in Chap. 1.

Additionally, the temporal issue described above is particularly critical when comparing satellite-derived estimates to ground-based irradiance measurements. The former are not irradiance measurements, but predictions based on a combination of radiation models (CSRM and cloud transmittance model) using cloud input data derived from satellite images obtained at fixed intervals, e.g., every 5 to 30 min for current geosynchronous satellites. An instantaneous irradiance is thus calculated for the precise snapshot time, and converted into, e.g., hourly irradiations. Depending on the satellite data source, the time stamp may refer to the specific time of the snapshot, or to the end of the accumulation period. Great care must thus be exerted when comparing such satellite-derived data to predictions from other models, including CSRMs.

Finally, it is stressed that irradiance predictions or measurements obtained for a specific site differ substantially in their spatial resolution from gridded irradiance estimates derived from reanalysis or satellite-based models. The results apply to a single point in the former case, and to an average over a grid cell in the latter case. This difference introduces a source of error in the comparison, particularly over complex terrain (due to, e.g., shading or elevation-induced effects) or wherever the atmospheric conditions are not homogeneous over the whole grid cell. Under clear conditions, fortunately, the conditions are spatially much more homogeneous than under most cloudy conditions, but may still introduce errors.

8.2 Sources of Error

Based on the information contained in previous sections, it is stressed that any discrepancy between modeled irradiance predictions and actual measurements can result from a number of factors:

1. Intrinsic model errors, caused by inadequacies in the simplified representation of physical phenomena attempted by the model.
2. Errors in the inputs to the model, which impact the predictions even if the model itself is "perfect". Some compensation or amplification of errors may occur, depending on the model algorithm and sources of data. Only rare studies have looked into this issue, e.g., (Polo et al. 2014).

3. Errors in the measured irradiance: Any bias in the measurement could be incorrectly interpreted as a bias of opposite sign in the modeled predictions. To limit experimental errors, the best measurement practices must be followed (Sengupta et al. 2017), appropriate corrections must be implemented (Gueymard and Myers 2009), good knowledge of each instrument's uncertainty is necessary (Gueymard and Myers 2010; Habte et al. 2016), and a thorough quality-control procedure must be implemented using, e.g., the BSRN protocol (Long and Shi 2008).

4. Errors caused by differences in the spatiotemporal references of the predictions and measurements; this is particularly to be expected when comparing gridded predictions to point-source measurements.

5. Errors induced by false identifications of clear-sky periods. Ideally, such periods would be determined using coincident observations of sky cover with a whole-sky camera or imager. Since such observations are usually not available at validation sites, clear-sky periods must be inferred from an analysis of the irradiance time series. Simple filters based solely on a threshold value of the clearness index, K_T, are too imprecise because, in case of cloud or albedo enhancement, K_T can be higher under cloudy conditions than under clear conditions (Gueymard 2017). More elaborate filters do exist (Inman et al. 2015; Larrañeta et al. 2017; Long and Ackerman 2000; Reno and Hansen 2016), but still lack extensive validation, particularly under highly turbid conditions. Due to the rapid irradiance fluctuations that occur under partly cloudy situations, these cloud screening methods can only work well for measurement time steps less than $\approx$3 min. Hence, it is not recommended to conduct a CSRM validation study using hourly data, for instance.

Note that, in the case of the "black-box" approach described in Sect. 5, the errors associated with points 1 and 2 above cannot be disentangled. The validation example below attempts to provide additional information on the magnitude of the errors to be specifically expected from point 2.

8.3 Example of Validation Study

A single location is selected for this example: Tamanrasset, Algeria (lat. 29.210°N, long. 47.060°E, elev. 1377 m). This location is particularly interesting because (i) it harbors a radiometric station of the Baseline Surface Radiation Network (BSRN[23]) that is collocated with a sunphotometric station of the AERONET network; (ii) it stands in the Sahara area at a relatively high elevation (Fig. 13); and (iii) it has been used in previous validation studies (Amillo et al. 2014; Gueymard 2014b; Gueymard and Ruiz-Arias 2015; Ineichen 2016; Kosmopoulos et al. 2018). The first reason is important from a validation perspective because it allows the ideal use

---

[23]http://bsrn.awi.de/.

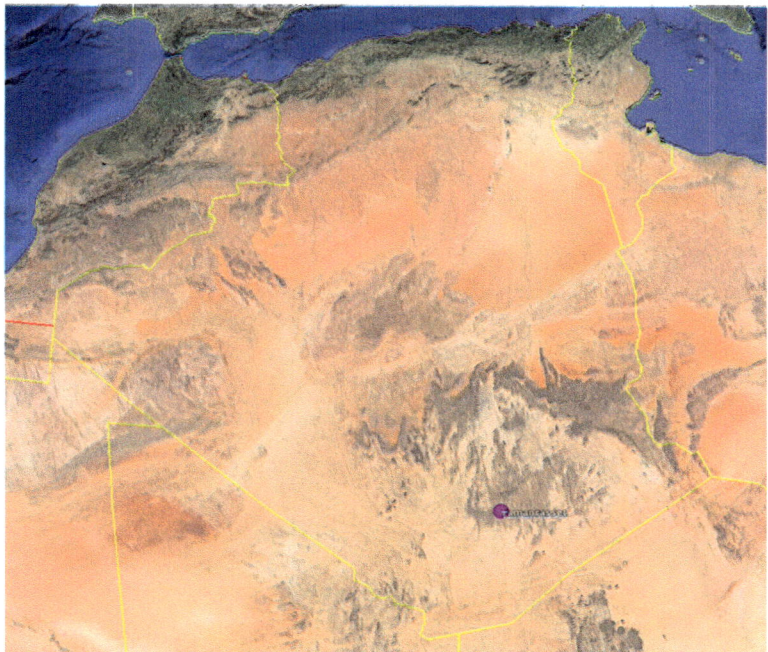

Fig. 13 Location of the Tamanrasset station in southern Algeria

of the best possible aerosol data as input to CSRMs, and the comparison of their output with the highest possible quality of the three irradiance components (DNI, DIF, and GHI), obtained here with separate thermopile radiometers at a 1-min time step. The second reason is significant because a recent study (Ruiz-Arias and Gueymard 2018b) showed that CSRMs tend to disagree more over elevated areas and wherever the atmospheric aerosol load is high, such as over or around the Sahara. The third reason indicates that, to gain perspective, the present results can be compared to those from existing validation studies.

The user-operable models selected here are those described in Sect. 5: Bird, Ineichen (2008, 2018), REST2 v5 and v9.1, and SMARTS v3.2. To compare the ideal scenario (ground-based aerosol observations) to the normal practice (modeled aerosol estimates), they are alternatively operated with aerosol data (AOD and α) from AERONET observations and MERRA-2 predictions. All other inputs are provided by MERRA-2. The McClear v3.1 black-box model, tied with CAMS reanalysis inputs, is also tested.

Each irradiance prediction is compared to the corresponding quality-controlled measurement after appropriate synchronization. Only those 1-min irradiance measurements whose center time is within ±2 min of the AERONET time stamp are considered. Similarly, the center time of the hourly MERRA-2 inputs is within ±30 min of the AERONET time stamp.

Table 4 Cumulative statistics describing the main atmospheric inputs and the irradiance observations at Tamanrasset during 2006–2014

Statistic	α M2	β M2	τ_{a550} M2	α AER	β AER	τ_{a550} AER	w M2	GHI	DNI	DIF
Mean	0.320	0.172	0.193	0.411	0.152	0.176	0.882	458.2	687.8	104.3
St. Dev.	0.244	0.146	0.149	0.366	0.168	0.178	0.494	267.3	265.5	64.7
Min	0.000	0.006	0.013	−1.410	0.002	0.006	0.054	59.3	6.0	17.0
Max	1.485	1.155	1.155	2.944	1.974	1.953	3.219	1141.2	1135.0	495.0

The alternate input data sources are MERRA-2 (M2) and AERONET (AER). Irradiances are expressed in W/m^2

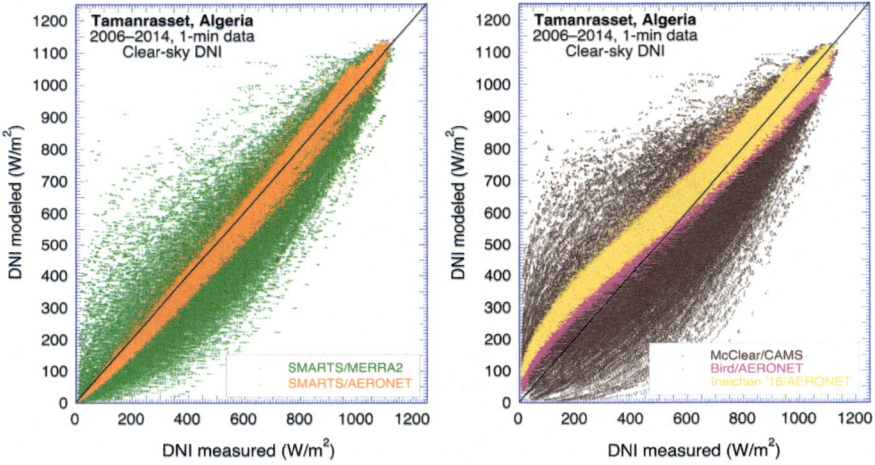

Fig. 14 Scatterplots of modeled versus measured DNI at Tamanrasset using various clear-sky radiation models and aerosol input combinations

The comparison of the predictions from all models is carried out during a 9-year period, 2006–2014, which is long enough to generate 257,090 valid 1-min data points. Because the AERONET observations are limited to air masses below 7, no comparison is done for $Z > 82.6°$. Some statistics about the main input data and the measured (reference) irradiance appear in Table 4.

A scatterplot of the SMARTS-predicted DNI versus its measured counterpart appears in Fig. 14, when using either the reference aerosol observations from AERONET or the modeled hourly MERRA-2 values. The much larger scatter that appears in the second case is striking, which confirms the importance of the propagation of aerosol-related errors. Similar results are obtained with the REST2 model. Another scatterplot is shown in Fig. 14, this time comparing the results of the McClear/CAMS black box to both the Bird and Ineichen-2018 models operated with AERONET data. The McClear/CAMS results appear even more scattered than those of SMARTS/MERRA-2, suggesting that the MERRA-2 aerosol information is more precise than that of CAMS, at least for the Tamanrasset area. The Bird and Ineichen-2018 results appear distorted, with strong overestimation of DNI when it is lower than ≈650 W/m², because of these models' difficulty of dealing with high-AOD situations created by dust aerosols. In particular, the Ineichen-2018 model does not use the Ångström exponent as an input, and excludes dust from its list of aerosol types, so that a default type—rural aerosols—must be used instead.

The results for DIF also show significant scatter when using aerosol inputs from reanalysis data in any model (only McClear/CAMS is shown here). The Bird and Ineichen-2018 models strongly underestimate DIF with AERONET data, whereas REST2 and SMARTS behave satisfactorily (Fig. 15).

Because of the expected compensations of errors between DNI and DIF, the results for GHI are normally much more precise, even when using aerosol inputs from reanalysis (Fig. 16). One exception is the Ineichen-2008 model, which generates a lot of random errors when using either source of aerosol inputs. In that regard, the updated Ineichen-2018 performs much better, despite its limitations under dusty conditions.

In addition to scatterplots, various statistical and visual tools can be used to assess the performance of CSRMs (Gueymard 2014b). Here, because of space limitations, only the most usual statistics, mean bias difference (MBD) and root mean square difference (RMSD), are provided in Table 5 and expressed in percent of the measured mean value (from Table 4). The increase in RMSD is significant when the aerosol inputs are provided by reanalysis rather than local observations. The concomitant change in MBD is significant too. In the present case, MERRA-2

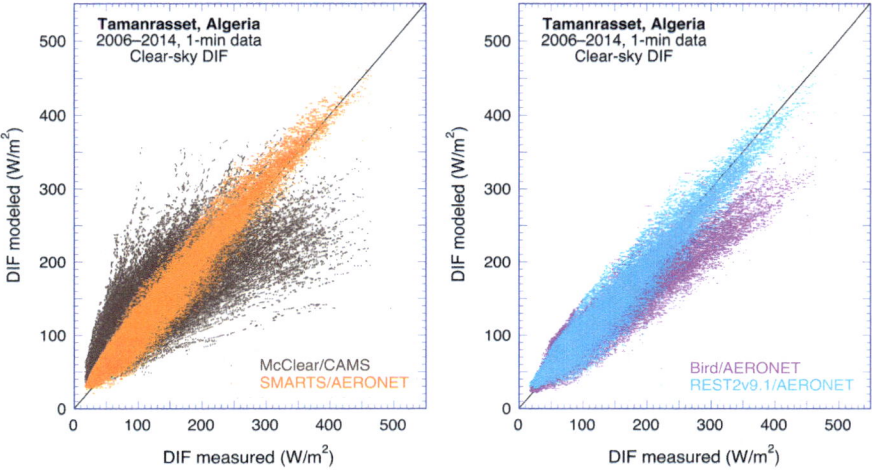

Fig. 15 Scatterplots of modeled versus measured DIF at Tamanrasset using various clear-sky radiation models and aerosol input combinations

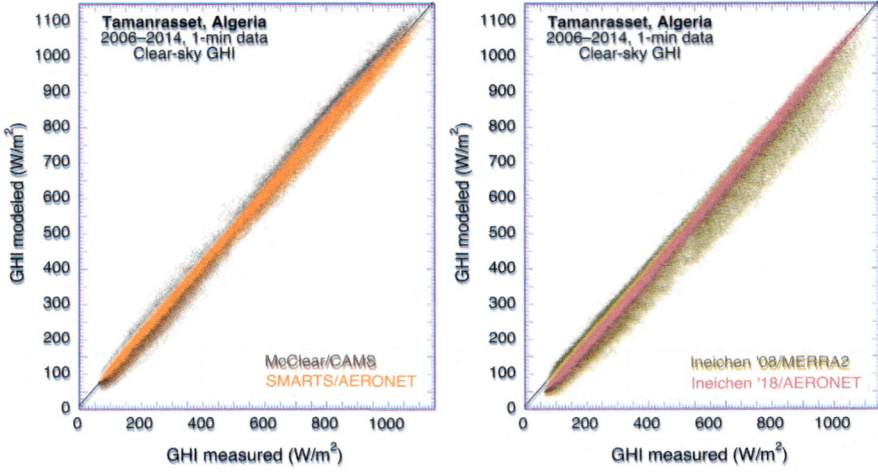

Fig. 16 Scatterplots of modeled versus measured GHI at Tamanrasset using various clear-sky radiation models and aerosol input combinations

tends to underestimate AOD around Tamanrasset, so that MBD increases for DNI and GHI, but decreases for DIF.

It is clear from the present results that the proper selection of a CRSM and of its optimum source of aerosol inputs is actually a difficult task, for which there is currently no definitive answer on a worldwide basis.

9 Conclusion and Future Developments

The calculation of clear-sky irradiance has made great progress in recent decades and continues to improve as the solar community now benefits from substantially more sources of data made possible by parallel developments in remote sensing, numerical weather prediction, and reanalysis models. Various radiation modeling approaches are possible to improve the accuracy of the predicted broadband irradiance while maintaining very fast speeds of execution. In general, the main difficulty is the correct modeling of aerosol attenuation, particularly under turbid conditions. It has been shown here that the accurate prediction of direct and diffuse irradiance is conditional to the use of a detailed radiation model and of sufficiently precise aerosol information. Using practical examples based on actual observations, it has been shown that the latter condition is often the most limiting factor.

In the future, the convergence of interests between the solar and atmospheric sciences communities is expected to continue. The use of more elaborate radiative models, providing specific spectral information for accuracy and more direct operability with photovoltaic applications, should become more prevalent. This will

Table 5 Performance of clear-sky radiation models under the conditions of Tamanrasset

Model	Bird	Ineichen (2008)	Ineichen (2018)	REST2 v5	REST2 v9.1	SMARTS v3.2	McClear v3.1
DNI							
MBD (AER.)	4.4	2.2	8.0	−2.2	−1.2	−0.8	–
MBD (Rean.)	1.6	−3.1	5.0	−7.1	−5.8	−6.9	−3.8
RMSD (AER.)	8.6	7.3	10.8	4.3	3.5	3.7	–
RMSD (Rean.)	14.1	13.8	14.7	15.1	14.3	15.0	18.0
DIF							
MBD (AER.)	−8.2	−21.2	−8.5	9.1	3.0	8.9	–
MBD (Rean.)	−2.5	−15.1	−1.9	19.4	11.9	20.7	17.6
RMSD (AER.)	23.2	37.6	21.8	17.0	12.9	15.0	–
RMSD (Rean.)	31.4	39.9	30.7	35.7	30.4	35.1	40.5
GHI							
MBD (AER.)	0.3	−4.5	−3.3	0.7	−0.1	1.6	–
MBD (Rean.)	−0.3	−6.5	−3.9	−0.2	−1.1	1.2	1.8
RMSD (AER)	3.3	8.4	4.6	3.9	3.2	3.2	–
RMSD (Rean.)	4.1	10.1	5.5	4.5	4.2	3.8	4.8

The MBD and RMSD statistics are expressed in percent of the mean measured irradiance. Separate statistics are provided depending on the source of aerosol inputs: from AERONET (AER.) or reanalysis (Rean.). The reanalyses are CAMS for McClear and MERRA-2 for all other cases

become possible thanks to computing speed improvements, including special technologies such as parallelization, clustering, or graphic card acceleration. The accuracy gains resulting from these modeling improvements will probably remain constrained by the availability of atmospheric data at high spatiotemporal resolution. Since more resolution implies more computing power and more data storage requirements, this limiting factor can only improve slowly. Furthermore, the access to, and easy manipulation of, large specialized atmospheric databases created by, e.g., reanalysis models, constitutes a serious limiting factor for most solar analysts. The development of open-source scripts for such operations would considerably improve the general accessibility to the best sources of data and, by way of consequence, would potentially improve the accuracy of all solar resource tools on a worldwide basis.

The validation of clear-sky radiation models is still difficult because of the limitations just described, and also because of the lack of reference method(s) for the screening of cloudy periods from irradiance time series. The development of such methods would benefit from the installation of all-sky imagers with appropriate cloud-detection software at primary radiometric stations.

Acknowledgements The AERONET and BSRN staff and participants are thanked for their successful effort in establishing and maintaining the various sites whose data were advantageously used in the present developments. The author also wishes to thank Dr. José Antonio Ruiz-Arias for his insightful comments and preparation of Fig. 4.

References

Amillo AG, Huld T, Müller R (2014) A new database of global and direct solar radiation using the Eastern Meteosat satellite, models and validation. Remote Sens 6:8165–8189

Ångström A (1929) On the atmospheric transmission of sun radiation and on dust in the air. Geografis Annal 2:156–166

Antonanzas-Torres F, Antonanzas J, Urraca R, Alia-Martinez M, Martinez-de-Pison FJ (2016) Impact of atmospheric components on solar clear-sky models at different elevation: case study Canary Islands. Energy Convers Manag 109:122–129

Badescu V (2013) Assessing the performance of solar radiation computing models and model selection procedures. J Atmos Sol-Terr Phys 105:119–134

Badescu V et al (2012a) Accuracy and sensitivity analysis for 54 models of computing hourly diffuse solar irradiation on clear sky. Theor Appl Climatol 111:379–399

Badescu V et al (2012b) Computing global and diffuse solar hourly irradiation on clear sky. Review and testing of 54 models. Renew Sust Energ Rev 16:1636–1656

Badescu V et al (2013) Accuracy analysis for fifty-four clear-sky solar radiation models using routine hourly global irradiance measurements in Romania. Renew Energy 55:85–103

Baig H, Fernández EF, Mallick TK (2016) Influence of spectrum and latitude on the annual optical performance of a dielectric based BICPV system. Sol Energy 124:268–277

Bird RE, Hulstrom RL (1980) Direct insolation models. Solar Energy Research Institute (now NREL) Rep. SERI/TR-335-344, Golden, CO. https://www.nrel.gov/docs/legosti/old/344.pdf

Bird RE, Hulstrom RL (1981a) Review, evaluation, and improvement of direct irradiance models. J Sol Energy Eng 103:182–192

Bird RE, Hulstrom RL (1981b) A simplified clear sky model for direct and diffuse insolation on horizontal surfaces. Solar Energy Research Institute (now NREL) Rep. SERI/TR-642–761, Golden, CO. http://www.nrel.gov/docs/legosti/old/761.pdf

Blanc P et al (2014) Direct normal irradiance related definitions and applications: the circumsolar issue. Sol Energy 110:561–577

Blanc P, Wald L (2016) On the effective solar zenith and azimuth angles to use with measurements of hourly irradiation. Adv Sci Res 13:1–6

Cachorro VE, Casanova JL, Frutos AMd (1987a) The influence of Angström parameters on calculated direct solar spectral irradiances at high turbidities. Sol Energy 39:399–407

Cachorro VE, de Frutos AM, Casanova JL (1987b) Determination of the Angström turbidity parameters. Appl Opt 26:3069–3076

Chauvin R, Nou J, Eynard J, Thil S, Grieu S (2018) A new approach to the real-time assessment and intraday forecasting of clear-sky direct normal irradiance. Sol Energy 167:35–51

Che H et al (2015) Ground-based aerosol climatology of China: aerosol optical depths from the China Aerosol Remote Sensing Network (CARSNET) 2002–2013. Atmos Chem Phys 15:7619–7652

d'Oliveira FA, de Melo FCL, Devezas TC (2016) High-altitude platforms—present situation and technology trends. J Aerosp Technol Manag 8:249–262

Eissa Y et al (2015) Validating surface downwelling solar irradiances estimated by the McClear model under cloud-free skies in the United Arab Emirates. Sol Energy 114:17–31

Eissa Y et al (2018) Prediction of the day-ahead clear-sky downwelling surface solar irradiances using the REST2 model and WRF-CHIMERE simulations over the Arabian Peninsula. Sol Energy 162:36–44

Engerer NA, Mills FP (2015) Validating nine clear sky radiation models in Australia. Sol Energy 120:9–24

Fernández EF, Almonacid F, Rodrigo P, Pérez-Higueras P (2013) Model for the prediction of the maximum power of a high concentrator photovoltaic module. Sol Energy 97:12–18

Gueymard CA (1989) A two-band model for the calculation of clear sky solar irradiance, illuminance, and photosynthetically active radiation at the Earth's surface. Sol Energy 43:253–265

Gueymard CA (1994) Analysis of monthly average atmospheric precipitable water and turbidity in Canada and Northern United States. Sol Energy 53:57–71

Gueymard CA (1995) SMARTS2, simple model of the atmospheric radiative transfer of sunshine: algorithms and performance assessment. Florida Solar Energy Center, Cocoa, FL, Rep. FSEC-PF-270-95

Gueymard CA (1996) Multilayer-weighted transmittance functions for use in broadband irradiance and turbidity calculations. In: Campbell-Howe R, Wilkins-Crowder B (eds) Proceeding of solar '96, American Solar Energy Society, Asheville, NC, pp 281–288

Gueymard CA (1998) Turbidity determination from broadband irradiance measurements: a detailed multicoefficient approach. J Appl Meteorol 37:414–435

Gueymard CA (2001) Parameterized transmittance model for direct beam and circumsolar spectral irradiance. Sol Energy 71:325–346

Gueymard CA (2003a) Direct solar transmittance and irradiance predictions with broadband models. Pt 1: detailed theoretical performance assessment. Solar Energy 74:355–379. Corrigendum: Solar Energy 376, 513 (2004)

Gueymard CA (2003b) Direct solar transmittance and irradiance predictions with broadband models. Pt 2: validation with high-quality measurements. Solar Energy 74:381–395. Corrigendum: Solar Energy 376, 515 (2004)

Gueymard CA (2005) Interdisciplinary applications of a versatile spectral solar irradiance model: a review. Energy 30:1551–1576

Gueymard CA (2008a) Prediction and validation of cloudless shortwave solar spectra incident on horizontal, tilted, or tracking surfaces. Sol Energy 82:260–271

Gueymard CA (2008b) REST2: high-performance solar radiation model for cloudless-sky irradiance, illuminance, and photosynthetically active radiation—validation with a benchmark dataset. Sol Energy 82:272–285

Gueymard CA (2009) Direct and indirect uncertainties in the prediction of tilted irradiance for solar engineering applications. Sol Energy 83:432–444

Gueymard CA (2011) Uncertainties in modeled direct irradiance around the Sahara as affected by aerosols: are current datasets of bankable quality? J Sol Energy Eng 133:031013–031024

Gueymard CA (2012a) Clear-sky irradiance predictions for solar resource mapping and large-scale applications: improved validation methodology and detailed performance analysis of 18 broadband radiative models. Sol Energy 86:2145–2169

Gueymard CA (2012b) Temporal variability in direct and global irradiance at various time scales as affected by aerosols. Sol Energy 86:2553–3544

Gueymard CA (2014a) Impact of on-site atmospheric water vapor estimation methods on the accuracy of local solar irradiance predictions. Sol Energy 101:74–82

Gueymard CA (2014b) A review of validation methodologies and statistical performance indicators for modeled solar radiation data: towards a better bankability of solar projects. Renew Sust Energ Rev 39:1024–1034

Gueymard CA (2017) Cloud and albedo enhancement impacts on solar irradiance using high-frequency measurements from thermopile and photodiode radiometers. Part 1: impacts on global horizontal irradiance. Sol Energy 153:755–765

Gueymard CA (2018) A reevaluation of the solar constant based on a 42-year total solar irradiance time series and a reconciliation of spaceborne observations. Sol Energy 168:2–9

Gueymard CA, Kambezidis HD (2004) Solar spectral radiation. In: Muneer T (ed) Solar radiation and daylight models. Elsevier

Gueymard CA, Myers D, Emery K (2002) Proposed reference irradiance spectra for solar energy systems testing. Sol Energy 73:443–467

Gueymard CA, Myers DR (2008) Validation and ranking methodologies for solar radiation models. In: Badescu V (ed) Modeling solar radiation at the earth surface. Springer

Gueymard CA, Myers DR (2009) Evaluation of conventional and high-performance routine solar radiation measurements for improved solar resource, climatological trends, and radiative modeling. Sol Energy 83:171–185

Gueymard CA, Myers DR (2010) Solar resource for space and terrestrial applications. In: Fraas LM, Partain LD (eds) Solar cells and their applications, 2nd ed. Wiley

Gueymard CA, Ruiz-Arias JA (2015) Validation of direct normal irradiance predictions under arid conditions: a review of radiative models and their turbidity-dependent performance. Renew Sust Energ Rev 45:379–396

Gueymard CA, Thevenard D (2009) Monthly average clear-sky broadband irradiance database for worldwide solar heat gain and building cooling load calculations. Sol Energy 83:1998–2018

Gueymard CA, Vignola F (1998) Determination of atmospheric turbidity from the diffuse-beam broadband irradiance ratio. Sol Energy 63:135–146

Habte A, Sengupta M, Andreas A, Wilcox S, Stoffel T (2016) Intercomparison of 51 radiometers for determining global horizontal irradiance and direct normal irradiance measurements. Sol Energy 133:372–393

Holben BN et al (1998) AERONET—a federated instrument network and data archive for aerosol characterization. Remote Sens Environ 66:1–16

Holben BN et al (2001) An emerging ground-based aerosol climatology: aerosol optical depth from AERONET. J Geophys Res D106:12067–12097

Iacono MJ, Delamere JS, Mlawer EJ, Clough SA, Morcrette JJ, Hou YT (2004) Development and evaluation of RRTMG_SW, a shortwave radiative transfer model for general circulation model applications. In: Proceeding of 14th ARM science meeting, Albuquerque, NM. http://www.arm.gov/publications/proceedings/conf14/extended_abs/iacono-mj.pdf

Ineichen P (2006) Comparison of eight clear sky broadband models against 16 independent data banks. Sol Energy 80:468–478

Ineichen P (2008) A broadband simplified version of the solis clear sky model. Sol Energy 82:758–762

Ineichen P (2016) Validation of models that estimate the clear sky global and beam solar irradiance. Sol Energy 132:332–344

Ineichen P (2018) High turbidity solis clear sky model: development and validation. Remote Sens 10:435

Inman RH, Edson JG, Coimbra CFM (2015) Impact of local broadband turbidity estimation on forecasting of clear sky direct normal irradiance. Sol Energy 117:125–138

Jessen W et al (2018) Proposal and evaluation of subordinate standard solar irradiance spectra for applications in solar energy systems. Sol Energy 168:30–43

Kosmopoulos PG, Kazadzis S, Taylor M, Raptis PI, Keramitsoglou I, Kiranoudis C, Bais AF (2018) Assessment of surface solar irradiance derived from real-time modelling techniques and verification with ground-based measurements. Atmos Meas Tech 11:907–924

Larrañeta M, Reno MJ, Lillo-Bravo I, Silva-Pérez MA (2017) Identifying periods of clear sky direct normal irradiance. Renew Energy 113:756–763

Lefèvre M et al (2013) McClear: a new model estimating downwelling solar radiation at ground level in clear-sky conditions. Atmos Meas Tech 6:2403–2418

Lefèvre M, Wald L (2016) Validation of the McClear clear-sky model in desert conditions with three stations in Israel. Adv Sci Res 13:21–26. https://doi.org/10.5194/asr-13-21-2016

Li J, Lv M, Sun K (2016) Optimum area of solar array for stratospheric solar-powered airship. Proc IMechE Part G J Aerospace Eng 231:2654–2665

Li ZQ et al (2018) Comprehensive study of optical, physical, chemical, and radiative properties of total columnar atmospheric aerosols over China: an overview of Sun-sky radiometer Observation Network (SONET) measurements. Bull Amer Meteorol Soc 99:739–755

Liu H, Aberle AG, Buonassisi T, Peters IM (2016) On the methodology of energy yield assessment for one-sun tandem solar cells. Sol Energy 135:598–604

Long CN, Ackerman TP (2000) Identification of clear skies from broadband pyranometer measurements and calculation of downwelling shortwave cloud effects. J Geophys Res 105D:15609–15626

Long CN, Shi Y (2008) An automated quality assessment and control algorithm for surface radiation measurements. Open Atmos Sci J 2:23–37

Martinez-Lozano JA, Utrillas MP, Tena F, Cachorro VE (1998) The parameterisation of the atmospheric aerosol optical depth using the Ångström power law. Sol Energy 63:303–311

Molineaux B, Ineichen P (1996) On the broad band transmittance of direct irradiance in a cloudless sky and its application to the parameterization of atmospheric turbidity. Sol Energy 56:553–563

Mueller RW et al (2004) Rethinking satellite-based solar irradiance modeling: the SOLIS clear-sky module. Remote Sens Environ 91:160–174

Oumbe A, Qu Z, Blanc P, Lefèvre M, Wald L, Cros S (2014) Decoupling the effects of clear atmosphere and clouds to simplify calculations of the broadband solar irradiance at ground level. Geosci Model Dev 7:1661–1669

Polo J, Alonso-Abella M, Ruiz-Arias JA, Balenzategui JL (2017) Worldwide analysis of spectral factors for seven photovoltaic technologies. Sol Energy 142:194–203

Polo J, Antonanzas-Torres F, Vindel JM, Ramirez L (2014) Sensitivity of satellite-based methods for deriving solar radiation to different choice of aerosol input and models. Renew Energy 68:785–792

Polo J, Estalayo G (2015) Impact of atmospheric aerosol loads on concentrating solar power production in arid-desert sites. Sol Energy 115:621–631

Reno MJ, Hansen CW (2016) Identification of periods of clear sky irradiance in time series of GHI measurements. Renew Energy 90:520–531

Rigollier C, Bauer O, Wald L (2000) On the clear sky model of the ESRA—European Solar Radiation Atlas—with respect to the Heliosat method. Sol Energy 68:33–48

Ruiz-Arias JA, Gueymard CA (2015) Solar resource for high-concentrator photovoltaic applications. In: Perez-Higueras PJ, Fernandez FE (eds) High concentrator photovoltaics: fundamentals, engineering and power plants. Springer

Ruiz-Arias JA, Gueymard CA (2018a) A multi-model benchmarking of direct and global clear-sky solar irradiance predictions at arid sites using a reference physical radiative transfer model. Sol Energy 171:447–465

Ruiz-Arias JA, Gueymard CA (2018b) Worldwide inter-comparison of clear-sky solar radiation models: consensus-based review of direct and global irradiance components simulated at the earth surface. Sol Energy 168:10–29

Ruiz-Arias JA, Gueymard CA, Quesada-Ruiz S, Santos-Alamillos FJ, Pozo-Vázquez D (2016) Bias induced by the AOD representation time scale in long-term solar radiation calculations. Part 1: sensitivity of the AOD distribution to the representation time scale. Sol Energy 137:608–620

Sengupta M, Habte A, Gueymard C, Wilbert S, Renné D (2017) Best practices handbook for the collection and use of solar resource data for solar energy applications, 2nd edn. National Renewable Energy Lab., Golden, CO. Rep. NREL/TP-5D00-68886, https://www.nrel.gov/docs/fy18osti/68886.pdf

Sengupta M, Xie Y, Lopez A, Habte A, Maclaurin G, Shelby J (2018) The National Solar Radiation Data Base (NSRDB). Renew Sust Energ Rev 89:51–60

Sun Q, Wang Z, Li Z, Erb A, Schaaf CB (2017) Evaluation of the global MODIS 30 arc-second spatially and temporally complete snow-free land surface albedo and reflectance anisotropy dataset. Int J Appl Earth Obs Geoinf 58:36–49

Theristis M, O'Donovan TS (2015) Electrical-thermal analysis of III–V triple-junction solar cells under variable spectra and ambient temperatures. Sol Energy 118:533–546

Vignola F, Grover C, Lemon N, McMahan A (2012) Building a bankable solar radiation dataset. Sol Energy 86:2218–2229

Xie Y, Sengupta M, Dudhia J (2016) A Fast All-sky Radiation Model for Solar applications (FARMS): algorithm and performance evaluation. Sol Energy 135:435–445

Xie Y, Sengupta M, Deline C (2017) Recent advancements in the numerical simulation of surface irradiance for solar energy applications. In: Proceeding of IEEE 44th photovoltaic specialists conference, Washington, DC. https://www.nrel.gov/docs/fy17osti/67804.pdf

Yang D (2016) Solar radiation on inclined surfaces: corrections and benchmarks. Sol Energy 136:288–302

Zhang H, Huang C, Yu S, Li L, Xin X, Liu Q (2018) A lookup-table-based approach to estimating surface solar irradiance from geostationary and polar-orbiting satellite data. Remote Sens 10:411

Zhang H, Zhang M, Cui Z, Wang Y, Xin J (2012) Simulation and validation of the aerosol optical thickness over China in 2006. Acta Meteorologica Sinica 26:330–344

Zhang T, Stackhouse PW, Chandler WS, Westberg DJ (2014) Application of a global-to-beam irradiance model to the NASA GEWEX SRB dataset: an extension of the NASA surface meteorology and solar energy datasets. Sol Energy 110:117–131

Zhang Y (2016) Simplified analytical model for investigating the output power of solar array on stratospheric airship. Int'l J Aeronautical Space Sci 17:432–441

Zhong X, Kleissl J (2015) Clear sky irradiances using REST2 and MODIS. Sol Energy 116: 144–164

Zhu W, Xu Y, Li J, Du H, Zhang L (2018) Research on optimal solar array layout for near-space airship with thermal effect. Sol Energy 170:1–13

Ziemke JR, Chandra S, Labow GJ, Bhartia PK, Froidevaux L, Witte JC (2011) A global climatology of tropospheric and stratospheric ozone derived from Aura OMI and MLS measurements. Atmos Chem Phys 11:9237–9251

Chapter 6
Solar Radiation Modeling from Satellite Imagery

Jesús Polo and Richard Perez

Abstract Satellites have been observing the earth-atmosphere system and delivering data since the 1960s. They constitute a crucial tool for observing, measuring, and understanding meteorological phenomena and radiative transfer budgets. Satellite observations are usually incorporated into numerical weather predictions (NWP) models, via data assimilation algorithms, to produce the best estimate of the atmospheric state and to improve weather forecasting. In the field of solar radiation, satellite imagery provides effective information since onboard sensors actually measure the incoming radiance form the earth-atmosphere system. The radiance received by satellites is related to solar radiation incident at the earth's surface since it results from the different interactions of the sun's radiation with the earth-atmosphere system—scattering, absorption, and reflection. Therefore, it is reasonable to design methods and algorithms to infer surface solar irradiance from the radiance received by the satellites' onboard instruments. The first algorithms were developed in the 1980s. They have significantly evolved since and have reached a high degree of maturity and accuracy thanks to continuous developments and improvements both in the methods and in onboard radiometric instrumentations (in particular, spectral and geographical resolution). Every solar energy project requires an accurate knowledge of the local solar resource. Although the number and geographical density of ground-based solar radiation sensors are continuously increasing, they can only supply the needed solar resource information in a handful of locations. In consequence, most solar energy projects rely on solar irradiance time series simulated from satellite imagery. Satellite-derived information can be processed into many forms useful to the solar community, including solar radiation

J. Polo (✉)
Photovoltaic Solar Energy Unit,
Renewable Energy Division (Energy Department) of CIEMAT,
Avda Complutense 40, 28040 Madrid, Spain
e-mail: jesus.polo@ciemat.es

R. Perez
Atmospheric Science Research Center, State University of New York,
251 Fuller Road, Albany, NY 12203, USA
e-mail: rperez@albany.edu

© Springer Nature Switzerland AG 2019
J. Polo et al. (eds.), *Solar Resources Mapping*, Green Energy and Technology,
https://doi.org/10.1007/978-3-319-97484-2_6

maps, typical meteorological (or representative) years, long-term characterization, and solar plant performance modeling. This chapter presents a state-of-the-art review of the modeling of surface solar irradiance from satellite images.

1 Geostationary and Polar-Orbiting Satellites

Meteorological satellites observing the earth-atmosphere system can be grouped into geostationary and polar-orbiting satellites. The geostationary satellites follow a circular orbit around the earth equator at a height of around 36,000 km. This orbit is a type of geosynchronous orbit, and thus geostationary satellites stand above the same location on earth. Figure 1 shows the positions of some current geostationary meteorological satellites. Polar-orbiting satellites circle around the earth over the north and south poles at around 850 km elevation. They orbit the earth in about 100 min and pass over the same region 1–3 times a day. The sensors onboard polar satellites have typically provided higher radiometric and spectral resolution than radiometers in geostationary satellites (Janssen and Huurneman 2001).

The instruments on board of geostationary weather satellites have been improving and increasing in capabilities and resolution. They usually measure the upwelling radiance in three main spectral bands: a visible band ($\sim$0.5–1.1 μm), a thermal infrared band ($\sim$10.5–12.5 μm), and a water vapor infrared band ($\sim$5.7–7.1 μm). As an example, the instrumentation onboard of the European geostationary satellites evolved from Meteosat First Generation sensor MVIRI that had three spectral channels to the meteosat Second Generation sensor SEVIRI with twelve spectral bands. The third generation will expand to four imager and two

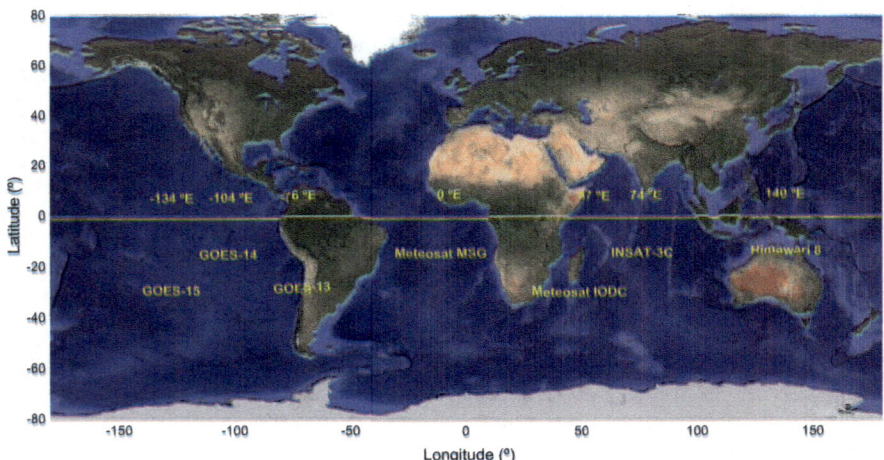

Fig. 1 Position of several geostationary weather satellites for earth-atmosphere observation

sounder satellites (www.eumetsat.int). Likewise, the last generation of the GOES satellite series becoming operational in early 2018 includes a sounder with a 19-spectral channel radiometer and an imager with a five-channel radiometer (http:// noaasis.noaa.gov/NOAASIS/). Similarly, the Japanese satellites Himawari-8 and -9 have Advanced Baseline Imager instrument with 16 spectral channels in different bands from ~ 0.43 to ~ 13.4 µm (www.data.jma.go.jp). In terms of temporal resolution, recent geostationary satellites collect a full-disk image every 15–30 min. Newer generation satellites will deliver full-disk data every five minutes. The geographical resolution of geostationary satellites ranges from 1 km at NADIR in the visible band to 2.5 km in the IR and water vapor bands. The new GOES satellites push the visible resolution down to 500 m.

Radiometers in polar-orbiting satellites have a higher spatial resolution thanks to their lower vantage point. For instance, MODIS and MISR instruments have spatial resolutions in the range of 250 m–1 km. The spectral resolution is variable and diverse in polar-orbiting satellites, ranging from multispectral sensors (MODIS) to hyperspectral sensors (OMI, AIRS).

The differences in radiometric, spectral, spatial, and temporal resolution between the geostationary and polar-orbiting platforms imply different data retrievals and methodologies to derive surface and atmospheric parameters. Nevertheless, there are several meteorological quantities that can be estimated by similar methodologies for both polar and geostationary satellites. Downwelling solar radiation at the earth's surface is one of these quantities.

2   Physical Principles for Estimating Solar Radiation from Satellite Imagery

Sensors on board of meteorological satellites receive the upwelling shortwave (solar) radiance from the earth-atmosphere system. This embeds both solar radiations reflected by the earth surface and/or by clouds and the backscattered radiation coming from the interaction of the incident solar radiation with the atmosphere Fig. 2. Therefore, it is possible to infer the downwelling solar radiation at the earth's surface from the components measured by the satellite radiometer if some atmospheric information is available or some assumptions are taken on the interaction with the atmosphere.

The first models to infer downwelling solar radiation (aka, global horizontal irradiance or GHI) were proposed in the early 1980s based on the strong negative correlation observed between the solar radiation reflected by clouds and the solar radiation reaching the earth's surface (Tarpley 1979; Moser and Raschke 1984; Pinker 1985; Cano et al. 1986; Johannes 1989; Pinker and Laszlo 1991). These first approaches were often based on statistical regressions between the digital count measured by the satellite radiometer and the simultaneous solar irradiance measured by a pyranometer. Consequently, they were referred to as statistical or empirical

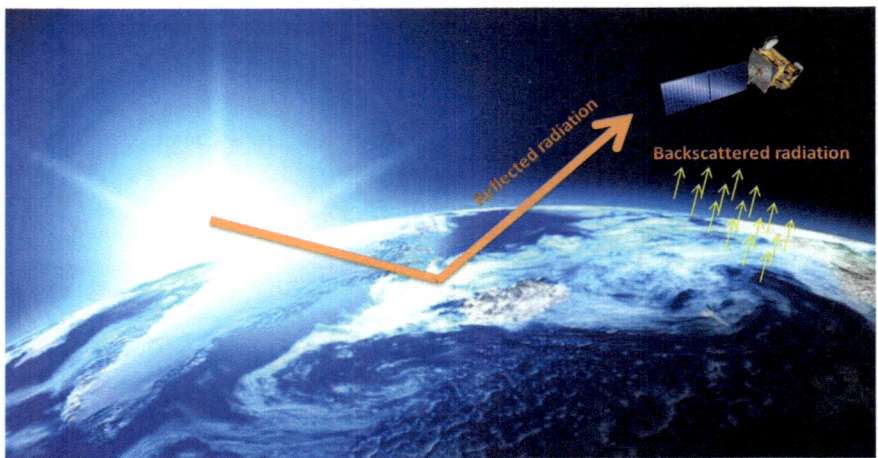

Fig. 2 Illustration of the solar radiation received by a satellite sensor

methods to distinguish from other methodologies based on an explicit radiative transfer modeling of the earth-atmosphere system (Noia et al. 1993a, b). Many of the recently proposed and currently used methodologies incorporate this statistical observation into physical transmittance models (Polo et al. 2008) and are denoted as semiempirical models.

2.1 Statistical and Semiempirical Models

The statistical methods for satellite-derived solar radiation are based on relationships between the satellite-sensed radiance and ground observations to compute global horizontal irradiance (GHI) and additional models to derive direct normal and diffuse horizontal irradiance (DNI and DIF, respectively) from the global component (Sengupta et al. 2017).

The earliest statistical models were purely empirical. Tarpley (1979) proposed a method by fitting locally measured GHI to the cosine-corrected count of the satellite.

The Cano's model was the origin of the Heliosat family of models (Cano et al. 1986). This method consisted of finding a linear fit of the clearness index (k_T) with the cloud cover index (n) using Meteosat visible images and hourly global irradiance for 27 French ground stations.

$$k_T = \frac{GHI}{G_0} = an + b \qquad (1)$$

The cloud cover index is the basic concept of several operational statistical and semiempirical models (Diabaté et al. 1987, 1989; Perez 2002; Perez et al. 2004; Rigollier et al. 2004). The cloud cover index is a relative parameter that normalizes the reflectivity measured by the satellite sensor within the dynamic range of observed radiances. In terms of both digital count at the sensor and reflectivity and albedo, the dynamic range establishes the variability from the darkest pixel or lowest albedo values (associated with the reflectivity of the ground or ground albedo ρ_g) and the brightest pixel or highest albedo values (representing the reflectivity of clouds or cloud albedo ρ_c). Therefore, for a specific pixel at the image with a measured albedo value denoted as planetary instantaneous albedo, ρ, the cloud index is defined by (Diabaté et al. 1987; Beyer et al. 1996; Ineichen and Perez 1999; Hammer et al. 2003),

$$n = \frac{\rho - \rho_g}{\rho_c - \rho_g} \qquad (2)$$

Thus, from Eq. (2) it can be straightforwardly observed that complete overcast conditions result in planetary albedo close to cloud albedo and then cloud cover index will tend to unity. Conversely, cloudless conditions result in albedo values close to ground albedo and the cloud cover index tends to zero. Figure 3 shows a time series of planetary albedo for several years of MSG images indicating the dynamic range.

The basic approach of statistical and semiempirical models is then to establish a relationship between a parameter that represents the atmospheric transmittance and the cloud cover index. The first statistical models used the clearness index as a

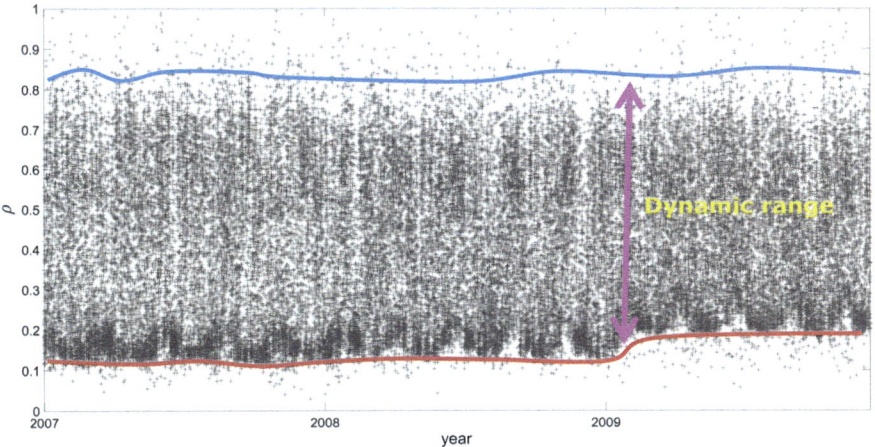

Fig. 3 Planetary albedo of for several years of MSG images and the corresponding dynamic range. The red curve corresponds to clear conditions, while the blue curve corresponds to fully overcast conditions. *Note* The change in dynamic range post 2009 is due to a change in location

representation of atmospheric transmittance for global irradiance. The clearness index is the ratio of global horizontal irradiance at the ground and extraterrestrial irradiance projected on a horizontal surface placed at the top of the atmosphere. However, the clearness index shows a dependence on the solar elevation angle, particularly for clear sky conditions, that limits its suitability for maximum irradiance conditions (Ineichen and Perez 1999). Therefore, statistical models evolved by the proposal of the clear sky index, k_c, as a better representation of solar radiation transmittance through the atmosphere. The clear sky index is the ratio of the global horizontal irradiance at the ground to the global irradiance at ground under complete cloudless conditions. The computation of the clear sky index thus requires the knowledge of irradiance under clear sky conditions, aka clear sky irradiance. This can be calculated with a rigorous radiative transfer model or approximated with simple a physical transmittance model.

Thus, the partial inclusion of physical equations in the calculation scheme of statistical models turned them into the category of semiempirical models. This is the case of Heliosat-2 and newer versions, and of the State University of New York (SUNY) models (Perez 2002; Perez et al. 2002; Hammer et al. 2003; Mueller et al. 2004; Rigollier et al. 2004).

The Heliosat-2 model proposed the following relationship between the clear sky index and the cloud cover index (Rigollier et al. 2004),

$$
k_c = \frac{G}{G_{clear}} =
\begin{cases}
1.2, & n < -0.2 \\
1 - n, & -0.2 \leq n < 0.8 \\
2.0667 - 3.6667n + 1.6667n^2, & 0.8 \leq n < 1.1 \\
0.05, & 1.1 \leq n
\end{cases}
\tag{3}
$$

Different modifications have subsequently been proposed years on both the Heliosat-2 method scheme (Schillings et al. 2004; Cros et al. 2006; Cebecauer and Suri 2010) and on the correlation between clear sky and cloud index (Zarzalejo et al. 2009). In addition, the Heliosat method, using MAGIC algorithm for clear sky calculations (Mueller et al. 2004), has also been adapted to work with polar-orbiting satellites (SCIAMACHY) to retrieve instantaneous surface solar irradiance (Wang et al. 2011).

New formulations for cloud index and clear sky transmittance calculations were proposed in the Heliosat-3 model as a revision and improvement of the earlier methods for adapting the methodology to the Meteosat Second Generation's visible channel (Dagestad 2004; Mueller et al. 2004; Dagestad and Olseth 2007). Some of the main novelties of the Heliosat-3 method were the computation of the instantaneous satellite albedo considering the backscattered radiation from the atmosphere as a function of the co-scattering angle (the angle subtended by the sun and satellite director vectors) and the algorithm for ground albedo computation. The Heliosat-3 method was later modified by proposing a dynamic model for estimating the ground albedo as a function of the co-scattering angle (thus improving the limitations of earlier versions for high reflective areas) and extending the application to other geostationary satellites (Polo et al. 2012, 2013). All these modifications and the possibility of using different clear sky models and atmospheric-derived input resulted in a model called Intisat Lib (Polo et al. 2015). The newest schemes proposed for the Heliosat family of models fall in the category of physical models.

The SUNY model uses a different way to compute the cloud index based directly on the pixel brightness that is considered proportional to the earth's radiance sensed by the satellite (Perez et al. 2002, 2004). The main difference with the Heliosat family models, in terms of cloud index concept, is that SUNY method works directly with the raw pixel brightness instead of estimate the satellite sensor reflectance. The raw pixel is first normalized by the cosine of solar zenith angle to account for first-order solar geometry effect. The relationship proposed for the clear sky index was

$$k_{\mathrm{c}} = \frac{G}{G_{\mathrm{clear}}} = 0.02 + 0.98(1 - n) \tag{4}$$

The second version of the SUNY model was used to produce the National Solar Resource Databases (NSRDB) of the National Renewable Energy Laboratory (NREL), and the model was also adapted to be used with Meteosat IODC images over the Indian Ocean (Perez et al. 2009). The SUNY model version 3 makes use of both visible and infrared channels imagery (Djebbar et al. 2012). The latest version is the fourth with notable performance improvement over the preceding versions, which have been improved in better source of aerosol optical depth, short-term forecast scheme integrated into the model and a better empirical method for estimating the dynamic range (Perez et al. 2015). The SUNY model is also the basis of the *SolarAnywhere* database (https://www.solaranywhere.com/). The SolarGIS model (https://solargis.info/) is based also on the earlier SUNY scheme, incorporating additional features for terrain effects, clear sky index calibration adapted to each satellite characteristics and dynamic computation of the dynamic range upper bound (Cebecauer and Suri 2010; Cebecauer et al. 2011; Perez et al. 2013).

The SARAH solar radiation dataset (Müller et al. 2015a; Pfeifroth et al. 2017) is the name of the retrieval supplied by Satellite Application Facility on Climate Monitoring (CM SAF, www.cmsaf.eu) covering from 1983 to 2015. The cloud cover index computation is based on the Heliosat method (Hammer et al. 2003) with some modifications in computing the ground and cloud albedo (Müller et al. 2015b). Clear sky irradiance is calculated by the SPECMAGIC model which is based on a Look-Up Table (LUT) derived from multiple rigorous radiative transfer calculations with libRadtran (Mueller et al. 2009). This methodology is also integrated into the PVGIS (http://re.jrc.ec.europa.eu/pvgis/) Web service offering solar resource and performance of photovoltaic technologies (Huld et al. 2012; Amillo et al. 2014). CM SAF solar radiation products have been widely evaluated and validated with ground measurements elsewhere (Posselt et al. 2012; Sanchez-Lorenzo et al. 2013; Riihelä et al. 2015; Žák et al. 2015).

2.2 Physical Models

Physical models differ conceptually from semiempirical models by solving the radiative transfer processes in the atmosphere. Satellite imagery is generally used for determining cloud properties, in particular, cloud optical depth. However, despite their different fundaments, physical models are not so significantly different from semiempirical models in practice. The first physical-based method was based on energy conservation within an earth-atmosphere column where the cloud effects were calculated from the measured satellite visible brightness (Gautier et al. 1980). Later a model based on the radiation budget for the tropical western Pacific ocean was proposed by (Nunez 1993), which served as a basis for a modified method to estimate solar radiation from satellite imagery in southeast Asia (Janjai et al. 2005, 2011, 2013).

Janjai's model is based on the consideration of multiple scattering, absorption, and reflections processes occurring in the atmosphere to compute the daily clearness index (kt_D) as (Janjai et al. 2005, 2013),

$$kt_D = \frac{\left(1 - \rho_A - \rho_{aer}\right)\left(1 - \alpha_w - \alpha_o - \alpha_{aer} - \alpha_g\right)}{1 - (\rho_A + \rho_{aer})\rho_G} \tag{5}$$

where α denotes the absorption coefficient of different gases (water vapor, ozone, aerosols, and mixed gases) and ρ indicates the different albedos (atmospheric, aerosol, and ground). This approach has been used for mapping several regions in southeast Asia.

BRASIL-SR is a physical model that employed the two-stream approach to solve the radiative transfer equation. The information on cloud optical depth is derived from satellite imagery. Clearness index is estimated by the transmittance of the cloudy and clear sky,

$$\frac{G}{G_0} = \left\{(\tau_{clear} - \tau_{cloud})(1 - n) + \tau_{cloud}\right\} \tag{6}$$

where G_0 is the irradiance at the top of the atmosphere, τ_{clear} is the clear sky transmittance, and τ_{cloud} is the overcast sky transmittance (Martins et al. 2003, 2007). This model has been recently modified to be adapted to the specific conditions of Chile creating a version named Chile-SR (Escobar et al. 2014).

The latest version of the Heliosat family, Heliosat-4, is a full physical model composed of two models mostly based on physically derived Look-Up Tables (Qu et al. 2017): the McClear model for solar irradiance under clear sky conditions and the McCloud model for determining the extinction of irradiance due to clouds. The McClear model is the result of multiple radiative transfer computations with libRadtran (Emde et al. 2015, 2016) for selected values of the inputs (Lefèvre et al. 2013). It basically consists of Look-Up Tables derived from these computations and interpolation functions. The input data regarding aerosol optical depth, aerosol type, ozone column, and water vapor is obtained from Copernicus Atmosphere Monitoring Service (CAMS, https://atmosphere.copernicus.eu/) that is a freely available operational service evolved from the former Monitoring Atmospheric Composition and Climate (MACC) project. The McCloud model is aimed at computing the clear sky indices for both global and beam irradiance as a function of cloud properties. The inputs to McCloud are essentially the bidirectional reflectance distribution function (BDRF) derived from MODIS (Blanc et al. 2014) and several cloud properties (cloud optical depth, cloud type, and cloud coverage) that are derived from satellite images using the AVHRR Processing scheme Over cLouds, Land and Ocean (APOLLO) scheme (Kriebel et al. 2003).

The Australian Bureau of Meteorology (BOM) has operated a physical model for deriving solar irradiation from GMS and MTSAT satellites. Instantaneous surface solar radiation at each grid point is calculated for the time of each satellite image with a physical model that parameterizes the important aspects of the radiative transfer in clear and cloudy atmospheres in two spectral bands, covering visible and near-infrared wavelengths, respectively (Weymouth and Le Marschall 2001). The physical parameterizations are adapted to the spectral response characteristics of each satellite. Total column water vapor amount, an ancillary input to the model, is taken from a numerical weather prediction (NWP) model field: the ERA and NCAR/NCEP reanalyses for archival processing.

3 Solar Products and Databases Based on Satellite Modeling

The high maturity and evolution of satellite methods for computing solar irradiance have resulted in the development of large data resources with considerable geographic coverage. Data products exist from both public and private sources. These databases are in continuous evolution and growth in terms of temporal coverage, spatial coverage, and quality/accuracy of the retrievals. Table 1 lists some of the best-known solar products based on satellite imagery.

The US National Renewable Energy Laboratory (NREL) has provided solar resource data for the USA for more than 25 years (https://nsrdb.nrel.gov). In addition, NREL recently updated the solar radiation database including data derived with SUNY semiempirical model for South Asia and Central America (https://nsrdb.nrel.gov/international-datasets).

Helioclim is a database developed by MINES-ParisTech, Armines, and Transvalor. It can be accessed via the SODA Web service (http://www.soda-pro.com/). At present, there are three versions of the database. These versions are based on the Heliosat-2 model applied to Meteosat first- and second-generation imagery, with differences also in the atmospheric information, used to input clear sky calculations. Time series, maps, and TMY are, among others, some of the products supplied by the SODA Web service. Solargis (https://solargis.info/) is another

Table 1 Solar radiation products and databases developed from satellite imagery

Name	Timestamp	Coverage	Web site
NASA SRB	3-hourly	World	http://gewex-srb.larc.nasa.gov/
DLR-ISIS	3-hourly	World	http://www.pa.op.dlr.de/ISIS/
HelioClim	hourly	Europe-Africa	http://www.soda-is.com/eng/helioclim/
SOLEMI	hourly	Europe-Africa-Asia	http://wdc.dlr.de/data_products/SERVICES/SOLARENERGY/
SolarGIS	30-min	World	http://solargis.info/
EnMetSol	hourly	Europe-Africa	https://www.uni-oldenburg.de/en/physics/research/ehf/energiemeteorology/enmetsol/
IrSOLaV	hourly	World	http://irsolav.com/
CM-SAF (SARAH)	hourly	Europe-Africa	http://www.cmsaf.eu/
SolarAnywhere	30-min	North America	http://www.solaranywhere.com/
CAMS	15-min	World	http://atmosphere.copernicus.eu/catalogue#/
PVGIS	hourly	Europe-Africa-Asia	http://re.jrc.ec.europa.eu/pvgis/
Vaisala	hourly	World	http://www.vaisala.com
Australian Bureau of Meteorology	hourly	Australia	http://www.bom.gov.au/climate/data-services/solar-information.shtml

well-known private service of satellite-derived solar radiation data products with worldwide coverage. Solaranywhere (https://www.solaranywhere.com/) is based on the SUNY model and covers North America and South-Central Asia. It is interesting to note that many of the semiempirical models (SUNY, Heliosat and their derivatives such as SolarGIS) have evolved into robust and reliable versions that have enabled commercial services for solar radiation data and solar resource assessment around the world.

In addition, there are also institutional services that offer access to satellite-derived solar resource information. This is the case of the CM SAF service, based on a modified version of Heliosat method. CM SAF is a joint activity of the National Meteorological Services of Belgium, Finland, the Netherlands, Sweden, Switzerland, UK, and Germany as leading entity, co-sponsored by EUMETSAT (http://www.cmsaf.eu/EN/Home/home_node.html). Through this service, we can access time series of solar radiation components at frequencies ranging from (15 min to annual) and from long periods. PVGIS is also a very well-known service of solar resource information based on satellite imagery which is targeted to the photovoltaic industry. It started as a geographical assessment of solar resource and performance of photovoltaic technology for Europe that was developed by the Joint Research Centre, Ispra (http://re.jrc.ec.europa.eu/pvgis/). The new version 5 includes data for Asia and America (Huld et al. 2012). One of the most recent institutional services is Copernicus

Atmosphere Monitoring Service (CAMS), implemented by the European Centre for Medium-Range Weather Forecasts ECMWF (http://www.copernicus.eu/main/atmosphere-monitoring), providing continuous data and information on atmospheric composition including solar resource data.

Both private and institutional services offer data on an archival historical basis as well. Some *also offer real-time and forecast data.*

4   Accuracy of Satellite-Derived Solar Irradiance

Satellite-based models and methods proposed for solar irradiance calculations are generally presented along with validation results against "ground-truth" experimental measurements. In addition to these direct comparisons against pinpoint ground measurements, there are evaluation studies worth to be mention because they attempt to characterize the uncertainty of satellite-derived data for geographically extended areas. Zelenka et al. (1999) compared the uncertainty of satellite estimations with the use of a neighboring station as a function of the distance. They concluded that for any application requiring site and time-specific data, the user should rely on satellite rather than on a ground station farther than 20–30 km from the site (Zelenka et al. 1999).

In the framework of Tasks 36 and 46 of International Energy Agency-Solar Heating and Cooling (IEA-SHC), Ineichen (2014) undertook a thorough validation of satellite models. He compared several well-known satellite-derived products against quality screened data at eighteen ground stations. He concluded that satellite-derived products are a reliable choice if no ground measurements are available in the vicinity of the considered location. As quantified by the relative root-mean-square error, the overall uncertainty of hourly satellite-derived irradiance was observed to be of the order of 17% for global horizontal and 34% for direct normal irradiance. Ineichen extended this validation study to the most recently developed satellite products—SolarGIS and Heliosat-4—to infer the impact of using MACC daily aerosol optical depth as input to the models, in contrast to previous models relying on climatological data often quantified by the Linke turbidity factor (Ineichen 2014). This updated study pointed out the improvement achieved in the estimation of direct beam component when using daily aerosol input data. Likewise a comparison among between different clear sky models and different daily aerosol products, MACC, MISR, and MODIS aerosol optical depth (Polo et al. 2014) pointed out also the higher accuracy achieved by the use of aerosols from MACC reanalysis and the REST2 clear sky model. The Helioclim-3 database of the direct beam was recently evaluated for Morocco conditions with acceptable results (Merrouni et al. 2017). A thorough validation of Heliosat and SPECMAGIC algorithms against 16 ground stations showed very low bias for global horizontal irradiance and less than 3% for the direct beam (Amillo et al. 2014). Extensive validations of other satellite products are found in the literature for lower temporal resolution (monthly or daily) due to the limitations of ground available data (Sanchez-Lorenzo et al. 2013). Other authors also have compared satellite data products against numerical model (Posselt et al. 2012).

The main source of uncertainty in satellite modeling is the treatment of clouds. Indeed radiative effect and optical properties of clouds cannot be completely defined, and models must rely on a degree of empiricism. Other sources of uncertainty are aerosol optical depth and clear atmospheric constituents (as noted above), as well as terrain effects and the high reflective albedo of deserts and snow, as well as spatial and temporal resolution—the higher the resolution in both time and space, the higher the uncertainty (Cebecauer and Suri 2010; Cebecauer et al. 2011). These sources of uncertainty produce systematic deviations, bias, or seasonal errors that cannot be avoided in most cases. To remedy this problem, several authors have recently proposed and developed methods for correcting bias or systematic errors in satellite retrievals at a given location with feedback from short-term measurements at the considered location (Polo et al. 2016). Thus, with the help of a short period of ground measurements, it is possible to de-bias satellite models, hence improve the accuracy (and consequently the bankability) of long-term time series of satellite-derived solar irradiance for a specific site. It is very important, however, to ensure that ground measurements used for this purpose are of the highest accuracy; indeed a recent study by Perez et al. showed that, on the contrary, satellite models were able to detect ground measurement bias at some of the world's most trusted reference stations (Perez et al. 2017).

References

Amillo A, Huld T, Müller R (2014) A new database of global and direct solar radiation using the eastern meteosat satellite, models and validation. Remote Sens 6:8165–8189. https://doi.org/10.3390/rs6098165

Beyer HG, Costanzo C, Heinemann D (1996) Modifications of the Heliosat procedure for irradiance estimates from satellite images. Sol Energy 56:207–212. https://doi.org/10.1016/0038-092X(95)00092-6

Blanc P, Gschwind B, Evre M, Wald L (2014) Twelve monthly maps of ground albedo parameters derived from MODIS data sets. IGARSS 2014, Jul 2014. Quebec, Canada, pp 3270–3272

Cano D, Monget JM, Albuisson M et al (1986) A method for the determination of the global solar radiation from meteorological satellite data. Sol Energy 37:31–39. https://doi.org/10.1016/0038-092X(86)90104-0

Cebecauer T, Suri M (2010) Accuracy improvements of satellite-derived solar resource based on GEMS re-analysis aerosols. In: Proceedings of the conference SolarPACES 2010 Perpignan, France, pp 1–4

Cebecauer T, Suri M, Gueymard CA (2011) Uncertainty sources in satellite-derived direct normal irradiance: how can prediction accuracy be improved globally. In Proceedings of the SolarPACES 2011 conference

Cros S, Albuisson M, Wald L (2006) Simulating Meteosat-7 broadband radiances using two visible channels of Meteosat-8. Sol Energy 80:361–367. https://doi.org/10.1016/j.solener.2005.01.012

Dagestad K-F (2004) Mean bias deviation of the Heliosat algorithm for varying cloud properties and sun-ground-satellite geometry. Theor Appl Climatol 79:215–224

Dagestad K-F, Olseth JA (2007) A modified algorithm for calculating the cloud index. Sol Energy 81:280–289. https://doi.org/10.1016/j.solener.2005.12.010

Diabaté L, Demarq H, Michaud-Regas N, Wald L (1987) Estimating incident solar radiation at the surface from images of the earth transmitted by geostationary satellites: the Heliosat project. Int J Sol Energy 5:261–278. https://doi.org/10.1080/01425918708914425

Diabaté L, Moussu G, Wald L (1989) Description of an operational tool for determining global solar radiation at ground using geostationary satellite images. Sol Energy 42:201–207. https://doi.org/10.1016/0038-092X(89)90012-1

Djebbar R, Morris R, Thevenard D et al (2012) Assessment of SUNY version 3 global horizontal and direct normal solar irradiance in Canada. Energy Procedia 30:1274–1283. https://doi.org/10.1016/j.egypro.2012.11.140

Emde C, Buras-Schnell R, Kylling A et al (2015) The libRadtran software package for radiative transfer calculations (Version 2.0). Geosci Model Dev Discuss 8:10237–10303. https://doi.org/10.5194/gmdd-8-10237-2015

Emde C, Buras-Schnell R, Kylling A, et al (2016) The libRadtran software package for radiative transfer calculations (version 2.0.1). Geosci Model Dev 9:1647–1672. https://doi.org/10.5194/gmd-9-1647-2016

Escobar RA, Cortés C, Pino A et al (2014) Solar energy resource assessment in Chile: satellite estimation and ground station measurements. Renew Energy 71:324–332

Gautier C, Diak G, Masse S et al (1980) A simple physical model to estimate incident solar radiation at the surface from GOES satellite data. J Appl Meteorol 19:1005–1012. https://doi.org/10.1175/1520-0450(1980)019%3c1005:ASPMTE%3e2.0.CO;2

Hammer A, Heinemann D, Hoyer C et al (2003) Solar energy assessment using remote sensing technologies. Remote Sens Environ 86:423–432. https://doi.org/10.1016/S0034-4257(03)00083-X

Huld T, Müller R, Gambardella A (2012) A new solar radiation database for estimating PV performance in Europe and Africa. https://doi.org/10.1016/j.solener.2012.03.006

Ineichen P (2014) Satellite irradiance based on MACC aerosols : Helioclim 4 and SolarGIS, global and beam components validation. In: EuroSun 2014 conference proceedings

Ineichen P, Perez R (1999) Derivation of cloud index from geostationary satellites and application to the production of solar irradiance and daylight illuminance data. Theor Appl Climatol 64:119–130

Janjai S, Laksanaboonsong J, Nunez M, Thongsathitya A (2005) Development of a method for generating operational solar radiation maps from satellite data for a tropical environment. Sol Energy 78:739–751. https://doi.org/10.1016/j.solener.2004.09.009

Janjai S, Masiri I, Laksanaboonsong J (2013) Satellite-derived solar resource maps for Myanmar. Renew Energy 53:132–140. https://doi.org/10.1016/j.renene.2012.11.014

Janjai S, Pankaew P, Laksanaboonsong J, Kitichantaropas P (2011) Estimation of solar radiation over Cambodia from long-term satellite data. Renew Energy 36:1214–1220. https://doi.org/10.1016/j.renene.2010.09.023

Janssen LLF, Huurneman GC (2001) Principles of remote sensing an introductory textbook. ISBN 90–6164–199–3 ITC, Enschede, The Netherlands

Johannes S (1989) Towards a surface radiation climatology: retrieval of downward irradiances from satellites. Atmos Res 23:287–321. https://doi.org/10.1016/0169-8095(89)90023-9

Kriebel KT, Gesell G, Kästner M, Mannstein H (2003) The cloud analysis tool APOLLO: improvements and validations. Int J Remote Sens 24:2389–2408. https://doi.org/10.1080/01431160210163065

Lefèvre M, Oumbe A, Blanc P et al (2013) McClear : a new model estimating downwelling solar radiation at ground level in clear-sky conditions. Atmos Meas Tech 6:2403–2418. https://doi.org/10.5194/amt-6-2403-2013

Martins FR, Pereira EB, Abreu SL (2007) Satellite-derived solar resource maps for Brazil under SWERA project. Sol Energy 81:517–528. https://doi.org/10.1016/j.solener.2006.07.009

Martins FR, Pereira EB, Abreu SL, et al (2003) Cross validation of satellite radiation transfer models during SWERA project in Brazil. In: Proceedings of the ISES solar world congress

Merrouni AA, Ghennioui A, Wolfertstetter F, Mezrhab A (2017) The uncertainty of the HelioClim-3 DNI data under Moroccan climate. In: AIP Conference proceedings 1850:140002–140011. https://doi.org/10.1063/1.4984519

Moser W, Raschke E (1984) Incident solar radiation over Europe estimated from METEOSAT data. J Clim Appl Meteorol 23:166–170

Mueller RW, Dagestad KF, Ineichen P et al (2004) Rethinking satellite-based solar irradiance modelling: the SOLIS clear-sky module. Remote Sens Environ 91:160–174. https://doi.org/10.1016/j.rse.2004.02.009

Mueller RW, Matsoukas C, Gratzki A et al (2009) The CM-SAF operational scheme for the satellite based retrieval of solar surface irradiance—a LUT based eigenvector hybrid approach. Remote Sens Environ 113:1012–1024. https://doi.org/10.1016/j.rse.2009.01.012

Müller R, Pfeifroth U, Träger-Chatterjee C et al (2015a) Surface solar radiation data set-heliosat (SARAH)-edition 1. Satellite application facility on climate monitoring

Müller R, Pfeifroth U, Träger-Chatterjee C et al (2015b) Digging the METEOSAT treasure—3 decades of solar surface radiation. Remote Sens 7:8067–8101. https://doi.org/10.3390/rs70608067

Noia M, Ratto CF, Festa R (1993a) Solar irradiance estimation from geostationary satellite data: II physical models. Sol Energy 51:457–465

Noia M, Ratto CF, Festa R (1993b) Solar irradiance estimation from geostationary satellite data: I statistical models. Sol Energy 51:449–456

Nunez M (1993) The development of a satellite-based insolation model for the tropical western Pacific Ocean. Int J Climatol 13:607–627. https://doi.org/10.1002/joc.3370130603

Perez R (2002) Time specific irradiances derived from geostationary satellite images. J Sol Energy Eng 124:2012

Perez R, Cebecauer T, Šúri M (2013) Chapter 2—Semi-empirical satellite models. In: Kleiss J (ed) Solar energy forecasting and resource assessment, pp 21–48

Perez R, Ineichen P, Kmiecik M et al (2004) Producing satellite-derived irradiances in complex arid terrain. Sol Energy 77:367–371. https://doi.org/10.1016/j.solener.2003.12.016

Perez R, Ineichen P, Moore K et al (2002) A new operational model for satellite-derived irradiances: description and validation. Sol Energy 73:307–317. https://doi.org/10.1016/S0038-092X(02)00122-6

Perez R, Schlemmer J, Hemker K et al (2015) Satellite-to-irradiance modeling–a new version of the SUNY model. In: Proceedings of the 42nd IEEE photovoltaic specialist conference, at New Orleans, LA

Perez R, Schlemmer J, Kankiewicz A et al (2017) Detecting calibration drift at reference ground truth stations-A demonstration of satellite irradiance model's accuracy. In: Proceedings of the IEEE PVSC-44, Washington, DC

Perez R, Schlemmer J, Renne D et al (2009) Validation of the suny satellite model in a meteosat environment. In: Proceedings of the American solar energy society annual conference ASES Annua:

Pfeifroth U, Kothe S, Müller R et al (2017) Surface radiation data set - Heliosat (SARAH)-Edition 2. https://doi.org/10.5676/eum_saf_cm/sarah/v002

Pinker RT (1985) Determination of surface albedo from satellites. Adv. Sp. Res. 5:333–343

Pinker RT, Laszlo I (1991) Effects of spatial sampling of satellite data on derived surface solar irradiance. J Atmos Ocean Technol 8:96–107

Polo J, Antonanzas-Torres F, Vindel JMM, Ramirez L (2014) Sensitivity of satellite-based methods for deriving solar radiation to different choice of aerosol input and models. Renew Energ 68:785–92. http://dx.doi.org/10.1016/j.renene.2014.03.022

Polo J, Martín L, Cony M (2012) Revision of ground albedo estimation in Heliosat scheme for deriving solar radiation from SEVIRI HRV channel of Meteosat satellite. Sol Energy 86:275–282. https://doi.org/10.1016/j.solener.2011.09.030

Polo J, Martín L, Vindel JM (2015) Correcting satellite derived DNI with systematic and seasonal deviations: application to India. Renew energy 80:238–243. https://doi.org/10.1016/j.renene.2015.02.031

Polo J, Vindel JM, Martín L (2013) Angular dependence of the albedo estimated in models for solar radiation derived from geostationary satellites. Sol Energy 93:256–266. https://doi.org/10.1016/j.solener.2013.04.019

Polo J, Wilbert S, Ruiz-Arias JA et al (2016) Preliminary survey on site-adaptation techniques for satellite-derived and reanalysis solar radiation datasets. Sol Energy 132:25–37. https://doi.org/10.1016/j.solener.2016.03.001

Polo J, Zarzalejo LF, Ramírez L (2008) Solar radiation derived from Satellite images. In: Badescu V (ed) Modeling solar radiation at the earth's surface: recent advances. Springer, Berlin Heidelberg, Berlin, Heidelberg, pp 449–461

Posselt R, Mueller RW, Stöckli R, Trentmann J (2012) Remote sensing of solar surface radiation for climate monitoring — the CM-SAF retrieval in international comparison. Remote Sens Environ 118:186–198. https://doi.org/10.1016/j.rse.2011.11.016

Qu Z, Oumbe A, Blanc P et al (2017) Fast radiative transfer parameterisation for assessing the surface solar irradiance: The Heliosat-4 method. Meteorol Zeitschrift 26:33–57. https://doi.org/10.1127/metz/2016/0781

Rigollier C, Lefèvre M, Wald L (2004) The method Heliosat-2 for deriving shortwave solar radiation from satellite images. Sol Energy 77:159–169. https://doi.org/10.1016/j.solener.2004.04.017

Riihelä A, Carlund T, Trentmann J et al (2015) Validation of CM SAF surface solar radiation datasets over Finland and Sweden. Remote Sens 7:6663–6682. https://doi.org/10.3390/rs70606663

Sanchez-Lorenzo A, Wild M, Trentmann J (2013) Validation and stability assessment of the monthly mean CM SAF surface solar radiation dataset over Europe against a homogenized surface dataset (1983–2005). Remote Sens Environ 134:355–366. https://doi.org/10.1016/j.rse.2013.03.012

Schillings C, Mannstein H, Meyer R, Mannstein H (2004) Validation of a method for deriving high resolution direct normal irradiance from satellite data and application for the Arabian Peninsula. Sol Energy 76:485–497. https://doi.org/10.1016/j.solener.2003.07.037

Sengupta M, Habte A, Gueymard C, Wilbert S, Renné D (2017) Best practices handbook for the collection and use of solar resource data for solar energy applications, 2nd edn. https://doi.org/10.18777/ieashc-task46-2015-0001

Tarpley JD (1979) Estimating incident solar radiation at the surface from geostationary satellite data. J Appl Meteorol 18:1172–1181. https://doi.org/10.1175/1520-0450(1979)018%3c1172:EISRAT%3e2.0.CO;2

Wang P, Stammes P, Mueller R (2011) Surface solar irradiance from SCIAMACHY measurements: algorithm and validation. Atmos Meas Tech 4:875–891. https://doi.org/10.5194/amt-4-875-2011

Weymouth GT, Le Marschall JF (2001) Estimate of daily surface solar exposure using GMS-5 stretched-VISSR observations: the system and basic results. Aust Meteorol Mag 50:263–278

Žák M, Mikšovský J, Pišoft P (2015) CMSAF radiation data: new possibilities for climatological applications in the Czech republic. Remote Sens 7:14445–14457. https://doi.org/10.3390/rs71114445

Zarzalejo LF, Polo J, Martín L et al (2009) A new statistical approach for deriving global solar radiation from satellite images. Sol Energy 83:480–484. https://doi.org/10.1016/j.solener.2008.09.006

Zelenka A, Perez R, Seals R, Renne D (1999) Effective accuracy of satellite-derived hourly irradiances. Theor Appl Climatol 62:199–207. https://doi.org/10.1007/s007040050084

Chapter 7
Solar Resource Evaluation with Numerical Weather Prediction Models

Pedro A. Jiménez, Jared A. Lee, Branko Kosovic and Sue Ellen Haupt

Abstract The use of numerical weather prediction (NWP) models for solar resource evaluation is examined. The theory behind NWP models is described highlighting relevant components for solar energy applications as well as how to use NWP models for mapping the solar resource at the regional scale. Future perspectives are briefly outlined.

1 Introduction

In 1904, Norwegian physicist Vilhelm Bjerknes envisioned weather forecasting through solving a system of nonlinear partial differential equations describing physical processes in the atmosphere. Following Bjerknes, in 1922 an English mathematician and physicist, Lewis Fry Richardson, published a paper with the title: "Weather Prediction by Numerical Process." In this paper Richardson, after whom the Richardson number (the non-dimensional ratio of the temperature gradient and the shear) is named, outlined a procedure for numerical weather prediction (NWP) carried out by a large number of human "computers." Implementation of Richardson's idea had to wait for the development of electronic programmable computers. In 1948 at the Institute for Advanced Study, John von Neumann formed a group of scientists headed by Jule Charney. This group was tasked with using the first programmable electronic computer, ENIAC, to demonstrate NWP by solving a simplified set of equations governing atmospheric dynamics.

While initially NWP considered only atmospheric dynamics, development of computational capabilities resulted in higher resolution simulations in recognition of a need to include cloud physics and radiative transfer effects in simulations. This led to the development of the first limited area models in the late 1960s and early

P. A. Jiménez (✉) · J. A. Lee · B. Kosovic · S. E. Haupt
Research Applications Laboratory, National Center for Atmospheric Research,
Boulder, CO 80307-3000, USA
e-mail: jimenez@ucar.edu

© Springer Nature Switzerland AG 2019
J. Polo et al. (eds.), *Solar Resources Mapping*, Green Energy and Technology,
https://doi.org/10.1007/978-3-319-97484-2_7

1970s. One of the first limited area models was the Mesoscale Model or MM, developed by Anthes and Warner (Anthes and Warner 1974, 1978). The MM was a predecessor to the Weather Research and Forecasting (WRF) model (Powers et al. 2017; Skamarock et al. 2008), a limited area model widely used for a range of applications including solar energy forecasting.

The use of NWP models to estimate the solar resource is an alternative to other traditional methodologies (Habte et al. 2017). A standard practice is to use satellite retrievals to estimate the solar resource. The resource is generally estimated using retrievals from geostationary satellites (e.g., Blanc et al. 2011; Martins et al. 2007). The estimations provided using these satellite methodologies are that NWP models must outperform.

Some efforts have been oriented toward quantifying the performance of NWP models in reproducing the solar resource (e.g., Charabi et al. 2016; Paquin-Ricard et al. 2010; Ruiz-Arias et al. 2016). These studies have identified a tendency of the models to overestimate the solar irradiance.

Model developments have been also introduced to improve the value of NWP models for solar energy applications. For example, the Sun4Cast® project (Haupt et al. 2018a) included developments of the WRF model specifically targeted to meet the needs of the solar industry. The main emphasis in developing the WRF-Solar® model (Jiménez et al. 2016b) was to improve the representation of the model physics. In particular, WRF-Solar provided a better representation of the cloud–aerosol–radiation feedbacks. The model has been shown to largely improve the clear-sky estimations of the WRF model (Jiménez et al. 2016b), as well as in all sky conditions, wherein including the radiative effects of unresolved clouds is necessary to largely reduce a positive bias in the model (Jiménez et al. 2016b).

A key component of NWP-based assessments of the solar resource is to employ data assimilation (DA). Due to the chaotic nature of the atmosphere, NWP model solutions eventually drift away from the true atmospheric state unless the NWP model is corrected by observations of atmospheric variables. DA is the process by which observations are blended with a model solution to generate the best estimate of the model state (Talagrand 1997). The resultant "analysis" can then serve as a more accurate initial state for a subsequent NWP model simulation, leading to more accurate simulations of the atmosphere (Fig. 1). Over the decades, DA has matured from relatively simple objective analysis to computationally intensive variational and ensemble Kalman filter approaches.

The present chapter provides an introduction to the evaluation of solar resources using NWP models. A brief outline of the main classes of modern DA schemes including concepts relevant for solar energy applications are discussed first (Sect. 2). Afterward, the theoretical basis of NWP models is explained (Sect. 3), and guidelines to efficiently use NWP models for solar resource evaluation are provided (Sect. 4). Potential enhancements via statistical post-processing of the model estimations are also discussed (Sect. 5). Finally, the chapter concludes by examining future perspectives (Sect. 6).

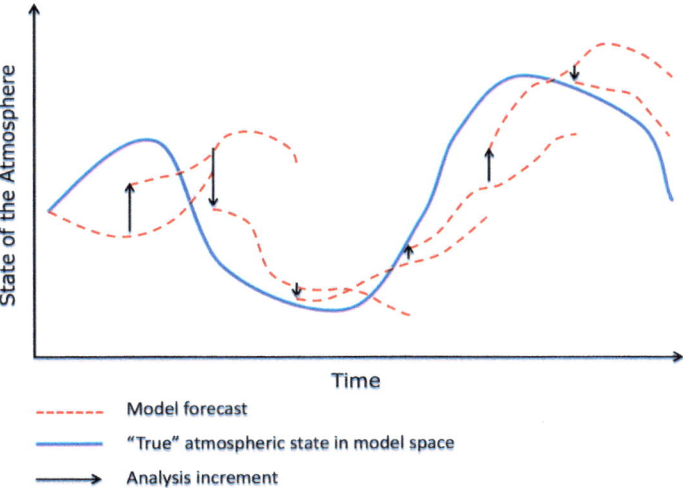

Fig. 1 Schematic diagram illustrating the basic DA procedure of blending a short-range forecast with observations to create an analysis state to initialize a new forecast

2 Data Assimilation

The summary of the various types of DA algorithms presented here in Sects. 2.1–2.4 is adapted from previous works by the authors (Haupt et al. 2017, 2018a) and is included here for completeness.

2.1 Nudging DA

Nudging, also called Newtonian relaxation, is a relatively simple optimal interpolation algorithm that builds on objective analysis techniques (Cressman 1959). In nudging schemes, small terms are added to the NWP model prognostic tendency equations in order to "nudge" the model state toward either a point observation or a gridded observation field (Hoke and Anthes 1976). Four-dimensional data assimilation (FDDA) is a modern nudging scheme in which the nudging terms are weighted in both time and space (Deng et al. 2007, 2009; Stauffer and Seaman 1990, 1994; Stauffer et al. 1991), to avoid contaminating the NWP model solution with artificial gravity waves. FDDA is a continuous DA algorithm, meaning that observations are assimilated continuously at every model time step, rather than only at specific analysis times. FDDA is often used in short-term case studies and longer-duration re-forecasts for weather and regional climate applications (e.g., Jonassen et al. 2012; Otte et al. 2012; Rogers et al. 2013), though it is also used for operational real-time forecasting applications (e.g., Liu et al. 2008), including for renewable energy forecasting (e.g., Mahoney et al. 2012).

2.2   *Variational DA*

The other main classes of DA algorithms are statistical interpolation techniques, which require estimates of the error covariance between background state variables and between the observed variables. Accurate computation or estimation of both of these error covariance matrices is often difficult (e.g., Kalnay 2003; Talagrand 1997).

One such class of techniques is variational data assimilation ("Var"). In Var DA, the analysis is obtained by combining observations with an initial state and model trajectory via minimizing a global cost function that accounts for the distance between the analysis and observations that are within the assimilation time window (Schlatter 2000; Talagrand and Courtier 1987). Three-dimensional variational DA (3D-Var) schemes vastly simplify the calculations by using a static climatology to estimate the error covariance matrices, at the expense of not being able to represent the true "errors of the day." Four-dimensional variational DA (4D-Var) schemes, by contrast, calculate flow-dependent error covariance matrices, and minimize the cost function over the entire period between analysis times, rather than only at the analysis time. These differences make 4D-Var schemes more accurate than 3D-Var, but also far more computationally intensive (e.g., Klinker et al. 2000; Yang et al. 2009).

2.3   *Ensemble Kalman Filters*

Like 3D-Var or 4D-Var, Kalman filters (Kalman 1960; Kalman and Bucy 1961) are statistical interpolation DA schemes and use forward operators to map observations of a number of quantities onto model variable space. Unlike 3D-Var or 4D-Var, however, KF algorithms update the model background state sequentially with each observation within the assimilation window, instead of solving a global cost function. To make the computational cost of estimating a flow-dependent background error covariance matrix for complex nonlinear geophysical models tractable, the ensemble Kalman filter (Bannister 2017; Evensen 1994) leverages a finite ensemble of model simulations. Each ensemble member has different initial conditions, boundary conditions, physics parameterization schemes, and/or key physical parameters and is assumed to be a random sample of the model state's true flow-dependent probability distribution. To reduce the impact of sampling error and its attendant noisy background error covariances and inadequate ensemble variances, several dozen ensemble members are typically used with EnKF. Integrating several dozen NWP ensemble members simultaneously is a large portion of the cost of an EnKF approach. Several different kinds of ensemble Kalman filters are in use by the NWP community (e.g., Anderson et al. 2009; Anderson 2001; Bishop et al. 2001; Schwartz et al. 2015; Whitaker and Hamill 2002; Yang et al. 2007). Houtekamer and Zhang (2016) contain a helpful, in-depth overview of several EnKF approaches in active use today by NWP researchers and practitioners.

2.4 Hybrid DA Approaches

Many applications have been developed that combine the most useful aspects of the DA approaches outlined above. Some approaches hybridize nudging with Var (Lei et al. 2012) and latent heat nudging with a type of EnKF for assimilating radar reflectivity data (Schraff et al. 2016). Most hybrid DA approaches, however, combine Var and EnKF algorithms in a variety of ways (e.g., Lorenc et al. 2015; Pan et al. 2014; Schwartz et al. 2014; Wang et al. 2008a, b; Zhang et al. 2013). Most operational national forecast centers today use some form of "EnVar" approach as the core of their DA system. For a detailed review of the myriad of hybrid DA approaches in use today, see (Bannister 2017).

2.5 Cloud DA for Solar Energy Applications

For the purposes of solar energy forecasting and solar resource modeling, generating accurate simulations of clouds and aerosols is crucial. Clouds are difficult to simulate with great accuracy, however, because the complex microphysical processes that are being simulated are sometimes not well known, and even if they are well known, they occur at such small scales that they cannot be simulated explicitly and must be parameterized. This parameterization introduces uncertainty and error into NWP simulations of clouds. Several studies have shown a general tendency of NWP models to over-predict global horizontal irradiance (GHI) in cloudy conditions (e.g., Charabi et al. 2016; Lee et al. 2017; Mathiesen and Kleissl 2011; Paquin-Ricard et al. 2010; Ruiz-Arias et al. 2016). Furthermore, unless an NWP model is specially initialized with all the requisite 3D microphysical fields (e.g., number concentration and mixing ratio for water vapor, rain, snow, graupel, etc.), it often takes a few hours for the NWP model to generate (or "spin up") fully formed cloud and precipitation fields (Warner 2011). In addition to improving cloud microphysics parameterizations, improving DA specifically for cloud forecasts is a key to improving solar irradiance forecasts.

The chief obstacle to use DA to initialize an NWP model with realistic cloud fields is that good observations of microphysical variables in three dimensions are generally not available. Radiosondes provide sparse vertical profiles of temperature, humidity, and wind, but not any other variables. From ground-based sensors or sky imagers, cloud base height can be determined with reasonable accuracy, but those are only point observations (e.g., Yang et al. 2014). Multiple sky imagers can be used to triangulate cloud base heights and motion over the combined field of view (e.g., Peng et al. 2015), but sky imagers are expensive and not widely deployed. Satellite radiances, in contrast, can provide information about cloud top height, cloud top motion, water vapor mixing ratio, and other important variables over a wide area. Visible and infrared bands are unable to penetrate clouds and remotely sense conditions below cloud top, however, while microwave bands can penetrate clouds but generally only give column-total information, rather than a high-resolution vertical profile. Thus, there are large regions of the atmosphere over which we have no direct or remotely sensed observations of key variables, meaning various inferences and assumptions about those fields must be made.

Several attempts have been made to assimilate cloud information into NWP models in the past two decades, with varying degrees of success in reducing errors in NWP model simulations. This includes assimilating upper-air cloud water mixing ratio observations from Geostationary Operational Environmental Satellite (GOES) platforms (Bayler et al. 2000), directly replacing model cloud cover with GOES-derived cloud cover (Yucel et al. 2002, 2003), assimilating Meteosat water vapor clear-sky radiance data with 4D-Var (Köpken et al. 2004; Munro et al. 2004), cloudy-sky radiances with 1D-Var to modify profiles of ice and liquid water (Martinet et al. 2013), cloudy-sky radiances with 4D-Var (Stengel et al. 2013, 2010), GOES cloud water path with EnKF (Jones et al. 2013), and cloud liquid and cloud ice water paths with 3D-Var (Chen et al. 2015). van der Veen and van der Veen (2013) and de Haan et al. (2014) combine Meteosat Second Generation (MSG) cloud height information with surface observations of cloud base height to insert clouds by modifying cloud fraction and virtual temperature profiles at the analysis time with 4D-Var and 3D-Var, respectively. In yet another approach, White et al. (2018) use FDDA to assimilate satellite-derived vertical velocity, in an effort to provide more dynamical support for cloudy and clear areas. Even with all these proposed approaches, it remains a matter of ongoing research to determine the best method to assimilate cloud information into NWP models and to maintain cloudy and clear areas in the right locations more than an hour or two into the simulation.

Yet another approach involves blending satellite cloud advection techniques with NWP models specifically for the purpose of predicting GHI. One such approach is the Cooperative Institute for Research in the Atmosphere (CIRA) solar forecasting system (CIRACast; Lee et al. 2017; Miller et al. 2018), which embeds satellite cloud information into the wind field of the Global Forecast System (GFS) model before running a radiative transfer code to calculate GHI. Another approach is the Multisensor Advection Diffusion Nowcast (MADCast; Auligné and Auligné 2014a, b; Descombes et al. 2014), which assimilates imager and profiler data from multiple satellites, converting that into tracers that represent cloud fraction, which is then advected and diffused in a dynamic-only version of the WRF model. These methods both perform best in the first 1–2 h, but have no inherent way to account for cloud formation and dissipation (Haupt et al. 2016b, 2018a, b; Lee et al. 2017). Research is currently ongoing at the National Center for Atmospheric Research (NCAR) to blend MADCast more fully with WRF-Solar ("MAD-WRF"; Haupt et al. 2018a, b) to update the dynamics and microphysical profiles in WRF and thus hopefully achieves longer-lasting error reduction in GHI forecasts.

DA is also a central feature in the creation of long-term retrospective analysis ("reanalysis") datasets. A reanalysis is typically a 20–30-year long (or more) record of the atmospheric state, as generated by a single, static version of an NWP model with all available observations assimilated. One such prominent reanalysis dataset that is useful for solar forecasting and resource assessment is NASA's Modern-Era Retrospective Analysis for Research and Applications, Version 2 (MERRA-2; Gelaro et al. 2017). The MERRA-2 reanalysis extends from the current day back to 1980. Using a 3D-Var system, the MERRA-2 assimilates all standard ground-based observations as well as a host of satellite-based observations. Among the positives of the satellite DA that goes into the MERRA-2 is a relatively high-resolution dataset of atmospheric aerosol concentration. Along with clouds, knowledge of the aerosol loading is crucial for accurately simulating GHI at the surface. Aerosol data from the MERRA-2 can also be used as input into other models like WRF-Solar that are more tailored for solar irradiance modeling applications and run at higher resolution.

3   Theoretical Basis of NWP Modeling

The main components of an NWP model are the dynamical core and the physics package. The dynamical core is responsible for a suitable discretization of the atmosphere to approximate a solution of the atmospheric equations of motion. The discretized set of equations typically accounts for advection, rotational effects, pressure gradient forces and gravity. Other relevant atmospheric processes are more difficult to represent explicitly. When this is the case, models include the effects of these processes in the parameterizations of the physical package.

The following sections describe the fundamentals of NWP models, namely the atmospheric equations of motion (Sect. 3.1), the dynamical core (Sect. 3.2), the physics (Sect. 3.3), and how the dynamical core and the physics interact (Sect. 3.4).

3.1   Atmospheric Equations of Motion

Flows in the atmosphere are governed by the Navier–Stokes equations. Atmospheric scales of motion span eight orders of magnitude. Since it is not possible to simultaneously resolve all the scales of motion in NWP models, it is necessary to represent the effect of small, unresolved scales of motion on larger resolved scales of motion. This is accomplished using Reynolds Averaged Navier–Stokes (RANS) equations. The averaging used to derive RANS equations is formally ensemble averaging; however, since forecasting deals with specific realizations of atmospheric flows that evolve in time, ensemble averaging is not appropriate. By invoking the ergodic hypothesis, an ensemble average can be replaced with a time average. However, due to dynamic forcing (i.e., solar irradiance) and the required time resolution, a time average is not appropriate either. Therefore, the governing equations for the resolved scales of motion are obtained using spatial averaging, with the smallest scales resolved in a simulation being proportional to the grid cell size. There are several possible simplifications of the

Navier–Stokes equations that can be used to simultaneously reduce computational complexity and preserve the main characteristics of the full system. A discretized form of the compressible Navier–Stokes equations is implemented in WRF as well as COSMO (Brdar et al. 2013). While the exact form of the equations implemented in these models is different, in both models fast acoustic pressure waves are filtered. Here we present the set of equations similar to those implemented in the COSMO model in coordinate independent notation. The set of equations includes mass conservation, prognostic equations for momentum and potential temperature, pressure and constituent tendency, and the ideal gas law:

$$\frac{D\bar{\rho}}{Dt} = -\bar{\rho}\nabla \cdot \tilde{v} \tag{1}$$

$$\frac{D\bar{\rho}\tilde{v}}{Dt} = -\nabla\bar{p} + \nabla \cdot \tilde{\sigma} + \bar{\rho}g - 2\Omega \times \bar{\rho}\tilde{v} \tag{2}$$

$$c_v \frac{D\bar{\rho}\tilde{\theta}}{Dt} = -\bar{p}\nabla \cdot \tilde{v} + Q_h + Q_m \tag{3}$$

$$\frac{D\bar{p}}{Dt} = -\frac{c_p}{c_v}\bar{p}\nabla \cdot \tilde{v} + \left(\frac{c_p}{c_v} - 1\right)Q_h + \frac{c_p}{c_v}Q_m \tag{4}$$

$$\frac{D\bar{\rho}\tilde{q}}{Dt} = -\nabla \cdot F \tag{5}$$

$$\bar{p} = \bar{\rho}R_d\tilde{T} \tag{6}$$

The potential temperature is defined as

$$\tilde{\theta} = \tilde{T}\left(\frac{p_0}{\bar{p}}\right)^{\frac{R}{c_p}} \tag{7}$$

All the fields are filtered using the Favre filter defined as

$$\tilde{\varphi} = \frac{\bar{\rho}\bar{\varphi}}{\bar{\rho}} \tag{8}$$

Here, D/Dt denotes the material derivative, v is the velocity vector, θ is potential temperature, the overbar denotes spatial average, and the tilde denotes the Favre.

3.2 Dynamics

Large-scale atmospheric circulation is forced by a differential heating between the equator and the poles, modulated by the effects of Earth's rotation and the curvature of Earth's surface. The differential heating results in poleward motion at the surface of the Earth and a return flow above, forming a Hadley cell at low latitudes. In the tropics, the effect of solar irradiance results in rising air near the equator forming the Intertropical Convergence Zone (ITCZ). In addition to the Hadley cell, the other circulations spanning the range of latitudes from the equator to the poles are the Ferrel cell and polar cell. Due to the apparent Coriolis force, the flow is deflected eastward (westward) on the northern (southern) hemisphere. In the northern hemisphere, surface flows would, therefore, be easterly; however, while easterly winds dominate over the tropical and subtropical region, due to the conservation of mass there needs to be westerly return flow at higher altitudes. Where the Hadley and Ferrel cell downdrafts meet there are semi-permanent high-pressure regions (e.g., the Bermuda–Azores High and North Pacific High in the northern hemisphere). The updrafts where Ferrel cells meet Polar cells create low-pressure regions (e.g., the Aleutian Low and Icelandic Low in the northern hemisphere). Variation of the Coriolis effect and the associated distribution of high and low pressure at higher latitudes results in Rossby waves. Rossby wave crests always have a westward component; collections of short waves move with group velocity eastward while long waves move westward. Rossby waves are meanders of the jet stream that can form the cyclones and anticyclones responsible for weather patterns at mid-latitudes.

3.3 Physics

The physics parameterizations in NWP models represent the effects of those processes that cannot be explicitly modeled. This could be either because we do not know how to represent the process, because of the necessity to use large computational resources for its explicit representation, or because the process occurs at the subgrid scale.

Typical processes that are parameterized in numerical weather prediction models include radiation, turbulent mixing, land–atmosphere interactions, cloud microphysics, and cumulus clouds (Stensrud 2011). The radiative effects consist of two parameterizations: one for the shortwave effects and the other one for the longwave effects. The parameterizations calculate atmospheric heating rates and the incoming radiation to the Earth surface. The net radiative balance at the surface is used by the land surface model to parameterize soil, vegetation, and hydrological processes in order to calculate surface turbulent fluxes (i.e., sensible heat flux and latent heat flux). The surface fluxes are used by planetary boundary layer schemes that parameterize the effects of turbulent mixing in the resolved atmospheric variables. The core set of parameterizations is completed by the cumulus and microphysics parameterizations. Both of them account for cloud effects. The cumulus scheme accounts for the effects on unresolved cumulus clouds, whereas the microphysics parameterization models the evolution of hydrometeors and its impact on the

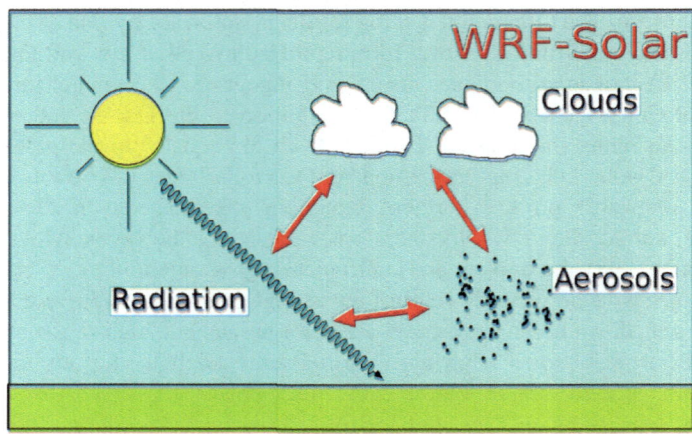

Fig. 2 Schematic diagram illustrating the feedbacks between the different components of the cloud–aerosol–radiation system that were improved in WRF-solar

resolved variables. Nonlinear feedbacks between the parameterizations further complicate their implementation in numerical weather prediction models.

Special efforts have been oriented to improve the physics packages for solar energy applications, e.g., the WRF-Solar model (Jiménez et al. 2016a, b). WRF-Solar efforts have focused on improving the cloud–aerosol–radiation feedbacks (Fig. 2). The model incorporates an efficient parameterization of the aerosol direct effect (Lee et al. 2017; Ruiz-Arias et al. 2014); improving the parameterization of the aerosol indirect effects (Thompson et al. 2016; Thompson and Eidhammer 2014); a shallow cumulus parameterization that accounts for radiative effects of unresolved clouds (Deng et al. 2014; Jiménez et al. 2016a; Lee et al. 2017); and a fast radiative transfer parameterization to update the surface irradiance every model time step (Xie et al. 2016). Recently, WRF-Solar incorporated a cloud fraction parameterization that uses relative humidity to estimate the horizontal cloud fraction (based on Sundqvist et al. 1989) and a dust model (Chin et al. 2000) to explicitly resolve dust transport and emission. These improvements to the WRF-Solar physical packages increase the value of WRF-Solar for solar irradiance estimations.

3.4 Coupling Dynamics and Physics

The approximations introduced in the dynamical core and the physical packages need to be integrated to predict the atmospheric evolution. This coupling is not straightforward (Caya et al. 1998; Gross et al. 2016, 2017; Lander and Hoskins 1997; Staniforth et al. 2002a, b; Williamson 2002). The challenge comes from splitting the contributions of the physical package into individual physical processes. The effects of these processes need to be coupled together and with the dynamical core to account for feedbacks between them. A suboptimal coupling combined with a large model time step may lead to processes that are isolated from the others for too long (Gross et al. 2017).

Different strategies have been proposed to couple the physics to the dynamical core. The basic approaches are process split and time split (e.g., Dubal et al. 2004; Williamson 2002). The process split calculates the tendencies of each process independently and then updates the atmospheric state. On the other hand, time splitting updates the atmospheric state each time a new tendency is available. This methodology requires a careful ordering of the physical processes, and it has been shown that this could lead to superior performance over the process split methodology (Beljaars et al. 2004). Hybrid approaches combining both methods have been also implemented in NWP models (Dubal et al. 2006).

In principle, reducing the time step of the model should alleviate problems with the coupling. However, one should ensure first that the numerical solutions will converge for sufficiently small time steps, and this still a topic of research (Gross et al. 2017). Fortunately, the coupling between the physics and dynamics has received more attention during recent years (Gross et al. 2016), which should contribute to reducing errors in this aspect of NWP models.

3.5 Chemistry

NWP models sometimes account for atmospheric chemistry. For example, the WRF model has the WRF-Chem extension (Fast et al. 2006; Grell et al. 2005) to simulate emissions, transport, mixing, and chemical reactions of gases and aerosols. The computational cost of these simulations increases as a result of accounting for the extra atmospheric processes and the need to advect the different gases and aerosols species. This can be a limitation for operational forecasting. However, accounting for these processes can enhance the value of NWP models for solar resource evaluations wherein the computational cost is not such a limiting factor. Previous studies suggest that accounting for atmospheric chemistry and aerosols can indeed enhance the performance of solar resource evaluations (e.g., Fountoukis et al. 2018). The modeler should consider the potential benefits of accounting for chemical and aerosol-related processes if it is possible to simulate them.

4 How to Use NWP for Solar Resource Evaluation

One should start by identifying previous solar resource assessments over the region. This can provide a preliminary idea of what the solar resource might be. If there are previous resource assessments based on NWP models, previous findings can point to challenges that the modeler should keep in mind when configuring the NWP model.

Before configuring and running the NWP model, it is desirable to gather available observations to aid with the design of the experimental setup. This includes both observations that can be used to decide the model configuration and observations that could be assimilated by the NWP model. The observations that will be used to configure the model should primarily consist of aerosols, irradiance, and cloud observations. These data will assist the modeler in deciding the physics packages that will be used in the solar resource assessment. The observations that may be assimilated can include any relevant atmospheric variable (e.g., winds, temperature, humidity, pressure, etc.) and should cover a wider region than the area of interest. This will help to properly simulate the atmospheric evolution over the area of interest.

A successful resource assessment requires considering the relevant physical mechanisms over the region of interest in order to configure the model. This will help the modeler select the physical parameterizations that should provide the best model performance over the region. For instance, the cloud parameterizations that produce the best performance in reproducing tropical convection may not be optimal over mid-latitudes. In addition, this analysis will help to identify relevant topographic features and surface heterogeneities that the model should represent. This inspection will help to identify an adequate spatial discretization of the atmosphere. Nesting strategies are frequently applied to progressively refine the vertical and horizontal grid spacing.

Another important decision is to select a dataset to create the initial and boundary conditions. Both reanalysis and analysis datasets assimilate observations to improve the estimation of the atmospheric state. Re-analyses present the advantage of using the same version of the model over a multi-year period. This is a desirable characteristic since changes in the model may introduce heterogeneities in the atmospheric state.

Independent of the dataset selected to create initial and boundary conditions, one can assimilate observations over the region to further improve the quality of the atmospheric state. This data assimilation is more important when assimilating observations that were not used by the (re)analysis dataset, or if the observations are assimilated at much finer spatial resolutions.

The modeler is at this point ready to decide a strategy to simulate the target temporal period. One can run the model continuously or perform shorter simulations that together span the target period. This second strategy ensures that the synoptic situation is close to the analysis due to the re-initializations of the model. Applying data assimilation also ensures a closer initial state to the observed one. This is not the case for a continuous simulation. However, different strategies can be used to minimize a potential drift of the synoptic situation. If one has decided to assimilate observations, one can run the simulations doing cycling every few hours (e.g., 6 h) wherein the model stops to perform assimilation before continuing the simulation. Alternatively, nudging strategies can be applied to reduce the synoptic drift. There are three nudging methodologies: observation nudging and analysis nudging, which were described in Sect. 2.1, and spectral nudging (Miguez-Macho et al. 2004, 2005). Observation nudging nudges the NWP model to point observations, while analysis nudging nudges the simulated atmospheric evolution to a gridded analysis. This may remove internal variability in the model. To overcome this limitation, spectral nudging only nudges the low-frequency modes, thereby allowing the NWP model to develop internal variability in the higher frequencies. Both nudging strategies also contribute to reducing errors in the lateral boundary conditions.

The selected experimental design should be tested and perhaps refined before simulating the complete target period. The testing period should cover at least a year in order to have a representation of the different synoptic patterns over the year. The simulation can be compared against the available observations to identify strengths and weaknesses in reproducing the solar resource. Hypotheses to explain the larger discrepancies should be formulated and potential solutions identified. One can then rerun the testing period to quantify any potential gain in the solar resource evaluation. This process should be repeated until a tolerable error in the solar resource estimation is obtained.

The final configuration can then be used to simulate the complete temporal period. The simulation should be validated against the available observations to quantify the limitations of the simulation. This will lead to a solar resource evaluation accompanied with a quantification of the modeled uncertainties.

5 Post-processing

As mentioned above, most current NWP models demonstrate biases in solar irradiance prediction. The older MM5 model tended to have a negative GHI bias (Armstrong 2000; Haupt et al. 2016a; Zamora et al. 2005) but WRF tends to have a positive GHI bias due to insufficient generation of cloudiness (Lara-Fanego et al. 2012; Lee et al. 2017; Mathiesen and Kleissl 2011; Ruiz-Arias et al. 2016). Although WRF-Solar partially alleviates that problem, the seasonal dependence of the bias remains (Jiménez et al. 2016a). If a bias is calculable, it can be removed via post-processing. Where there are seasonal or cloud type dependencies, more complex statistical and artificial intelligence (AI) methods may provide better corrections.

More of the post-processing work has been applied on the forecasting time scales where it is most important to obtain highly accurate real-time predictions of solar irradiance (Tuohy et al. 2015). Many of these same techniques are applicable to solar resource assessment. Some of these techniques include multi-linear corrections, typically known as Model Output Statistics (MOS; Glahn and Lowry 1972) applied to solar NWP predictions (Lorenz et al. 2014); applying a dynamic version of MOS to multiple individual models and blending those model outputs with optimized weights (Haupt et al. 2016b, 2018a, b); applying a Kalman filter to the NWP forecasts (Diagne et al. 2014); and using artificial neural networks to post-process NWP output (Marquez and Coimbra 2011).

Similar techniques have been applied for resource assessment for wind energy. Clifton et al. (2018) points out that machine learning methods are promising for improving the accuracy of such assessments while reducing the computational costs and sometimes providing probabilistic information. Kay and MacGill (2014) demonstrate a basic bias correction methodology for improving wind resource assessments. Zhang et al. (2015) demonstrate the efficacy of using an analog ensemble approach to post-processing wind resource data. They show that using the NWP output paired with historical observations, they can find analogous cases that form an ensemble that both improves the deterministic forecast and provides probabilistic information.

Finally, for long-term solar resource assessment, it is becoming more important to take into account the impacts of climate change. To do this requires judiciously interpreting and correcting climate model output. Pryor et al. (2005, 2009) studied the impact of projected climate change on the wind resource. Dutton et al. (2018) combined climate model data with a quantitative business model to estimate future plant performance. Haupt et al. (2016a) studied the projected changes in both the solar and wind resource over the contiguous USA using both a baseline reanalysis and combinations of global climate models and regional climate models. To best project the patterns, that study employed self-organizing maps and Monte Carlo selection to build a proxy future climate. Thus, multiple post-processing methods can help correct the NWP model output and project the solar resource into the future, even under a changing climate.

6 Future Perspectives

Historically, the tradeoff in satellite DA has been using geostationary or polar-orbiting satellites. Geostationary satellites have provided hemispheric coverage, but their spatial and temporal resolution has often been inadequate, while polar-orbiting satellites have offered a much higher spatial resolution, though with a limited field of view and infrequent overpasses. The most recent geostationary satellites that have been launched, however, contain imagers with high spatial and temporal resolution. For instance, the GOES-16 Advanced Baseline Imager has 16 bands with a spatial resolution of 0.5–2 km, while taking hemispheric scans every 15 min, and scans of the conterminous US every 5 min (GOES-R 2018). Such high spatial and temporal resolution from geostationary satellites will prove tremendously beneficial to satellite DA and solar resource assessment and forecasting in coming years.

Improvements in physics parameterizations should also lead to better solar resource evaluations. In particular, the effects that aerosols and clouds exert on the shortwave irradiance should be improved. For instance, including the radiative effects of unresolved clouds is an ongoing topic of research (Jiménez et al. 2016a). Historically, the effects of unresolved clouds affected temperature and moisture profiles. Only recently the radiative effects of the subgrid clouds have been considered, and improvements in the solar irradiance simulation have been identified (Alapaty et al. 2012; Berg et al. 2013; Deng et al. 2014; Herwehe et al. 2014; Jiménez et al. 2016a). Future research may identify the optimal methodology to represent the subgrid-scale clouds valid at the range of grid spacing used in NWP simulations.

Future efforts in aerosol modeling should be directed to identify the best way to represent the aerosol effects. There should be a balance between the amount of physical processes accounted for and the accuracy of the simulation. The recent availability of aerosol reanalysis (e.g., MERRA-2) should contribute toward making progress in this direction.

References

Alapaty K, Herwehe JA, Otte TL, Nolte CG, Bullock OR, Mallard MS, Dudhia J (2012) Introducing subgrid-scale cloud feedbacks to radiation for regional meteorological and climate modeling. Geophys Res Lett. https://doi.org/10.1029/2012GL054031

Anderson JL (2001) An ensemble adjustment Kalman Filter for data assimilation. Mon Weather Rev. https://doi.org/10.1175/1520-0493(2001)129%3c2884:AEAKFF%3e2.0.CO;2

Anderson J, Hoar T, Raeder K, Liu H, Collins N, Torn R, Avellano A (2009) The data assimilation research testbed a community facility. Bull Am Meteorol Soc. https://doi.org/10.1175/2009BAMS2618.1

Anthes RA, Warner TT (1974) Prediction of mesoscale flows over complex terrain. Retrieved from http://www.dtic.mil/docs/citations/AD0776333

Anthes RA, Warner TT (1978) Development of hydrodynamic models suitable for air pollution and other mesometeorological studies. Mon Weather Rev. https://doi.org/10.1175/1520-0493(1978)106%3c1045:DOHMSF%3e2.0.CO;2

Armstrong MA (2000) Comparison of MM5 forecast shortwave radiation with data obtained from the atmospheric radiation measurement program. University of Maryland. Retrieved from https://www.atmos.umd.edu/~bobe/downloads/armstrong.pdf

Auligné T, Auligné T (2014a) Multivariate minimum residual method for cloud retrieval. Part I: theoretical aspects and simulated observation experiments. Mon Weather Rev 142(12):4383–4398 https://doi.org/10.1175/MWR-D-13-00172.1

Auligné T, Auligné T (2014b) Multivariate minimum residual method for cloud retrieval. Part II: real observations experiments. Mon Weather Rev 142(12):4399–4415 https://doi.org/10.1175/MWR-D-13-00173.1

Bannister RN (2017) A review of operational methods of variational and ensemble-variational data assimilation. Q J R Meteorol Soc. https://doi.org/10.1002/qj.2982

Bayler GM, Aune RM, Raymond WH, Bayler GM, Aune RM, Raymond WH (2000) NWP cloud initialization using GOES sounder data and improved modeling of nonprecipitating clouds. Mon Weather Rev 128(11):3911–3920. https://doi.org/10.1175/1520-0493(2001)129%3c3911:NCIUGS%3e2.0.CO;2

Beljaars A, Bechtold P, Köhler M, Morcrette J-J, Tompkins AM, Viterbo P, Wedi N (2004) The numerics of physical parameterization. In: Seminar on recent developments in numerical methods for atmospheric and ocean modelling, 6–10 September 2004. ECMWF, Shinfield Park, Reading, pp 113–134

Berg LK, Gustafson WI, Kassianov EI, Deng L (2013) Evaluation of a modified scheme for shallow convection: implementation of CuP and case studies. Mon Weather Rev. https://doi.org/10.1175/MWR-D-12-00136.1

Bishop CH, Etherton BJ, Majumdar SJ (2001) Adaptive sampling with the ensemble transform Kalman Filter. Part I: theoretical aspects. Mon Weather Rev. https://doi.org/10.1175/1520-0493(2001)129%3c0420:ASWTET%3e2.0.CO;2

Blanc P, Gschwind B, Lefèvre M, Wald L (2011) The HelioClim project: surface solar irradiance data for climate applications. Remote Sens. https://doi.org/10.3390/rs3020343

Brdar S, Baldauf M, Dedner A, Klöfkorn R (2013) Comparison of dynamical cores for NWP models: comparison of COSMO and Dune. Theor Comput Fluid Dyn. https://doi.org/10.1007/s00162-012-0264-z

Caya A, Laprise R, Zwack P, Caya A, Laprise R, Zwack P (1998) Consequences of using the splitting method for implementing physical forcings in a semi-implicit semi-Lagrangian model. Mon Weather Rev 126(6):1707–1713. https://doi.org/10.1175/1520-0493(1998)126%3c1707:COUTSM%3e2.0.CO;2

Charabi Y, Gastli A, Al-Yahyai S (2016) Production of solar radiation bankable datasets from high-resolution solar irradiance derived with dynamical downscaling Numerical Weather prediction model. Energy Rep. https://doi.org/10.1016/j.egyr.2016.05.001

Chen Y, Wang H, Min J, Huang X-Y, Minnis P, Zhang R, Palikonda R (2015) Variational assimilation of cloud liquid/ice water path and its impact on NWP. J Appl Meteorol Climatol 54(8):1809–1825. https://doi.org/10.1175/JAMC-D-14-0243.1

Chin M, Rood RB, Lin SJ, Müller JF and Thompson AM (2000) Atmospheric sulfur cycle simulated in the global model GOCART: model description and global properties. J Geophys Res Atmos. https://doi.org/10.1029/2000JD900384

Clifton A, Hodge B-M, Draxl C, Badger J, Habte A (2018) Wind and solar resource data sets. Wiley Interdisc Rev Energy Environ 7(2):e276. https://doi.org/10.1002/wene.276

Cressman GP (1959) An operational objective analysis system. Mon Weather Rev. https://doi.org/10.1175/1520-0493(1959)087%3c0367:AOOAS%3e2.0.CO;2

de Haan S, van der Veen SH, de Haan S, van der Veen SH (2014) Cloud initialization in the rapid update cycle of HIRLAM. Weather Forecast 29(5):1120–1133. https://doi.org/10.1175/WAF-D-13-00071.1

Deng A, Stauffer DR, Dudhia J, Otte TL, Hunter GK (2007) Update on analysis nudging Fdda in Wrf-arw. National Center for Atmospheric Research, Boulder, CO. 4.8: 8th WRF users' workshop. Retrieved from http://www2.mmm.ucar.edu/wrf/users/workshops/WS2007/abstracts/4-8_Deng.pdf

Deng A, Stauffer D, Gaudet B, Dudhia J, Bruyere C, Wu W, Bourgeois A (2009) Update on WRF-ARW end-to-end multi-scale FDDA system. National Center for Atmospheric Research, Boulder, CO. 1.9: 10th WRF users' workshop. Retrieved from http://www2.mmm.ucar.edu/wrf/users/workshops/WS2009/WorkshopPapers.php

Deng A, Gaudet BJ, Dudhia J, Alapaty K (2014) Implementation and evaluation of a new shallow convection scheme in WRF. In: 26th Conference on weather analysis and forecasting/22nd conference numerical weather prediction. American Society, Atlanta, GA, p 25. Retrieved from https://ams.confex.com/ams/94Annual/webprogram/Paper236925.html

Descombes G, Auligné T, Lin H-C, Xu D, Schwartz C, Vandenberghe F (2014) NCAR technical notes multi-sensor advection diffusion nowCast (MADCast) for cloud analysis and short-term prediction. Retrieved from http://library.ucar.edu/research/publish-technote

Diagne M, David M, Boland J, Schmutz N, Lauret P (2014) Post-processing of solar irradiance forecasts from WRF model at Reunion Island. Sol Energy 105(0):99–108. https://doi.org/10.1016/j.solener.2014.03.016

Dubal M, Wood N, Staniforth A (2004) Analysis of parallel versus sequential splittings for time-stepping physical parameterizations. Mon Weather Rev. https://doi.org/10.1175/1520-0493(2004)131%3c0121:AOPVSS%3e2.0.CO;2

Dubal M, Wood N, Staniforth A (2006) Some numerical properties of approaches to physics-dynamics coupling for NWP. Q J R Meteorol Soc. https://doi.org/10.1256/qj.05.49

Dutton JA, James RP, Ross JD (2018) Probabilistic forecasts for energy: weeks to a century or more. In: Troccoli A (ed) Weather & climate services for the energy industry, UK Palgrave Macmillan, Cham, London, pp 161–177. https://doi.org/10.1007/978-3-319-68418-5_12

Evensen G (1994) Sequential data assimilation with a nonlinear quasi-geostrophic model using Monte Carlo methods to forecast error statistics. J Geophys Res. https://doi.org/10.1029/94JC00572

Fast JD, Gustafson WI, Easter RC, Zaveri RA, Barnard JC, Chapman EG, Peckham SE (2006) Evolution of ozone, particulates, and aerosol direct radiative forcing in the vicinity of Houston using a fully coupled meteorology-chemistry-aerosol model. J Geophys Res Atmos. https://doi.org/10.1029/2005JD006721

Fountoukis C, Martín-pomares L, Perez-astudillo D, Bachour D, Gladich I (2018) Simulating global horizontal irradiance in the Arabian Peninsula: sensitivity to explicit treatment of aerosols. Sol Energy 163:347–355. https://doi.org/10.1016/j.solener.2018.02.001

Gelaro R, McCarty W, Suárez MJ, Todling R, Molod A, Takacs L, Zhao B (2017) The modern-era retrospective analysis for research and applications, version 2 (MERRA-2). J Clim. https://doi.org/10.1175/JCLI-D-16-0758.1

Glahn HR, Lowry DA (1972) The use of model output statistics (MOS) in objective weather forecasting. J Appl Meteorol 11:1203–1211

GOES-R (2018) Instruments: advanced baseline imager (ABI). Retrieved 14 April 2018, from https://www.goes-r.gov/spacesegment/abi.html

Grell GA, Peckham SE, Schmitz R, McKeen SA, Frost G, Skamarock WC, Eder B (2005) Fully coupled "online" chemistry within the WRF model. Atmos Environ. https://doi.org/10.1016/j.atmosenv.2005.04.027

Gross M, Malardel S, Jablonowski C, Wood N, Gross M, Malardel S, Wood N (2016) Bridging the (knowledge) gap between physics and dynamics. Bull Am Meteor Soc 97(1):137–142. https://doi.org/10.1175/BAMS-D-15-00103.1

Gross M, Wan H, Rasch PJ, Caldwell PM, Williamson DL, Klocke D, Leung R (2017) Recent progress and review of issues related to physics dynamics coupling in geophysical models. https://doi.org/10.1175/MWR-D-17-0345.1

Habte A, Sengupta M, Lopez A (2017) Evaluation of the National Solar Radiation Database (NSRDB): 1998–2015. Technical Report TP-5D00-67722 (April), pp 1–38. Retrieved from http://www.nrel.gov/docs/fy17osti/67722.pdf

Haupt SE, Copeland J, Cheng WYY, Zhang Y, Ammann C, Sullivan P, Sullivan P (2016a) A method to assess the wind and solar resource and to quantify interannual variability over the united states under current and projected future climate. J Appl Meteorol Climatol 55(2):345–363. https://doi.org/10.1175/JAMC-D-15-0011.1

Haupt SE, Kosovic B, Jensen L, Lee J, Jimenez Munoz P, Lazo K, Hinkleman L (2016b) The Sun4Cast® solar power forecasting system: the result of the public-private-academic partnership to advance solar power forecasting. https://doi.org/10.5065/D6N58JR2

Haupt SE, Jiménez PA, Lee JA, Kosović B (2017) Principles of meteorology and numerical weather prediction. In: Kariniotakis G (ed) Renewable energy forecasting: from models to applications, Woodhead Publishing (Elsevier), Cambridge, MA, pp 3–28. https://doi.org/10.1016/B978-0-08-100504-0.00001-9

Haupt SE, Kosović B, Jensen T, Lazo JK, Lee JA, Jiménez PA, Heiser J (2018) Building the Sun4Cast system: improvements in solar power forecasting. Bull Amer Meteorol Soc 99:121–135. https://doi.org/10.1175/BAMS-D-16-0221.1

Haupt SE, Kosović B, Lee JA, Jiménez PA (2018b) Mesoscale modeling of the atmosphere. In: Veers P (ed) Wind power modelling: atmosphere and wind plant flow, IET Publishing, Stevenage, UK

Herwehe JA, Alapaty K, Spero TL, Nolte CG (2014) Increasing the credibility of regional climate simulations by introducing subgrid-scale cloud-radiation interactions. J Geophys Res. https://doi.org/10.1002/2014JD021504

Hoke JE, Anthes RA (1976) The initialization of numerical models by a dynamic-initialization technique. Mon Weather Rev. https://doi.org/10.1175/1520-0493(1976)104%3c1551:TIONMB%3e2.0.CO;2

Houtekamer PL, Zhang F (2016) Review of the ensemble Kalman Filter for atmospheric data assimilation. Mon Weather Rev. https://doi.org/10.1175/MWR-D-15-0440.1

Jiménez PA, Alessandrini S, Haupt SE, Deng A, Kosovic B, Lee JA, Delle Monache L (2016a) The role of unresolved clouds on short-range global horizontal irradiance predictability. Mon Weather Rev. https://doi.org/10.1175/MWR-D-16-0104.1

Jiménez PA, Hacker JP, Dudhia J, Haupt SE, Ruiz-Arias JA, Gueymard CA, Deng A (2016b) WRF-SOLAR: description and clear-sky assessment of an augmented NWP model for solar power prediction. Bull Am Meteorol Soc. https://doi.org/10.1175/BAMS-D-14-00279.1

Jonassen MO, Ólafsson H, Ágústsson H, Rögnvaldsson Ó, Reuder J (2012) improving high-resolution numerical weather simulations by assimilating data from an unmanned aerial system. Mon Weather Rev. https://doi.org/10.1175/MWR-D-11-00344.1

Jones TA, Stensrud DJ, Minnis P, Palikonda R (2013) Evaluation of a forward operator to assimilate cloud water path into WRF-DART. Mon Weather Rev. https://doi.org/10.1175/MWR-D-12-00238.1

Kalman RE (1960) A new approach to linear filtering and prediction problems. J Basic Eng. https://doi.org/10.1115/1.3662552

Kalman RE, Bucy RS (1961) New results in linear filtering and prediction theory. J Basic Eng. https://doi.org/10.1115/1.3658902

Kalnay E (2003) Atmospheric modeling, data assimilation, and predictability. Ann Phys. https://doi.org/10.1256/00359000360683511

Kay M, MacGill I (2014) Improving NWP forecasts for the wind energy sector. In: Troccoli A, Dubus L, Haupt SE (eds) Weather matters for energy, Springer, New York, NY, pp 413–428. https://doi.org/10.1007/978-1-4614-9221-4_20

Klinker E, Rabier F, Kelly G, Mahfouf JF (2000) The ECMWF operational implementation of four-dimensional variational assimilation. III: experimental results and diagnostics with operational configuration. Q J R Meteorol Soc. https://doi.org/10.1256/smsqj.56416

Köpken C, Kelly G, Thépaut J-N (2004) Assimilation of meteosat radiance data within the 4D-Var system at ECMWF: assimilation experiments and forecast impact. Q J R Meteorol Soc 130 (601):2277–2292. https://doi.org/10.1256/qj.02.230

Lander J, Hoskins BJ (1997) Believable scales and parameterizations in a spectral transform model. Mon Weather Rev. https://doi.org/10.1175/1520-0493(1997)125%3c0292:BSAPIA%3e2.0.CO;2

Lara-Fanego V, Ruiz-Arias JA, Pozo-Vázquez D, Santos-Alamillos FJ, Tovar-Pescador J (2012) Evaluation of the WRF model solar irradiance forecasts in Andalusia (Southern Spain). Sol Energy 86(8):2200–2217. https://doi.org/10.1016/j.solener.2011.02.014

Lee JA, Haupt SE, Jiménez PA, Rogers MA, Miller SD, McCandless TC (2017) Solar irradiance nowcasting case studies near sacramento. J Appl Meteorol Climatol. https://doi.org/10.1175/JAMC-D-16-0183.1

Lei L, Stauffer DR, Deng A (2012) A hybrid nudging-ensemble Kalman filter approach to data assimilation in WRF/DART. Q J R Meteorol Soc 138(669):2066–2078. https://doi.org/10.1002/qj.1939

Liu Y, Warner TT, Bowers JF, Carson LP, Chen F, Clough CA, Weingarten DS (2008) The operational mesogamma-scale analysis and forecast system of the U.S. army test and evaluation command. Part I: overview of the modeling system, the forecast products, and how the products are used. J Appl Meteorol Climatol. https://doi.org/10.1175/2007JAMC1653.1

Lorenc AC, Bowler NE, Clayton AM, Pring SR, Fairbairn D (2015) Comparison of hybrid-4DEnVar and hybrid-4DVar data assimilation methods for global NWP. Mon Weather Rev. https://doi.org/10.1175/MWR-D-14-00195.1

Lorenz E, Kuhnert J, Heinemann D (2014) Overview on irradiance and photovoltaic power prediction. In: Troccoli A, Dubus L, Haupt SE (eds) Weather matters for energy, Springer, New York, NY, pp 429–454. https://doi.org/10.1007/978-1-4614-9221-4_21

Mahoney WP, Parks K, Wiener G, Liu Y, Myers WL, Sun J, Haupt SE (2012) A wind power forecasting system to optimize grid integration. In: IEEE transactions on sustain energy. https://doi.org/10.1109/TSTE.2012.2201758

Marquez R, Coimbra CFM (2011) Forecasting of global and direct solar irradiance using stochastic learning methods, ground experiments and the NWS database. Solar Energy 85(5): 746–756. https://doi.org/10.1016/j.solener.2011.01.007

Martinet P, Fourrié N, Guidard V, Rabier F, Montmerle T, Brunel P (2013) Towards the use of microphysical variables for the assimilation of cloud-affected infrared radiances. Q J R Mseteorol Soc. https://doi.org/10.1002/qj.2046

Martins FR, Pereira EB, Abreu SL (2007) Satellite-derived solar resource maps for Brazil under SWERA project. Sol Energy. https://doi.org/10.1016/j.solener.2006.07.009

Mathiesen P, Kleissl J (2011) Evaluation of numerical weather prediction for intra-day solar forecasting in the continental United States. Sol Energy. https://doi.org/10.1016/j.solener.2011.02.013

Miguez-Macho G, Stenchikov GL, Robock A (2004) Spectral nudging to eliminate the effects of domain position and geometry in regional climate model simulations. J Geophys Res D Atmos 109(13):1–15. https://doi.org/10.1029/2003JD004495

Miguez-Macho G, Stenchikov GL, Robock A, Miguez-Macho G, Stenchikov GL, Robock A (2005) Regional climate simulations over North America: interaction of local processes with improved large-scale flow. J Clim 18(8):1227–1246. https://doi.org/10.1175/JCLI3369.1

Miller SD, Rogers MA, Haynes JM, Sengupta M, Heidinger AK (2018) Short-term solar irradiance forecasting via satellite/model coupling. Sol Energy 168:102–117. https://doi.org/10.1016/J.SOLENER.2017.11.049

Munro R, Köpken C, Kelly G, Thépaut JN, Saunders R (2004) Assimilation of Meteosat radiance data within the 4D-Var system at ECMWF: data quality monitoring, bias correction and single-cycle experiments. Q J R Meteorol Soc. https://doi.org/10.1256/qj.02.229

Otte TL, Nolte CG, Otte MJ, Bowden JH (2012) Does nudging squelch the extremes in regional climate modeling? J Clim. https://doi.org/10.1175/JCLI-D-12-00048.1

Pan Y, Zhu K, Xue M, Wang X, Hu M, Benjamin SG, Whitaker JS (2014) A GSI-Based coupled EnSRF–En3DVar hybrid data assimilation system for the operational rapid refresh model: tests at a reduced resolution. Mon Weather Rev 142(10):3756–3780. https://doi.org/10.1175/MWR-D-13-00242.1

Paquin-Ricard D, Jones C, Vaillancourt PA (2010) Using ARM observations to evaluate cloud and clear-sky radiation processes as simulated by the canadian regional climate model GEM. Mon Weather Rev. https://doi.org/10.1175/2009MWR2745.1

Peng Z, Yu D, Huang D, Heiser J, Yoo S, Kalb P (2015) 3D cloud detection and tracking system for solar forecast using multiple sky imagers. Sol Energy 118(0):496–519. https://doi.org/10.1016/j.solener.2015.05.037

Powers JG, Klemp JB, Skamarock WC, Davis CA, Dudhia J, Gill DO, Duda MG (2017) The weather research and forecasting model: overview, system efforts, and future directions. Bull Am Meteorol Soc. https://doi.org/10.1175/BAMS-D-15-00308.1

Pryor SC, Barthelmie RJ, Kjellström E (2005) Potential climate change impact on wind energy resources in northern Europe: analyses using a regional climate model. Clim Dyn. https://doi.org/10.1007/s00382-005-0072-x

Pryor SC, Barthelmie RJ, Young DT, Takle ES, Arritt RW, Flory D, Roads J (2009) Wind speed trends over the contiguous United States. J Geophys Res. https://doi.org/10.1029/2008JD011416

Rogers RE, Deng A, Stauffer DR, Gaudet BJ, Jia Y, Soong ST, Tanrikulu S (2013) Application of the weather research and forecasting model for air quality modeling in the San Francisco bay area. J Appl Meteorol Clim. https://doi.org/10.1175/JAMC-D-12-0280.1

Ruiz-Arias JA, Dudhia J, Gueymard CA (2014) A simple parameterization of the short-wave aerosol optical properties for surface direct and diffuse irradiances assessment in a numerical weather model. Geoscientific Model Dev 7(3):1159–1174. https://doi.org/10.5194/gmd-7-1159-2014

Ruiz-Arias JA, Arbizu-Barrena C, Santos-Alamillos FJ, Tovar-Pescador J, Pozo-Vázquez D (2016) Assessing the surface solar radiation budget in the WRF model: a spatiotemporal analysis of the bias and its causes. Mon Weather Rev. https://doi.org/10.1175/MWR-D-15-0262.1

Schlatter TW (2000) Variational assimilation of meteorological observations in the lower atmosphere: a tutorial on how it works. J Atmos Sol Terr Phys. https://doi.org/10.1016/S1364-6826(00)00096-1

Schraff C, Reich H, Rhodin A, Schomburg A, Stephan, K, Periáñez A, Potthast R (2016) Kilometre-scale ensemble data assimilation for the COSMO model (KENDA). Q J R Meteorol Soc. https://doi.org/10.1002/qj.2748

Schwartz CS, Liu Z, Lin H-C, Cetola JD (2014) Assimilating aerosol observations with a "hybrid" variational-ensemble data assimilation system. J Geophys Res Atmos 119(7):4043–4069. https://doi.org/10.1002/2013JD020937

Schwartz CS, Romine GS, Sobash RA, Fossell KR, Weisman ML (2015) NCAR's experimental real-time convection-allowing ensemble prediction system. Weather Forecast. https://doi.org/10.1175/WAF-D-15-0103.1

Skamarock WC, Klemp JB, Dudhi J, Gill DO, Barker DM, Duda MG, Powers JG (2008) A description of the advanced research WRF version 3. Technical report (June), p 113. https://doi.org/10.5065/D6DZ069T

Staniforth A, Wood N, Côté J (2002a) Analysis of the numerics of physics–dynamics coupling. Q J R Meteorol Soc 128(586):2779–2799. https://doi.org/10.1256/qj.02.25

Staniforth A, Wood N, Côté J, Staniforth A, Wood N, Côté J (2002b) A simple comparison of four physics–dynamics coupling schemes. Mon Weather Rev. https://doi.org/10.1175/1520-0493(2002)130%3c3129:ASCOFP%3e2.0.CO;2

Stauffer DR, Seaman NL (1990) Use of four-dimensional data assimilation in a limited-area mesoscale model. Part I: experiments with synoptic-scale data. Mon Weather Rev

Stauffer DR, Seaman NL (1994) Multiscale four-dimensional data assimilation. J Appl Meteorol. https://doi.org/10.1175/1520-0450(1994)033%3c0416:MFDDA%3e2.0.CO;2

Stauffer DR, Seaman NL, Binkowski FS (1991) Use of four-dimensional data assimilation in a limited-area mesoscale model. Part II: effects of data assimilation within the planetary boundary layer. Mon Weather Rev. https://doi.org/10.1175/1520-0493(1991)119%3c0734:UOFDDA%3e2.0.CO;2

Stengel M, Lindskog M, Undén P, Gustafsson N, Bennartz R (2010) An extended observation operator in HIRLAM 4D-VAR for the assimilation of cloud-affected satellite radiances. Q J R Meteorol Soc. https://doi.org/10.1002/qj.621

Stengel M, Lindskog M, Undén P, Gustafsson N (2013) The impact of cloud-affected IR radiances on forecast accuracy of a limited-area NWP model. Q J R Meteorol Soc. https://doi.org/10.1002/qj.2102

Stensrud DJ (2011). Parameterization schemes: keys to understanding numerical weather prediction models. https://doi.org/10.1017/CBO9780511812590

Sundqvist H, Berge E, Kristjánsson JE (1989) Condensation and cloud parameterization studies with a mesoscale numerical weather prediction model. Mon Weather Rev. https://doi.org/10.1175/1520-0493(1989)117%3c1641:CACPSW%3e2.0.CO;2

Talagrand O (1997) Assimilation of observations, an introduction. J Meteorol Soc Jpn. https://doi.org/10.1256/qj.02.132

Talagrand O, Courtier P (1987) Variational assimilation of meteorological observations with the adjoint vorticity equation. I: theory. Q J R Meteorol Soc. https://doi.org/10.1002/qj.49711347812

Thompson G, Eidhammer T (2014) A study of aerosol impacts on clouds and precipitation development in a large winter cyclone. J Atmos Sci. https://doi.org/10.1175/JAS-D-13-0305.1

Thompson G, Tewari M, Ikeda K, Tessendorf S, Weeks C, Otkin J, Kong F (2016) Explicitly-coupled cloud physics and radiation parameterizations and subsequent evaluation in WRF high-resolution convective forecasts. Atmos Res. https://doi.org/10.1016/j.atmosres.2015.09.005

Tuohy A, Zack J, Haupt SE, Sharp J, Ahlstrom M, Dise S, Collier C (2015) Solar forecasting: methods, challenges, and performance. IEEE Power Energy Magaz. https://doi.org/10.1109/MPE.2015.2461351

van der Veen SH, van der Veen SH (2013) Improving NWP model cloud forecasts using meteosat second-generation imagery. Mon Weather Rev 141(5):1545–1557. https://doi.org/10.1175/MWR-D-12-00021.1

Wang X, Barker DM, Snyder C, Hamill TM (2008a) A hybrid ETKF–3dvar data assimilation scheme for the WRF model. Part I: observing system simulation experiment. Mon Weather Rev. https://doi.org/10.1175/2008MWR2444.1

Wang X, Barker DM, Snyder C, Hamill TM (2008b) A hybrid ETKF–3DVAR data assimilation scheme for the WRF model. Part II: real observation experiments. Mon Weather Rev. https://doi.org/10.1175/2008MWR2445.1

Warner TT (2011) Quality assurance in atmospheric modeling. Bull Am Meteorol Soc. https://doi.org/10.1175/BAMS-D-11-00054.1

Whitaker JS, Hamill TM (2002) Ensemble data assimilation without perturbed observations. Mon Weather Rev. https://doi.org/10.1175/1520-0493(2002)130%3c1913:EDAWPO%3e2.0.CO;2

White AT, Pour-Biazar A, Doty K, Dornblaser B, McNider RT, White AT, McNider RT (2018) Improving cloud simulation for air quality studies through assimilation of geostationary satellite observations in retrospective meteorological modeling. Mon Weather Rev 146(1):29–48. https://doi.org/10.1175/MWR-D-17-0139.1

Williamson DL (2002) Time-split versus process-split coupling of parameterizations and dynamical core. Mon Weather Rev. https://doi.org/10.1175/1520-0493(2002)130%3c2024:TSVPSC%3e2.0.CO;2

Xie Y, Sengupta M, Dudhia J (2016) A fast all-sky radiation model for solar applications (FARMS): algorithm and performance evaluation. Eur Photovolt Sol Energy Conf 135:435–445. https://doi.org/10.1016/j.solener.2016.06.003

Yang L, Lam JC, Liu J (2007) Analysis of typical meteorological years in different climates of China. Energy Convers Manage 48(2):654–668. https://doi.org/10.1016/j.enconman.2006.05.016

Yang S-C, Corazza M, Carrassi A, Kalnay E, Miyoshi T (2009) Comparison of local ensemble transform Kalman Filter, 3DVAR, and 4DVAR in a Quasigeostrophic model. Mon Weather Rev. https://doi.org/10.1175/2008MWR2396.1

Yang H, Kurtz B, Nguyen D, Urquhart B, Chow CW, Ghonima M, Kleissl J (2014) Solar irradiance forecasting using a ground-based sky imager developed at UC San Diego. Sol Energy 103(0):502–524. https://doi.org/10.1016/j.solener.2014.02.044

Yucel I, Shuttleworth WJ, Pinker RT, Lu L, Sorooshian S, Yucel I, Sorooshian S (2002) Impact of ingesting satellite-derived cloud cover into the regional atmospheric modeling system. Mon Weather Rev 130(3):610–628. https://doi.org/10.1175/1520-0493(2002)130%3c0610:IOISDC%3e2.0.CO;2

Yucel I, James Shuttleworth W, Gao X, Sorooshian S (2003) Short-term performance of MM5 with cloud-cover assimilation from satellite observations. Mon Weather Rev. https://doi.org/10.1175//2565.1

Zamora RJ, Dutton EG, Trainer M, McKeen SA, Wilczak JM, Hou Y-T (2005) The accuracy of solar irradiance calculations used in mesoscale numerical weather prediction. Mon Weather Rev. https://doi.org/10.1175/MWR2886.1

Zhang F, Zhang M, Poterjoy J, Zhang F, Zhang M, Poterjoy J (2013) E3DVar: coupling an ensemble Kalman Filter with three-dimensional variational data assimilation in a limited-area weather prediction model and comparison to E4DVar. Mon Weather Rev 141(3):900–917. https://doi.org/10.1175/MWR-D-12-00075.1

Zhang J, Draxl C, Hopson T, Monache LD, Vanvyve E, Hodge BM (2015) Comparison of numerical weather prediction based deterministic and probabilistic wind resource assessment methods. Appl Energy. https://doi.org/10.1016/j.apenergy.2015.07.059

Chapter 8
Solar Radiation Interpolation

Ana M. Martín and Javier Dominguez

Abstract Geographic information systems provide different options to analyze and represent the spatial heterogeneity of solar radiation incident on a certain area. This chapter presents a description of the main and well-known methods for determining interpolation surfaces from a data sample. Moreover, using 3D model of the analyzed area, computer models of spatial analysis are precise techniques to adjust the results to the variability of surfaces in a geographic area. Both alternatives offer a great analysis capacity. The selection of a procedure will depend on the objective of the study and the available information.

1 Introduction

Studies about the amount of solar radiation that reaches a surface are of great importance in various areas such as agriculture, ecology, hydrology, biology, meteorology, architecture or the use of solar energy as an alternative for energy supply. The use of solar energy is conditioned by the intensity of the incident solar radiation, so that it is essentially an adequate knowledge of the solar resource distribution.

Although there are solar radiation maps for large geographic areas, in some detailed project solar radiation data are required. In this cases, where there are not measuring stations, will be needed to approximate the data. Sometimes, they can assume as a valid value, the nearest station data and others can use these measurements, using spatial interpolation techniques, to analyze and model the solar resource in no data locations within the same area.

A. M. Martín · J. Dominguez (✉)
Renewable Energy Division (Energy Department) of CIEMAT,
Avda Complutense 40, 28040 Madrid, Spain
e-mail: javier.dominguez@ciemat.es

A. M. Martín
e-mail: AnaMaria.Martin@externos.ciemat.es

© Springer Nature Switzerland AG 2019
J. Polo et al. (eds.), *Solar Resources Mapping*, Green Energy and Technology,
https://doi.org/10.1007/978-3-319-97484-2_8

If there are few measuring stations or the installed network is not dense enough, the addition of data from remote sensors, such as satellite images will improve the adjustment of the estimations. Another alternative to calculate the solar radiation is the use of models included in geographic information systems (GIS) software. These systems incorporate the option of evaluating the influence of elevation, considering the geospatial variations of solar radiation between areas of more and less complex relief.

The purpose of this chapter is to feature the capacity of this type of tools in the determination of solar radiation for a specific geographical area. First, the main interpolation methods will be summarized and then, one of the existing GIS models will be described for the estimation of solar radiation considering the influence of terrain topography.

2   Interpolation of Solar Radiation Data Using GIS

The values obtained by interpolation process would depend on the characteristics of the studied geographical variables, the available sample, factors associated with the distribution of the data, the required spatial resolution and the chosen predictor model (Burrough and McDonnell 1998). In general, interpolation techniques are classified as deterministic and geostatistical (Johnston et al. 2001; Santos Preciado and García Lázaro 2008).

The *deterministic techniques* create interpolated surfaces based on the adjustment of mathematical functions to the measured points. They define a set of explanatory variables so that the errors were minimal.

The *geostatistical techniques* generate the prediction surfaces using statistical models. These methods quantify the spatial correlation of the data and evaluate the uncertainty of the obtained results.

In addition, interpolation methods are classified as exact and inexact. They are considered exact when the interpolated values for a location correspond to the measured data (Fig. 1). Finally, they are divided into global and local methods. Global interpolations use a single function to create the continuous surface from all the sample values and, local interpolations adjust the function to different small areas of the sample points. Local methods are more appropriate when the total trend of the analyzed data is unknown (Fig. 2).

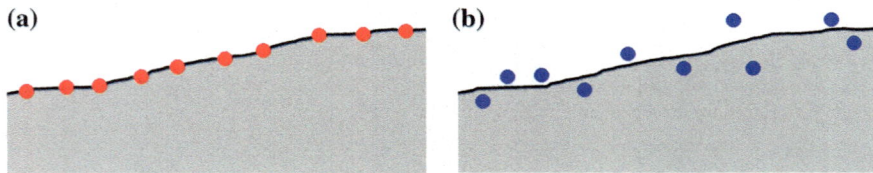

Fig. 1 a Exact and **b** inexact interpolation methods

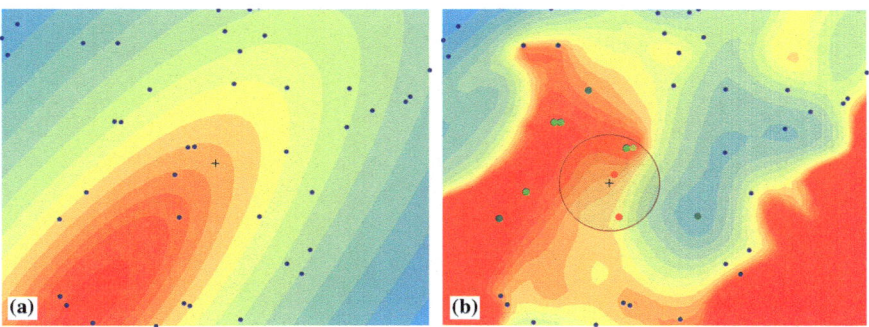

Fig. 2 a Global and **b** local interpolation methods

Table 1 Spatial interpolation methods (Burrough and McDonnell 1998; Johnston et al. 2001)

Method	Interpolation type	Local/global	Exact interpolation
Global polynomial	Deterministic	Global	No
Local polynomial	Deterministic	Local	No
Inverse distance weighted	Deterministic	Local	Yes
Radial basis functions	Deterministic	Local	Yes
Kriging	Geostatistical	Local with global trend	Yes
Cokriging	Geostatistical	Local with global trend	Yes

Table 1 presents a brief summary of the main methods that can be applied in the interpolation of solar radiation data.

The amount of interpolation methods available is quite broad, as well as the parameters susceptible of using for the adjustment in each of them. To calculate solar radiation values, some studies use other variables such as temperature, humidity, cloudiness, precipitation or elevation (Evrendilek and Ertekin 2008; Jolly et al. 2005). Comparative studies have been carried out to establish which is the optimal method for each variable (Apaydin et al. 2004; Vicente-Serrano et al. 2003), although the kriging method is one of the most applied (Antonanzas et al. 2015; Perea-Moreno and Hernandez-Escobedo 2016; Righini et al. 2005).

In order to illustrate the interpolation methods, we provide daily solar radiation data (summer solstice) from 45 radiometric stations in Spain (AEMET 2011) (Fig. 3).

Fig. 3 Distribution of radiometric stations from Spain, located in the Iberian Peninsula

2.1 Deterministic Methods

2.1.1 Global Polynomial Interpolation

When the analyzed parameter varies continuously in a certain area, this interpolation method defines the mathematical function that fits the values of the input sample points. The global polynomial interpolation models a smooth surface that best represents the trend in the data points, in order to differences of observed errors are minimal.

This method uses all data available of the study area to produce a prediction surface. The interpolated values are computed from their geographical location by multiple regressions using a least squares regression fit. The correlation between the variable to interpolate z and its coordinates (x, y) is defined with the function:

$$z(x, y) = \sum_{r+s \leq p} \left(b_{rs} \cdot x^r \cdot y^s \right) \tag{1}$$

The number of coefficients b_{rs} to be determined would depend on the order p of the trend surface. First- to third-order polynomial functions are the most common:

$$z(x, y) = b_0 + b_1 \cdot x + b_2 \cdot y \tag{2}$$

$$z(x, y) = b_0 + b_1 \cdot x + b_2 \cdot y + b_3 \cdot x^2 + b_4 \cdot xy + b_5 \cdot y^2 \tag{3}$$

$$z(x, y) = b_0 + b_1 \cdot x + b_2 \cdot y + b_3 \cdot x^2 + b_4 \cdot x \cdot y + b_5 \cdot y^2 + b_6 \cdot x^3$$
$$+ b_7 \cdot x^2 \cdot y + b_8 \cdot x \cdot y^2 + b_9 \cdot y^3 \tag{4}$$

Being a global interpolation method, it is likely to appear outliers at the edges of the surface, generally related to exceptionally high or low values. The resulting surfaces modeled with low-order polynomials may be suitable to represent certain processes. However, using high-order polynomials, a properly description of the trend surface becomes more complex. In addition, global interpolation is a complementary technique used to identify a general trend that influences on data when a local analysis is implemented.

2.1.2 Local Polynomial Interpolation

With the local polynomial interpolation, instead of using the points of the entire surface, the mathematical function is evaluated exclusively on the surface near each point of estimation. This method fits the function repeatedly to small sections of the sample points, defined by a window, to cover the whole area. The least squares procedure is used by minimizing the expression:

$$\sum_{i=1}^{n} \omega_i (z(x, y) - \mu_0(x, y))^2 \tag{5}$$

where n is the number of points into the window and the weight ω_i when, for example, the window is a circle is defined as:

$$\omega_i = \left(1 - \frac{d_i}{R}\right)^p \tag{6}$$

where d_i is the distance between the estimated point and a sample point within the window and R is the ratio of the window.

The value of the polynomial $\mu_0(x_i, y_i)$ for first- and second-order functions are:

$$\mu_0(x, y) = b_0 + b_1 \cdot x + b_2 \cdot y \tag{7}$$

$$\mu_0(x, y) = b_0 + b_1 \cdot x + b_2 \cdot y + b_3 \cdot x^2 + b_4 \cdot xy + b_5 \cdot y^2 \tag{8}$$

Local polynomial functions are suitable for evaluating data which have small variations in the nearest region.

2.1.3 Inverse Distance Weighting

Interpolation method based on the premise that considers, the points closest to a location are more similar than those further away. Inverse distance weighting method calculates the value of an unknown point by means of a combining weighted of the values in a sample of points. Using this method, greater weight is given to points located in the nearby position, decreasing their influence as a function of distance. The general formula is the following:

$$\hat{z}(x_0) = \frac{\sum_{i=1}^{n} \frac{Z(x_i)}{d_{i0}^p}}{\sum_{i=1}^{n} \frac{1}{d_{i0}^p}} \tag{9}$$

where $\hat{z}(x_o)$ is the figured value for the location x_o; n is the number of locations where a value has been measured; d_{i0} represents the distance between the sample locations x_i and the prediction location x_0; and, $z(x_i)$ is the value of the location x_i (Slocum et al. 2014).

The power parameter p is the main factor that affecting on the interpolated values, due to it controls assigned weights to the measured points. Weights are proportional to the inverse distance d_{i0} raised to the power p. When the parameter p increases, the weights of the furthest values diminish and nearby points will have a greater influence on the estimated values.

2.1.4 Radial Basis Functions

Radial basis functions are techniques in which the values are determined by different mathematical functions that force the surface to pass through all the measured points. These methods generate continuous flexible surfaces by adjusting the interpolated values to minimize, as much as possible, their total curvature. These functions are quite appropriate when it is necessary to create large smooth surfaces without many variations between an area and the contiguous one.

The interpolated result is defined by a linear combination of the basic functions:

$$\hat{z}(x_0) = \sum_{i=1}^{n} \omega_i \cdot \varnothing(r) + \beta \tag{10}$$

where n is the number of sample points; ω_i are weights to be estimated; β is a bias parameter; r is de Euclidean distance between the estimated point and each data location; and, $\phi(r)$ is the radial basis function.

Some of the types of radial basis functions commonly used are:
Thin-plate spline function:

$$\emptyset(r) = (\sigma \cdot r)^2 \cdot \ln(\sigma \cdot r) \tag{11}$$

Multiquadric function:

$$\emptyset(r) = \sqrt{r^2 + \sigma^2} \tag{12}$$

Inverse multiquadric function:

$$\emptyset(r) = \frac{1}{\sqrt{r^2 + \sigma^2}} \tag{13}$$

Spline with tension function:

$$\emptyset(r) = \ln\left(\sigma \cdot \frac{r}{2}\right) + K_0(\sigma \cdot r) + C_E \tag{14}$$

Completely regularized spline function (with tension and smoothing):

$$\emptyset(r) = -\left[\ln\left(\sigma \cdot \frac{r}{2}\right)^2 + E_1\left(\sigma \cdot \frac{r}{2}\right)^2 + C_E\right] \tag{15}$$

The parameter σ controls the smoothness of the function: K_0 is the modified Bessel function (Abramowitz and Stegun 1974); C_E is the Euler constant; and, E_1 is the exponential integral function (Mitášová and Mitáš 1993) (Fig. 4).

2.2 Geostatistical Methods

2.2.1 Kriging

Kriging interpolation methods are characterized by creating a surface applying statistical models and providing information about the accuracy of the results, including the spatial correlation of the data. They are based on the weighted average calculation of the sample measurements. The weights are defined with the distance between the measured points and the location of the prediction, in addition to the spatial structure of the sample points (Slocum et al. 2014).

The kriging procedure consists first, in examining the spatial distribution of the data (autocorrelation) and next, generating the interpolation surface with the most appropriate estimation method.

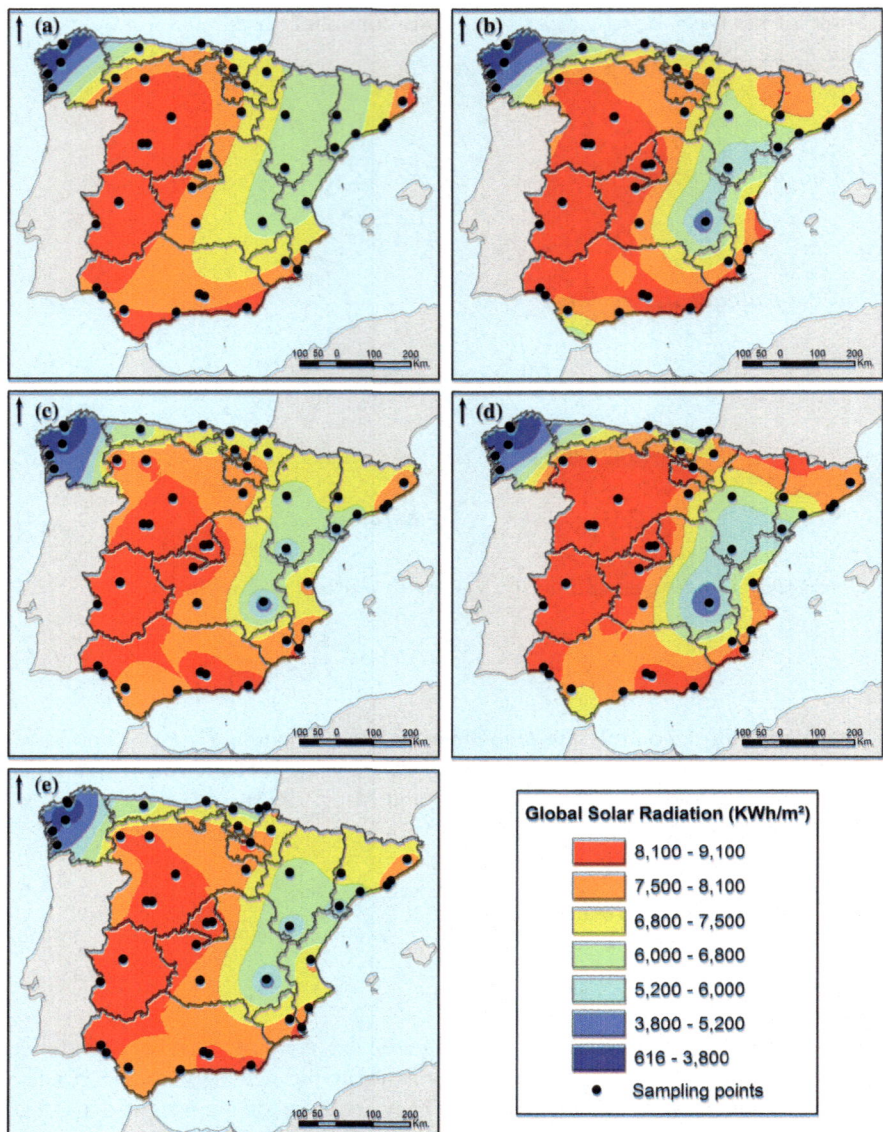

Fig. 4 Estimated global solar radiation for summer solstice (2011) from 45 stations in Spain. Deterministic interpolation methods: **a** global polynomial; **b** local polynomial; **c** inverse distance weighted; **d** thin-plane spline function; and, **e** completely regularized spline function

Autocorrelation

Spatial autocorrelation is analyzed by variance, represented graphically with the variogram, where the spatial variability of a phenomenon is shown according to the sampling points are further away. Semivariance $\gamma(h)$ can be estimated using:

$$\gamma(h) = \frac{1}{2 \cdot n} \sum_{i=1}^{n} [z(x_i) - z(x_i + h)]^2 \tag{16}$$

where n is the number of sample points separated by a distance interval h; $z(x_i)$ is the sample value in a location x_i; and, $z(x_i + h)$ is the value at a distance h from x_i (Burrough and McDonnell 1998).

However, to quantify the scale of spatial variation, it is necessary to adjust the variogram to a theoretical function. This may help to the extraction of a series of parameters which will be used in the kriging interpolation.

In the graph of the variogram (Fig. 5), the variance versus distance is represented. When the distance between points is zero, the semivariance should be zero, but the curve at this point has a value close to zero. This unexplained semivariance is the *nugget* effect, and it indicates measurement errors and variability at a lower scale than the sample. At high values of distance, there is a point at which the semivariance between pairs of points does not increase. The distance at which the semivariance levels off is the *range* and the *sill* is the height reached by the variogram at that point.

The equations of the models to adjust the semivariance are summarized in Table 2.

Types of kriging methods

In general, the predictions of the model for the variable in a location are variants of the equation:

$$\hat{Z}(x_0) - \mu = \sum_{i=1}^{n} \lambda_i [Z(x_i) - \mu(s_0)] \tag{21}$$

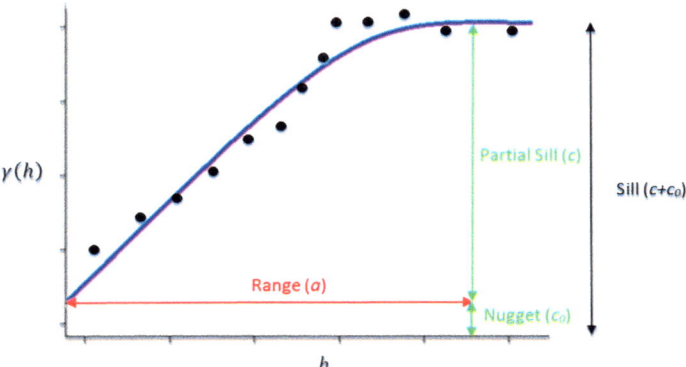

Fig. 5 Example of variogram

Table 2 Semivariance models (Burrough and McDonnell 1998)

Model	Equations	
Linear 	$\gamma(h) = c_0 + b \cdot h$ (b is the slope of the line)	(17)
Exponential 	$\gamma(h) = c_0 + c \cdot \left(1 - \exp\left(\frac{-h}{a}\right)\right)$	(18)
Gaussian 	$\gamma(h) = c_0 + c \cdot \left(1 - \exp\left(-\frac{h^2}{a^2}\right)\right)$	(19)
Spherical 	$\gamma(h) = \begin{cases} c_0 + c \cdot \left(\frac{3h}{2a} - \frac{1}{2} \cdot \left(\frac{h}{a}\right)^3\right) & 0 < h < a \\ c_0 + c & h \geq a \end{cases}$	(20)

where μ is a known stationary mean value; n is the number of point for the estimation; λ_i is the kriging weight; and $\mu(x_0)$ is the mean of sample data in the search window.

To carry out a prediction when mapping variables, there are different types of kriging methods that can be divided into linear and nonlinear (Cressie 2015; Chilès and Delfiner 2012; Goovaerts 1997; Wackernagel 2003):

Linear methods: The estimates are weighted linear combinations of the data.

- *Ordinary kriging* is the most general and widely used kriging methods. The estimated points are allocated values with a weighted linear combination using sample values. The method assumes that the mean value is constant and unknown over the search neighborhood.
- *Simple kriging* uses the average of the total set of data. It considers the premise that both, the mean value and the semivariance of the process, are known and remain constant in all locations.

- *Universal kriging* is a method that performs the estimation of the variable in the presence of a trend or drift. The analyzed phenomenon consists of a deterministic component and the corresponding residue. The variable is determined as the sum of the deterministic function of the drift and a random function with zero mean, which represents the fluctuation or residual error.

Nonlinear methods: The estimates depend on the statistical distribution of the variables.

- *Indicator kriging* is a method in which continuous data are transformed to a binary scale. A binary indicator variable is obtained by establishing a threshold and assigning value 1 to those that are less or equal than it and 0 to the others. In the resulting interpolation, the distribution of the data will reflect being in a class, depending on whether the data exceeds or falls below the specified threshold.
- *Probability kriging* is a technique based on indicator kriging that uses binary data (0 or 1) and then, applies cokriging to perform a better estimation of the resulted probability. The indicator values are used as the primary variable and the original sample data as the secondary one in the cokriging.
- *Disjunctive kriging* is also a nonlinear method in which the global distributions of the data are normalized by Hermite polynomials. The estimate of the variable is developed with a linear combination of the estimated polynomial values. In addition, this method estimates the probability that a random variable shows of exceeding, or not exceeding, a predetermined level in the analyzed area.

2.2.2 Cokriging

Sometimes the phenomena analyzed depend on the values of an analyzed variable, but other times several phenomena are related to each other. Cokriging offers the option of identifying the characteristics of a primary variable from the data of another variable. This method considers secondary information that can be obtained about the variable investigated, referring to other attributes related to the main one (Goovaerts 1997). The general equation that shows the estimated interpolation of the combination between the primary and secondary variables is:

$$\hat{Z}_1(x_0) - \mu_1 = \sum_{i_1=1}^{n_1} \lambda_{i_1}[Z_1(x_{i_1}) - \mu_1(x_{i_1})] + \sum_{j=2}^{n_v} \sum_{i_j=1}^{n_j} \lambda_{i_j}[Z_j(x_{i_j}) - \mu_j(x_{i_j})] \quad (22)$$

where μ_1 is a known stationary mean value of the primary variable; n_1 is the number of points for the estimation in the search window; λ_{i1} is the weight of the primary variable; $Z_1(x_{i1})$ is the data of primary variable; $\mu_1(x_{i1})$ is the mean of sampled data in the search window; n_v is the number of secondary variables; n_j number of j secondary variable in the search window; λ_{ij} is the weight of secondary variable;

$Z_j(x_{ij})$ is the data of secondary variable; and, $\mu_j(x_{ij})$ is the mean of sample secondary variable in the search window.

In cokriging, the spatial dependency relationships are specified by the autocorrelation of the different variables and the cross-correlation between the data. To define the coherence between the variables, a crossed variogram is elaborated, where the variance represented will no longer be between points of the same variable, but between the values of one variable in relation to the other. To verify that there is a covariation, between the primary and secondary variable, a semi-variogram can be estimated from the following equation:

$$\gamma_{12} = \frac{1}{2 \cdot n} \sum_{i=1}^{n} [z_1(x_i) - z_1(x_i + h)] \cdot [z_2(x_i) - z_2(x_i + h)] \tag{23}$$

where n is the number of pairs of sampled points of variable z_1 and z_2 in locations x_i and $x_i + h$ separated by a distance h (Burrough and McDonnell 1998).

There are several cokriging methods that include ordinary cokriging, simple cokriging, universal cokriging, indicator cokriging, probability cokriging and disjunctive cokriging (Cressie 2015; Chilès and Delfiner 2012; Goovaerts 1997; Isaaks and Srivastava 1989) (Fig. 6) (Table 3).

3 Modeling Solar Radiation Using GIS

When the objective is to determine more precisely the distribution of the incident solar radiation in a region or a specific location, establishing a single value for an area that is too wide may be insufficient. At regional and local scales, altitude, orientation, slope, and shading can generate microclimates and, a more or less homogeneous distribution of solar radiation. In these cases, the topography of the area helps to incorporate these factors into the analysis, improving the estimation of solar radiation when the variations caused by the terrain effect are considered.

There are different GIS software packages that have models for estimating solar radiation. The r.sun model designed for the free software GRASS GIS (GRASS Development Team 2018) calculates the three components of solar radiation (direct, diffuse and reflected) with clear sky conditions. It incorporates the possibility of including the effect of shading due to topography and a cloud attenuation factor (Hofierka et al. 2007; Šúri and Hofierka 2004).

In addition, the models developed for the ArcGIS software by ESRI (Environmental Systems Research Institute) (Esri 2018), provided the tool 'Area Solar Radiation' available with the extension 'Spatial Analyst'. This tool of ArcGIS represents and analyzes the insolation for a period of time in a geographical area that is represented by terrain raster file (Digital Surface Model—DSM). The analysis result is the global solar radiation for each location of a surface and is calculated as the addition of the direct and diffuse solar radiation (Esri 2017a). It is

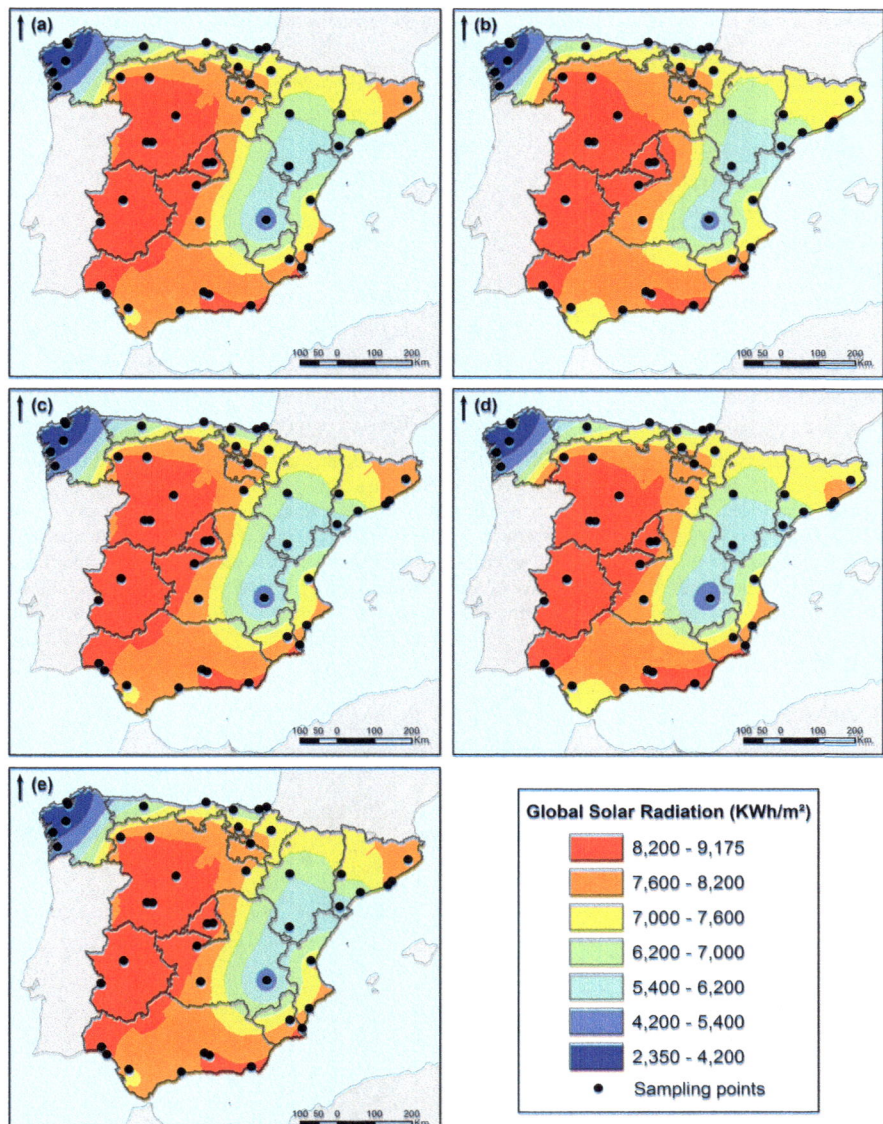

Fig. 6 Estimated global solar radiation for summer solstice (2011) from 45 stations in Spain. Geostatistical interpolation methods: **a** ordinary kriging; **b** simple kriging; **c** universal kriging; **d** disjunctive kriging; and, **e** ordinary cokriging

designed to work on local scales because it only defines a latitude value for the entire area. It can also be used for national and continental scales if the input DSM is divided into small areas (Fig. 7).

Table 3 Errors statistic of the prediction values of global solar radiation interpolation (summer solstice of 2011) from 45 stations in Spain

Interpolation method	ME	RMAE	MSE	RMSSE	ASE
Global polynomial	10.26	727.53	–	–	–
Local polynomial	71.28	790.75	–	–	–
Inverse distance weighted	17.23	820.78	–	–	–
Thin-plate spline function	5.34	829.69	–	–	–
Completely regularized spline function	2.52	774.18	–	–	–
Ordinary kriging	13.85	759.95	0.004	0.98	793.58
Simple kriging	5.38	763.89	0.006	0.93	841.38
Universal kriging	14.09	759.88	0.004	1.003	783.74
Disjunctive kriging	18.05	752.12	0.008	0.87	821.30
Ordinary cokriging	9.63	782.54	0.00004	1.02	777.97

ME Mean Error, *RMSE* Root Mean Square Error, *MSE* Mean Standardized Error, *RMSSE* Root Mean Square Standardized Error, *ASE* Average Standard Error (*Mean Error* (ME) shows the average difference value between the measured data and the prediction. *Root Mean Square Error* (RMSE) indicates the grade of bias from the predictions with the measured values. The smaller this error, the prediction is better. *Mean Standardized Error* (MSE) is the average of the standardized error whose value should be nearby 0. *Root Mean Square Standardized Error* (RMSSE) should be close to 1. If the error is greater than 1, the prediction is underestimated and if the error is less than 1, the prediction is overestimated. *Average Standard Error* (ASE) is the mean of the prediction standard error.) (Esri 2017b)

Fig. 7 Digital surface model (pixel 25 m). Area of Guadalajara and Madrid provinces (Spain). Coordinates: 40° 51′ 59.56″N, 3° 21′ 15.18″O (IGN 2015)

3.1 Calculation of Solar Radiation with ArcGIS

'Area Solar Radiation' tool is based on algorithms developed by Fu and Rich (1999) that determine the hemispheric viewshed, the sunmap and the skymap that calculate the amount of solar radiation on a location:

1. The *viewshed* shows, by searching in a series of directions, those areas of the sky that are visible or hidden due to topography or nearby structures, when they are observed from a certain point.

   In the raster representation of the viewshed, each cell is assigned a value relative to the visibility of the direction of the sky and its location (row and column) is represented for the zenith and azimuth angles.

2. The *sunmap* determines the way of the Sun in the sky over a period of time. In the resulting raster map, the apparent position of the Sun is calculated with the latitude of the area, and it is represented with intervals that vary during the periods of the day (hours) and of the year (days or months). Each sector of the sunmap is assigned an identifying value together with the zenith and azimuthal angles.

3. A *skymap* consists of the division of the sky into a series of sectors in which diffuse radiation can be originated. The sectors that shape the map of the sky are also assigned an identifying value. They defined by the zenith and azimuthal angles, which calculate the diffuse solar radiation in each sky sector (Fig. 8).

Then viewshed is overlaid with the sunmap and the skymap to calculate, respectively, the direct and diffuse solar radiation that are originated from each direction of the sky (Esri 2017a; Kodysh et al. 2013) (Fig. 9).

Analyzing solar radiation in a specific area requires taking into account factors that are responsible for attenuating the amount of radiation that finally reaches the surface. Topography, atmospheric agents, and seasonal variation of insolation are major factors that affect the spatial distribution of solar radiation.

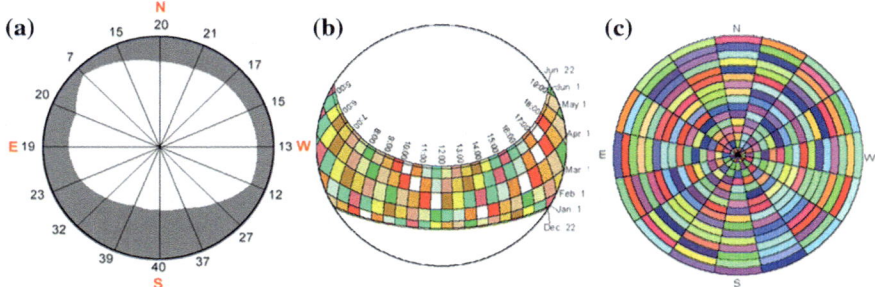

Fig. 8 a Viewshed, **b** sunmap for winter to summer solstices y **c** a skymap with sky sectors defined by 16 zenith and azimuth divisions (Fu and Rich 2000)

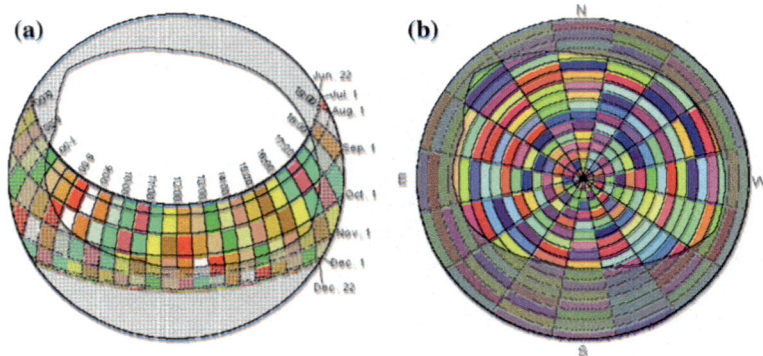

Fig. 9 **a** Overlay of viewshed with sunmap y **b** overlay of viewshed with skymap (Fu and Rich 2000)

Shading limits the amount of insolation in a specific location and the characteristics of the surfaces, their slope, and orientation determine the angle of incidence of solar radiation. In addition, the weather conditions and the effect of the atmosphere also influence the attenuation of the final values of insolation. These factors are considered by estimating atmospheric transparency (transmissivity) and diffuse proportion of solar radiation.

'Area Solar Radiation' tool has options to establish parameters to fit the variables that define the study area and influence on the insolation levels. The inclusion of atmospheric parameters to adjust the calculation of solar radiation with ArcGIS can be done using different methodologies. Sun et al. (2013) apply an annual value for these factors. However, other studies use monthly data provide by agencies such as NASA or the PVGIS databases (Brito et al. 2012; Fogl and Moudrý 2016), which adapt the results better to the monthly and seasonal variations of the solar radiation. Cloud coverage data from meteorological stations and weather databases are also utilized in the parameter calculation for ArcGIS (Oloo et al. 2015; Wong et al. 2016).

Some works use models to calibrate and obtain adequate data for each month of the year. With the solar radiation values get in different measuring stations, Tooke et al. (2011) determine the atmospheric transparency index, from which the diffuse proportion derives, using a first-order model proposed by Orgill and Hollands (1977). Mavromatidis et al. (2015) calculate the global radiation for all possible combinations of transmissivity and diffuse proportion, selecting a set of monthly values that give a result closer to the values calculated with the Meteonorm software (Meteotest 2017).

We propose to estimate the monthly values of atmospheric parameters, using a reference day for each month of the year and the horizontal radiation data of a location. First, analytically the transmissivity is determined and subsequently the diffuse proportion is derived with a linear correlation. To show the temporal variation of the insolation, global solar radiation map is calculated for all the months of the year and then, adds the results of the twelve maps to obtain the total annual value (Figs. 10 and 11).

3.1.1   Transmissivity

Transmittivity is the proportion of the solar radiation that goes through the atmosphere and reaches the surface of the Earth with respect to the solar radiation received outside the atmosphere (extraterrestrial). The values that this parameter can take values between 0 (without transmission) and 1 (complete transmission), considering that a value of 0.5 corresponds to a generally clear sky.

To establish the monthly transmissivity values, the monthly average clearness index K_T is calculated. This parameter is defined as the ratio between is the monthly average daily radiation on a horizontal surface H_h and the extraterrestrial solar radiation incident on a horizontal plane H_0:

$$K_T = H_h/H_0 \tag{24}$$

The global radiation on a horizontal surface values H_h for each month is obtained from the PVGIS database by selecting a location on the interactive map (European Commission 2012).

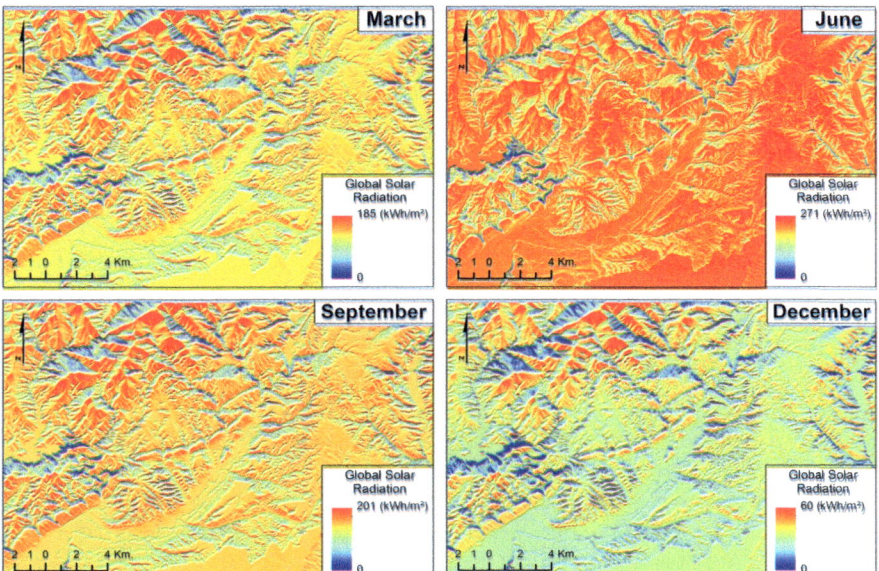

Fig. 10 Global solar radiation for some months (March, June, September, and December). Area of Guadalajara and Madrid provinces (Spain)

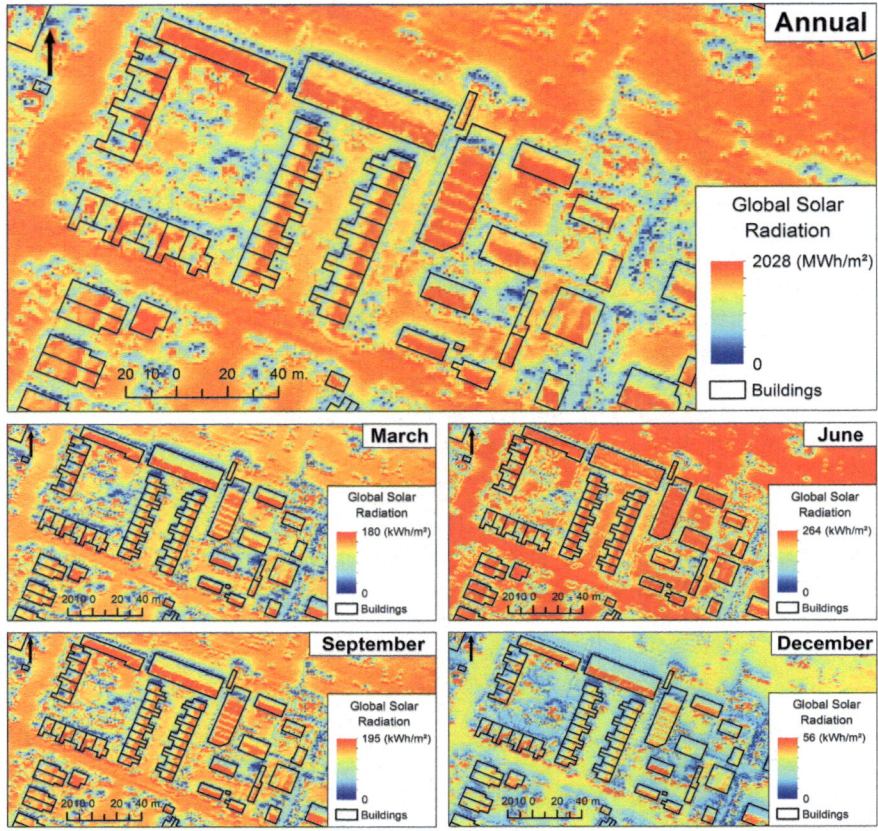

Fig. 11 Global solar radiation for some months (March, June, September, and December) and annual. Set of buildings in the town of Alpedrete (Madrid)

Analytically, the extraterrestrial solar radiation incident on a horizontal plane H_0 can be calculated by the expression:

$$H_0 = (24/\pi) \cdot I_{SC} \cdot [1 + 0.033 \cdot (360 \cdot n/365)] \\ \cdot [\cos\phi \cdot \cos\delta \cdot \sin\omega_s + (\pi \cdot \omega_s/180) \cdot \sin\phi \cdot \sin\delta] \tag{25}$$

where I_{SC} is the solar constant (1367 W/m^2 day); n is the selected day for each month (Table 4); and, ϕ is the latitude.

There are different approaches for the determination of declination angle δ. According to the equation of Cooper (1969), for a Julian day n of year, the declination is expressed as:

Table 4 Recommended average day for each month (Klein 1977)

Month	Day of the year	Date	Month	Day of the year	Date
January	17	17 Jan.	July	198	17 Jul.
February	47	16 Feb.	August	228	16 Aug.
March	75	16 Mar.	September	258	15 Sep.
April	105	15 Apr.	October	288	15 Oct.
May	135	15 May.	November	318	14 Nov.
June	162	11 Jun.	December	344	10 Dec.

$$\delta = 23.45 \cdot \sin\left(360 \cdot \frac{284 + n}{365}\right) \tag{26}$$

Sunset hour angle ω_s is calculated by the equation (Duffie and Beckman 2013):

$$\cos \omega_s = -\tan \phi \cdot \tan \delta \tag{27}$$

3.1.2   Diffuse Proportion

Diffuse proportion is the fraction of the global solar radiation that is diffuse. The values of this parameter vary from 0 to 1, establishing as a value of 0.3 for generally clear sky conditions.

The coefficient of monthly average diffuse ratio K_D, which represents the relation between the global solar radiation and the diffuse component of the radiation, is calculated to determine the diffuse proportion. This parameter is based on the monthly average clearness index K_T and is calculated using the correlation of Gopinathan and Soler (1995) (Table 5).

$$K_D = 0.91138 - 0.6225 \cdot K_T \tag{28}$$

Table 5 Values of transmissivity and diffuse proportion

Month	H_h (Wh/m^2)	Transmissivity	Diffuse proportion
January	1930	0.45	0.47
February	2980	0.39	0.54
March	4430	0.34	0.59
April	5260	0.38	0.55
May	6330	0.36	0.57
June	7430	0.29	0.64
July	7800	0.25	0.69
August	6710	0.27	0.66
September	5090	0.32	0.62
October	3510	0.36	0.57
November	2230	0.42	0.50
December	1770	0.45	0.48

Area of Guadalajara and Madrid provinces (Spain)

4 Conclusions

The purpose of this chapter has been to describe some of the available options for estimating the distribution of solar radiation in geographic areas. Several methods of interpolation and approximation were developed to predict the values of a spatial phenomenon in a location. In the interpolation examples shown previously, although the polynomial global method has the smallest root mean quadratic error that results in interpolations closer to the real value, it is observed that the geo-statistical methods, in general show very low quadratic errors. Within this group, the universal kriging and ordinary Cokriging methods are the ones that show the greatest adjustment in the results. The mean standardized error close to 0 and the root mean square standardized error nearby 1 (Table 3).

Although all interpolation techniques are valid, it is important to analyze whether the interpolation method and the selected criteria are the most appropriate. A method that conforms well to a data set may not be the most appropriate for a different data set. The application of each method will consider the objective of the interpolation, the properties of the available data and the distribution of the sample.

Sometimes, there are not enough measurement points or the solar radiation distribution changes on a very short spatial scale, such as mountainous regions or urban areas with a complex morphology. An alternative to interpolation procedures are the modeling techniques developed for GIS software that manage large amounts of geo-referenced data. Models for estimating solar radiation are mainly based on a three-dimensional surface model and, using a series of zonal parameters, they are adjusted as much as possible to the characteristics of the geographical area under study.

GIS analysis tools for the prediction and mapping of the solar resource are increasingly powerful. The objective is to highlight the potential presented by this type of tools for the representation of this phenomenon without forgetting the purpose of carrying out a specific study. Users will have to choose the technology that best suits each one and analyze the results according to the method and parameters that defines it.

References

Abramowitz M, Stegun IA (1974) Handbook of mathematical functions, with formulas, graphs, and mathematical tables. Dover Publications Inc, New York

Agencia Estatal de Meteorología AEMET (2011) Agencia Estatal de Meteorología. http://www.aemet.es/es/portada. Accessed 29 Mar 2018

Antonanzas J, Urraca R, Martinez-de-Pison FJ, Antonanzas-Torres F (2015) Solar irradiation mapping with exogenous data from support vector regression machines estimations. Energy Convers Manage 100:380–390. https://doi.org/10.1016/j.enconman.2015.05.028

Apaydin H, Kemal SF, Yildirim YE (2004) Spatial interpolation techniques for climate data in the GAP region in Turkey. Clim Res 28:31–40

Brito MC, Gomes N, Santos T, Tenedório JA (2012) Photovoltaic potential in a Lisbon suburb using LiDAR data. Sol Energy 86:283–288

Burrough PA, McDonnell RA (1998) Principles of geographical information systems. Oxford University Press, Oxford

Chilès J-P, Delfiner P (2012) Geostatistics: modeling spatial uncertainty. Willey Series in Probability and Statistics. Wiley, Hoboken

Cooper PI (1969) The absorption of radiation in solar stills. Sol Energy 12:333–346. https://doi.org/10.1016/0038-092X(69)90047-4

Cressie N (2015) Statistics for spatial data, revised edition. Wiley Classic Library. Wiley, Hoboken

Duffie JA, Beckman WA (2013) Solar engineering of thermal processes. Wiley-Interscience, Hoboken

Esri (2017a) ArcGIS desktop. Tools: an overview of the solar radiation toolset. http://desktop.arcgis.com/en/arcmap/10.3/tools/spatial-analyst-toolbox/an-overview-of-the-spatial-analyst-toolbox.htm. Accessed 29 Mar 2018

Esri (2017b) ArcGIS Pro. Tool reference: cross validation. https://pro.arcgis.com/en/pro-app/tool-reference/geostatistical-analyst/cross-validation.htm. Accessed 29 Mar 2018

Esri (2018) Esri: GIS mapping software, spatial data analytics and location plarform. https://www.esri.com/es-es/home. Accessed 29 Mar 2018

European Commission (2012) Photovoltaic geographical information system interactive maps (PVGIS). http://re.jrc.ec.europa.eu/pvgis/apps4/pvest.php?lang=es&map=europe. Accessed 29 Mar 2018

Evrendilek F, Ertekin C (2008) Assessing solar radiation models using multiple variables over Turkey. Clim Dyn 31:131–149. https://doi.org/10.1007/s00382-007-0338-6

Fogl M, Moudrý V (2016) Influence of vegetation canopies on solar potential in urban environments. Appl Geogr 66:73–80. https://doi.org/10.1016/j.apgeog.2015.11.011

Fu P, Rich PM (1999) Design and implementation of the solar analyst: an ArcView extension for modeling solar radiation at landscape scales. In: Proceedings of the 19th annual ESRI user conference, San Diego, USA

Fu P, Rich PM (2000) The solar analyst 1.0 user manual. Helios Environmental Modeling Institute (HEMI), USA

Goovaerts P (1997) Geostatistics for natural resources evaluation. Applied geostatistics series. Oxford University Press, New York

Gopinathan KK, Soler A (1995) Diffuse radiation models and monthly-average, daily, diffuse data for a wide latitude range. Energy 20:657–667. https://doi.org/10.1016/0360-5442(95)00004-Z

GRASS Development Team (2018) GRASS GIS. https://grass.osgeo.org/. Accessed 29 Mar 2018

Hofierka J, Šúri M, Huld T (2007) r.sun. Solar irradiance and irradiation model. https://grass.osgeo.org/grass75/manuals/r.sun.html. Accessed 29 Mar 2018

Instituto Geográfico Nacional IGN (2015) Modelos Digital del Terreno - MDT25. Hoja MTN50 0485. http://www.ign.es/web/ign/portal. Accessed 29 Mar 2018

Isaaks EH, Srivastava RM (1989) An introduction to applied geostatistics. Oxford University Press, New York

Johnston K, Ver Hoef JM, Krivoruchko K, Lucas N (2001) ArcGIS 9. Using ArcGIS geostatistical analyst. ESRI, Redlands, CA, USA

Jolly WM, Graham JM, Michaelis A, Nemani R, Running SW (2005) A flexible, integrated system for generating meteorological surfaces derived from point sources across multiple geographic scales. Environ Model Softw 20:873–882

Klein SA (1977) Calculation of monthly average insolation on tilted surfaces. Sol Energy 19:325–329. https://doi.org/10.1016/0038-092X(77)90001-9

Kodysh JB, Omitaomu OA, Bhaduri BL, Neish BS (2013) Methodology for estimating solar potential on multiple building rooftops for photovoltaic systems. Sustain Cities Soc 8:31–41

Mavromatidis G, Orehounig K, Carmeliet J (2015) Evaluation of photovoltaic integration potential in a village. Sol Energy 121:152–168. https://doi.org/10.1016/j.solener.2015.03.044

Meteotest (2017) Meteonorm. Global irradiation and climate data. https://meteotest.ch/en/. Accessed 29 Mar 2018

Mitášová H, Mitáš L (1993) Interpolation by regularized spline with tension: I. Theory and implementation. Math Geol 25:641–655. https://doi.org/10.1007/BF00893171

Oloo FO, Olang L, Strobl J (2015) Spatial modelling of solar energy potential in Kenya. Int J Sustain Energy Plann Manage 6:17–30. https://doi.org/10.5278/ijsepm.2015.6.3

Orgill JF, Hollands KGT (1977) Correlation equation for hourly diffuse radiation on a horizontal surface. Sol Energy 19:357–359. https://doi.org/10.1016/0038-092X(77)90006-8

Perea-Moreno A-J, Hernandez-Escobedo Q (2016) Solar resource for urban communities in the Baja California Peninsula, Mexico. Energies 9 https://doi.org/10.3390/en9110911

Righini R, Grossi Gallegos H, Raichijk C (2005) Approach to drawing new global solar irradiation contour maps for Argentina. Renew Energy 30:1241–1255. https://doi.org/10.1016/j.renene.2004.10.010

Santos Preciado JM, García Lázaro FJ (2008) Análisis estadístico de la información geográfica. Cuadernos UNED. Universidad Nacional de Educación a Distancia, UNED, Madrid

Slocum TA, McMaster RB, Kessler FC, Howard HH (2014) Thematic cartography and geovisualization: pearson new international edition. Thematic cartography and geovisualization. Pearson Education Limited, London

Sun YW, Hof A, Wang R, Liu J, Lin YJ, Yang DW (2013) GIS-based approach for potential analysis of solar PV generation at the regional scale: a case study of Fujian Province. Energy Policy 58:248–259

Šúri M, Hofierka J (2004) A new GIS-based solar radiation model and its application to photovoltaic assessments. Trans GIS 8:175–190. https://doi.org/10.1111/j.1467-9671.2004.00174.x

Tooke TR, Coops NC, Voogt JA, Meitner MJ (2011) Tree structure influences on rooftop-received solar radiation. Landscape Urban Planning 102:73–81

Vicente-Serrano SM, Saz-Sánchez MA, Cuadrat JM (2003) Comparative analysis of interpolation methods in the middle Ebro Valley (Spain): application to annual precipitation and temperature. Climate Res 24:161–180

Wackernagel H (2003) Multivariate geostatistics: an introduction with applications. Springer, Berlin

Wong MS et al (2016) Estimation of Hong Kong's solar energy potential using GIS and remote sensing technologies. Renew Energy 99:325–335. https://doi.org/10.1016/j.renene.2016.07.003

Chapter 9
Basics on Mapping Solar Radiation Gridded Data

Jesús Polo and Luis Martín-Pomares

Abstract There is a lot of gridded information on meteorological variables elsewhere. Numerical weather prediction models and satellite-derived models deliver time series and aggregates of main meteorological variables with global coverage that can be finally used to create maps offering information on the spatial variability of those magnitudes. This chapter intends to give a very simple overview of mapping solar radiation data or any other gridded variable using QGIS open-source Geographic Information System.

1 Spatial Data and Geographic Information Systems

Spatial data is nowadays used in many disciplines of very different nature from social sciences to engineering. Basically, spatial data is information geographically referenced on some specific variables with the associated spatial reference; that is, spatial data needs to have geographic coordinates (latitude and longitude or UTM coordinates) and a system of reference. Spatial information can be found in several data types: point (single location of a city), line (border of a country), polygon (area of a country) and grid (rectangular matrix of points). Points, lines and polygons are denoted as vector data, and the gridded information is named raster data. Systems of reference are structures used to define unique positions in the space. The coordinate system that is most commonly used to define locations on the three-dimensional earth is called the geographic coordinate system (GCS), and it is based on a sphere or spheroid. Locations in the GCS are defined by their respective latitude and

J. Polo (✉)
Photovoltaic Solar Energy Unit, Renewable Energy Division (Energy Department),
CIEMAT, Avda Computense 40, 28040 Madrid, Spain
e-mail: jesus.polo@ciemat.es

L. Martín-Pomares
Qatar Environment & Energy Research Institute, Hamad Bin Khalifa University,
5825, Doha, Qatar
e-mail: lpomares@hbku.edu.qa

© Springer Nature Switzerland AG 2019
J. Polo et al. (eds.), *Solar Resources Mapping*, Green Energy and Technology,
https://doi.org/10.1007/978-3-319-97484-2_9

longitude within the GCS. One of the most used systems of reference is the WGS84 projection (World Geodetic Survey 1984) which uses the centre of the earth as the origin of the GCS and is used for defining locations across the globe.

Spatial data is conveniently exploited through the Geographic Information Systems (GIS). A GIS is a collection of computer tools for visualization, handling, processing and analysis of spatial data. GIS can be presented in a commercial or free software tool, or as a toolbox (collection of functions and methods) in programming packages frequently used in engineering like MATLAB®, R or Phyton (Bivand et al. 2008; Trauth 2010). The latter is the case of the Mapping Toolbox of MATLAB®, the sp package in R or the GeoPandas package in Python.

The most well-known commercial GIS software is probably ArcGIS, professional software containing the state of the art in geographic information systems (https://www.arcgis.com/index.html). Nevertheless, there is open-source alternative software that can be useful for many applications. One of the most widely used is QGIS (https://qgis.org/es/site/), whose community is actively and continuously developing new features and updating the program. QGIS was developed as a project of the Open Source Geospatial Foundation (OSGeo), a non-profit organization for fostering and supporting broader open geospatial technologies (https://www.osgeo.org/). Figure 1 shows the QGIS 3.2 Desktop.

Very basically the spatial data is presented in a GIS as a set of layers. Each layer can be a vector or a raster layer, and the GIS software usually provides specific methods that operate with vector or with raster layers. A GIS is able to show in one single map multiple layers of information. This requires that all the layers use the same system of reference in order to be visualized together. In addition, there are also methods for data restructuration and conversion into different formats, such as raster-to-vector translation. The layer has also attributes and metadata associated.

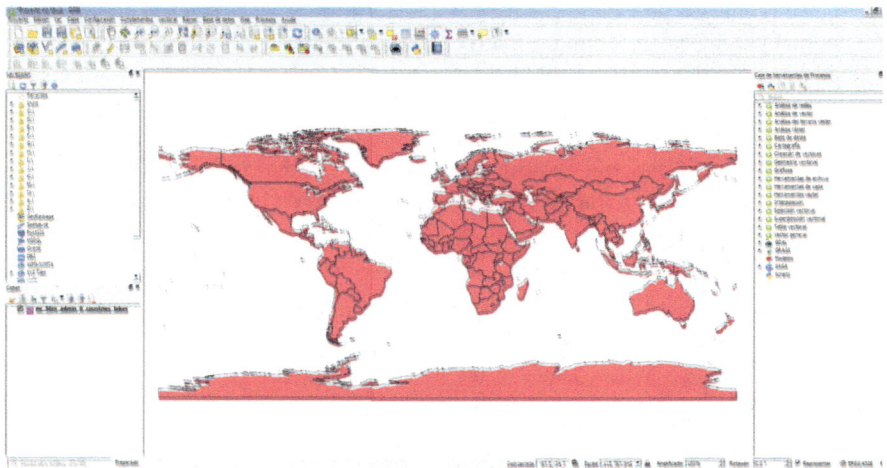

Fig. 1 QGIS Bonn desktop

Every analysis, visualization or mapping activity to be done in a GIS requires of the availability of several basic spatial data (world, countries borders, coastline, etc.) that usually act as a background layer or map. There is a lot of free available spatial data on the Internet that can be used for mapping gridded solar radiation and other meteorological gridded data. Natural Earth is a good example of public domain with vector and raster data for being used with GIS software (https://www. naturalearthdata.com/). Data themes are available at three levels of detail (scales 1: 10, 1:50 and 1:110 m).

1.1 Shapefiles

The shapefile is very common and widely used format for vector data that was developed by Esri (ESRI 1998). A shapefile consists actually of three files with the extension .shp, .shx and.dbf. The first one stores the geometry of the digital features as a set of vector coordinates. The second one contains an index for allowing quick access to the spatial features. The third required file stores attribute data. All of them must present in the same folder in order to be usable by a GIS. For instance, Natural Earth is a public domain offering free shapefiles of the countries, boundary lines, coastlines and polygons of the world with at least first-order administrative attributes (names of countries, provinces, states, etc.) that can be very useful in mapping gridded information. Figure 2 shows several shapefiles (administrative level 1, roads and water areas and labels on provinces) of east Spain obtained from the DIVA-GIS free spatial data repository (http://diva-gis.org/Data). This figure was created with QGIS 3.2 loading three shapefiles and selecting the attributes to be shown of each layer. GADM is also a free repository with spatial data for all countries and their subdivisions (https://gadm.org/).

Shapefiles can contain also polygons that represent specific geographic areas that share a common characteristic not only countries boundaries and regions. For instance, Fig. 3 shows an updated version of the Köppen–Geiger climatic zones attending exclusively to precipitation criterion (Peel et al. 2007). In addition, QGIS can be used to convert categorized raster information (land cover classification) into shapefile.

1.2 Raster Data

Raster data is information on a specific feature represented as a rectangular surface divided into a regular grid of cells. In a raster file, every cell or pixel is associated with a particular latitude and longitude. Therefore, raster data is made up of pixels (also referred to as grid cells) and although they are usually regularly spaced and square, they can also be irregular. Satellite image and satellite-derived meteorological information are the most typical examples of raster spatial data. Raster files

Fig. 2 Illustration of basic vector data

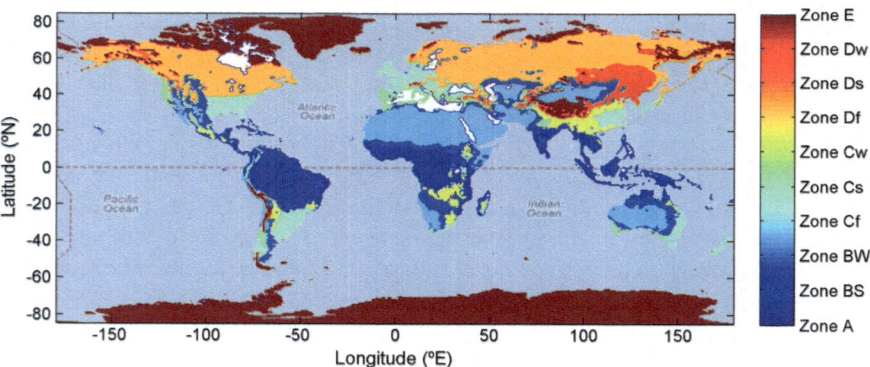

Fig. 3 Köppen–Geiger climatic zones considering only precipitation

contain basically a 2D matrix of data, and thus, map algebra can be easily performed with any raster file in most of the GIS tools. Raster data may be discrete data as occurs in land-use maps or continuous data as elevation maps (DEM—digital elevation map). QGIS uses the GDAL library to read and write raster data formats and is compatible with most common raster formats (ArcInfo ASCII, GeoTIFF, EDRAS, etc.).

Typical raster datasets include remote sensing data, such as aerial photography or satellite imagery data. Unlike vector data, raster data is geocoded by pixel resolution and the x/y-coordinate of a corner pixel of the raster layer. This allows QGIS or any other GIS to position the data correctly in the map canvas. Figure 4 shows an image of the elevation map of Spain from the raster information available from DIVA-GIS website.

QGIS includes most of the common methods for raster data incorporating procedures from GDAL, GRASS and SAGA libraries. In mapping solar radiation and in particular solar potential, it is required frequently to deal with the slope of the terrain. Loading a DEM raster in QGIS, the slope can be estimated directly by the slope process of GDAL or *r.slope* method in GRASS. In solar potential, estimation is very common to establish a threshold for the slope that makes null the solar potential; one of the most widely used criteria of slope considers that terrains with a slope beyond 3% are not available for solar power deployment (Navarro et al. 2016). The estimation of the slope can be used to generate a raster mask or exclusion layer that multiplies the raster of solar irradiation to generate the corresponding solar potential. A raster mask or exclusion layer is a raster of the same size and resolution than the target which contains only 1 or 0 values according to the exclusion criteria specified. Figure 5 illustrates the process of generating an exclusion map for the slope criteria of the 3% used in the determination of the solar potential for Vietnam (Polo et al. 2015a, b). Different exclusion criteria are used depending on the solar power technology for determining the availability of a geographic area for deployment solar energy systems (Omitaomu et al. 2012; Martín-Chivelet 2016). Every exclusion criteria can be determined by binary exclusion masks that result in the final solar potential map using raster algebra.

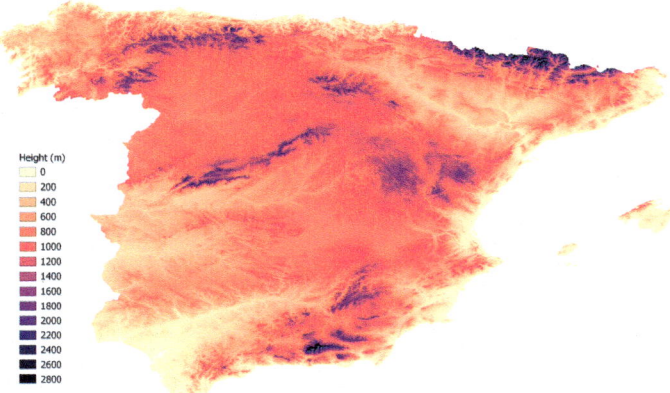

Fig. 4 Elevation map of Spain

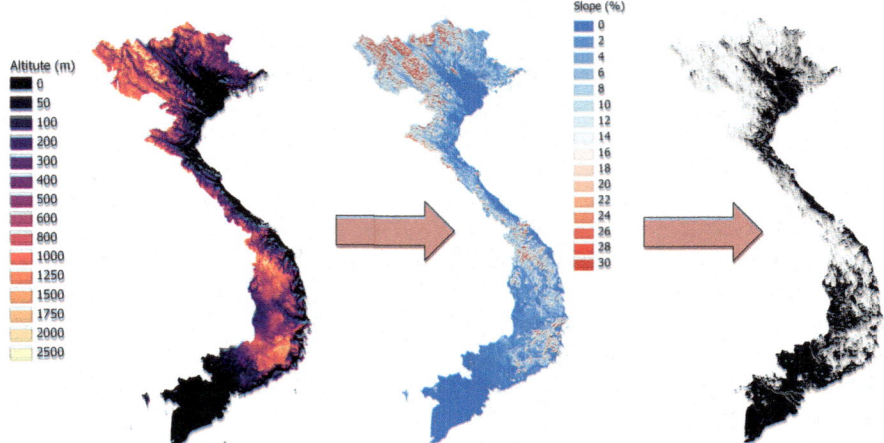

Fig. 5 Process for the slope exclusion mask; left, elevation map; middle, slope map; right, exclusion mask

2 Simple Exercise of Solar Radiation Mapping

In this section, a simple step-by-step procedure for creating a solar radiation map with QGIS is described. Let us start with the rectangular raster data, a regular matrix of daily mean solar global irradiation for the month of February derived from Meteosat satellite imagery from 10 years. The matrix consists of 300 columns and 150 rows of data that represent the domain covering −10–5°E longitude and 35–44°N latitude at a spatial resolution of 0.05° (Polo 2015).

The raster data is stored in an ASCII file in the ESRI ASCII format which allows a very easy handling a transformation of the data. ESRI ASCII format is basically a matrix of data in ASCII file with a header as follows:

NCOLS 300
NROWS 180
XLLCORNER-10
YLLCORNER 35
CELLSIZE 0.05
NODATA_VALUE-9999.

The header gives information on the number of rows and columns of the matrix, the spatial resolution, the geographic coordinates of the lower left corner and the flag used in those cells where there is no data.

The second step is to download the geographic boundary of the region that is going to be mapped. In this case, the Iberian Peninsula and Balearic Islands can be obtained from Natural Earth free spatial data as a shapefile.

Finally, the raster must be cut to fit the boundaries of the selected region. QGIS has a method for intersecting and clipping operations with at least two layers: a raster layer and a vector layer. In QGIS Bonn version, this can be found in the *Raster/Extract/Cut* menu. Figure 6 illustrates all the process and shows the final map.

The raster data is commonly used also in Web services through what is named as geoportal. A geoportal is a Web-based geospatial resource that allows users to access geospatial information and services through the Internet. For that purpose, it is needed a Web map, a Web server for publishing the map online and a set of methods (java-script, PHP, etc.) for interacting with the map. QGIS includes a plugin for creating a Web visor map in a very straightforward way, called *qgis2map*. Figure 7 shows a very simple example of a Web map where uses *OpenStreetMap* as a background map for showing the solar radiation map of February in Spain.

There are many examples of solar radiation Web maps and services through the Internet provided by both public and private organizations. As illustrative mention of a few of them, some of the most well-known and used free services are:

– PVGIS, developed and supplied by the European Commission—Joint Research Centre (JRC), focused on the performance of the photovoltaic technology. PVGIS offers Web-based maps and information covering Europe, Africa and Asia based on Meteosat satellite imagery (http://re.jrc.ec.europa.eu/pvgis_v2.html)
– IRENA Global Atlas 3.0 is a browser that compiles Web maps from several suppliers on renewable energy resources including solar radiation data (https://irena.masdar.ac.ae/gallery/#gallery).
– National Solar Radiation Database viewer from NREL offering solar radiation maps for USA, Central America, part of South America and part of Asia (https://maps.nrel.gov/nsrdb-viewer).

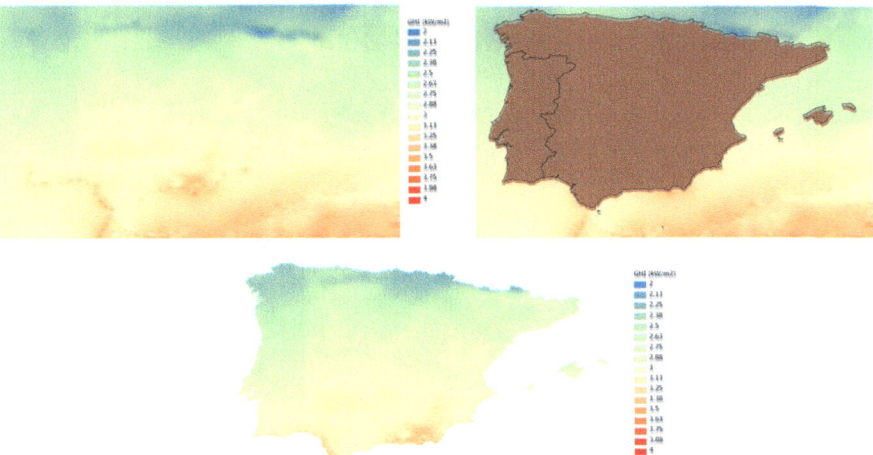

Fig. 6 Illustration of the intersection between a regular raster of gridded data and a vector shapefile to generate a map of solar radiation

Fig. 7 Illustration of Web map visor using *OpenStreetMap* as background

– The "Solar Atlas for the Mediterranean" is a portal for global horizontal and direct normal irradiance data for the southern and eastern Mediterranean region (http://www.solar-med-atlas.org/solarmed-atlas).

3 Interpolating Tables of Gridded Data

In some cases, the gridded information of solar radiation data can be regularly distributed with a coarser spatial resolution or can be formed by a discrete set of points. Interpolation is then a commonly used GIS technique to create a continuous surface. Interpolation of solar radiation is described in detail in Chap. 8. Therefore, this section illustrates a few examples of interpolation of discrete points using QGIS tools.

Let us consider as an example monthly means of global solar irradiation in three Morocco regions that were derived from satellite imagery at a spatial resolution of $0.1° \times 0.1°$. The data was stored in a simple ASCII or .csv file containing three columns of data: longitude, latitude and GHI in kW m^{-2}. QGIS can load a layer of delimited text data that is viewed in QGIS canvas as a regular grid of discrete points. Interpolation of this regular grid can be done in QGIS using the GDAL library, which contains around twenty interpolation methods. The list can range from methods that do not perform interpolation or smoothing, like the nearest neighbour, to interpolation techniques like inverse distance weighted, ordinary or

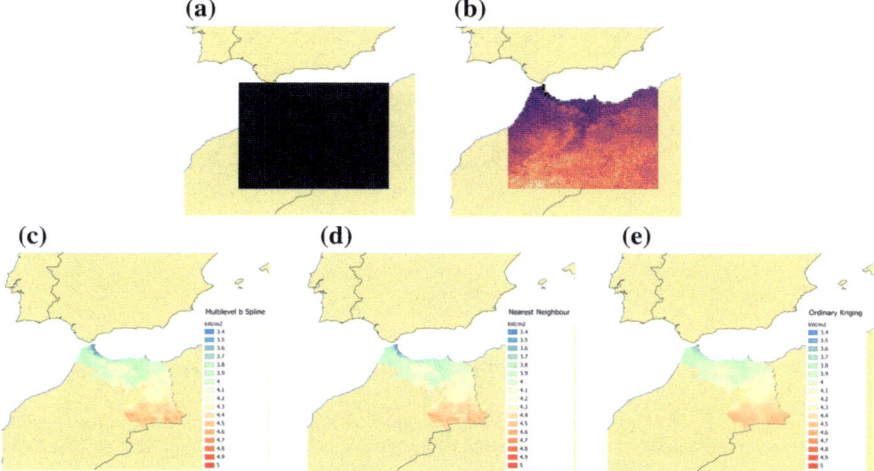

Fig. 8 Examples of spatial interpolation of discrete points of solar radiation data. **a** Regular grid of satellite-derived discrete points, **b** regular grid of categorized discrete points, **c** Interpolation with multilevel B Spline, **d** Nearest neighbour method for creating a continuous surface, **e** Ordinary kriging interpolation

regression-kriging. Figure 8 illustrates some examples of interpolation. The choice of one or another interpolation method depends on the purpose or visualizing needs and on the characteristics of the discrete point set as well (Chelbi et al. 2015; Hofierka et al. 2017):

- Nearest neighbour consists in assigning to each cell the value of the nearest cell in the original grid; it can be used for categorical data or for classifying data.
- Spline interpolation is a method for smoothing the data, since estimate values using a mathematical function that minimizes the surface curvature among points.
- Kriging is an advanced geostatistical procedure that uses the semivariogram to express the spatial continuity (autocorrelation). The semivariogram measures the strength of the statistical correlation as a function of distance.

4 Sharing Geospatial Data

Nowadays, it is very common to find geospatial information that is being shared through the Internet. GeoServer (http://geoserver.org/) is an open-source code, written in Java, that allows to display and share spatial data using open standards set by the Open Geospatial Consortium (OGC). GeoServer has become an easy method to connect Web maps such as OpenLayers, Google maps or Bing Maps. There are two basic service sets—the Web Feature Services (WFS) and the Web Map Services (WMS). The WFS is concerned with direct access to data—reading, writing and updating features. The WMS is concerned with transforming your data into a map (image). QGIS versions 2.x are compatible with a plugin called GeoServer Explorer that makes rather easy to share a QGIS project or maps through GeoServer. Likewise, QGIS allows loading easily WMS and WFS services available on the Internet.

References

Bivand RS, Pebesma EJ, Gómez-Rubio V (2008) Applied spatial data. Analysis with R. Springer, New York

Chelbi M, Gagnon Y, Waewsak J (2015) Solar radiation mapping using sunshine duration-based models and interpolation techniques: application to Tunisia. Energy Convers Manag 101:203–215. https://doi.org/10.1016/J.ENCONMAN.2015.04.052

ESRI (1998) ESRI shapefile technical description

Hofierka J, Lacko M, Zubal S (2017) Parallelization of interpolation, solar radiation and water flow simulation modules in GRASS GIS using OpenMP. Comput Geosci 107:20–27. https://doi.org/10.1016/J.CAGEO.2017.07.007

Martín-Chivelet N (2016) Photovoltaic potential and land-use estimation methodology. Energy 94:233–242. https://doi.org/10.1016/j.energy.2015.10.108

Navarro AA, Ramírez L, Domínguez P et al (2016) Review and validation of Solar Thermal Electricity potential methodologies. Energy Convers Manag 126:42–50. https://doi.org/10.1016/j.enconman.2016.07.070

NIMA (1997) Department of defense world geodetic system 1984, its definition and relationships with local geodetic systems. Technical Report TR8350.2 3rd edn

Omitaomu OA, Blevins BR, Jochem WC et al (2012) Adapting a GIS-based multicriteria decision analysis approach for evaluating new power generating sites. Appl Energy 96:292–301

Peel MC, Finlayson BL, McMahon TA (2007) Updated world map of the Köppen-Geiger climate classification. Hydrol Earth Syst Sci 11:1633–1644. https://doi.org/10.5194/hess-11-1633-2007

Polo J (2015) Solar global horizontal and direct normal irradiation maps in Spain derived from geostationary satellites. J Atmos Solar Terr Phys 130–131:81–88. https://doi.org/10.1016/j.jastp.2015.05.015

Polo J, Bernardos A, Navarro AAA et al (2015a) Solar resources and power potential mapping in Vietnam using satellite-derived and GIS-based information. Energy Convers Manag 98:348–358. https://doi.org/10.1016/j.enconman.2015.04.016

Polo J, Gastón M, Vindel JMM, Pagola I (2015b) Spatial variability and clustering of global solar irradiation in Vietnam from sunshine duration measurements. Renew Sustain Energy Rev 42:1326–1334. https://doi.org/10.1016/j.rser.2014.11.014

Trauth MH (2010) MATLAB® Recipes for earth sciences, 3rd edn. Springer

Chapter 10
Sampling Design Optimization of Ground Radiometric Stations

Luis Martín-Pomares, Martín Gastón Romeo, Jesús Polo, Laura Frías-Paredes and Carlos Fernández-Peruchena

Abstract This chapter presents a methodology to identify optimal site locations to establish a surface radiometric monitoring network once the raw solar estimations are produced from satellite images or numerical models. The site selection is done considering the long-term solar resource, its spatial distribution, variability and technical and logistics aspects. The methodology presented here is an adaptive sampling strategy under an assumed population model derived from satellite images or numerical models. The objective is to install the radiometric stations in optimal locations to correct the systematic biases of the modelled solar radiation, improving the estimates and minimizing the number of stations needed. To achieve that, we need to identify the area with a similar dynamic in terms of solar radiation. Inside the areas identified, the most optimal locations will be used to place the radiometric stations. The methodology is divided into three phases. The first phase divides the geographical extension under study to identify the areas with a similar dynamic in terms of monthly solar radiation. The selection of the number of cluster/areas is done with an information criteria technique. Once the optimal number of clusters

L. Martín-Pomares (✉)
Qatar Environment & Energy Research Institute, Hamad Bin Khalifa University,
P.O. Box 5825, Doha, Qatar
e-mail: lpomares@hbku.edu.qa

M. Gastón Romeo · L. Frías-Paredes
National Renewable Energy Centre (CENER), C/Ciudad de La Innovación,
7, 31621 Sarriguren (Navarra), Spain
e-mail: mgaston@cener.com

L. Frías-Paredes
e-mail: lfrias@cener.com

J. Polo
Photovoltaic Solar Energy Unit. Renewable Energy Division (Energy Department),
Ciemat, Avda Complutense40, 28040 Madrid, Spain
e-mail: jesus.polo@ciemat.es

C. Fernández-Peruchena
National Renewable Energy Centre (CENER), C/Isaac
Newton n 4- Pabellón de Italia, 41092 Seville, Spain
e-mail: cfernandez@cener.com

© Springer Nature Switzerland AG 2019
J. Polo et al. (eds.), *Solar Resources Mapping*, Green Energy and Technology,
https://doi.org/10.1007/978-3-319-97484-2_10

and the extension of each area is defined, the second phase is the production of a long list of candidate sampling sites with GIS techniques based on constraints to identify the best locations in each area to place the radiometric stations. The third phase is based on ranking the long list of candidate locations with site visits and a checklist criterion based on BSRN recommendations to produce a final shortlist of optimal sites to measure solar radiation in each area. Finally, two tiers of radiometric stations are proposed to place in each area depending on the ranking level of each location.

1 Introduction

A detailed knowledge about the resource availability in a region appears as a key start point to the development and advance of renewable energies. This information should describe the temporal and spatial variability and will be a key tool for governments and investors in the decision-making process. To support stakeholders, the implementation of solar resource maps is a mandatory and very useful practice to plan the installation of solar technologies in a country. Any photovoltaic (PV) or concentrated solar power (CSP) project plan needs precise information on the long-term solar resources and its variability over time as the first step in the development. This information is very sensitive and requires the most accurate estimations since any small error on the total expected energy, or the inter-annual and intra-annual variability, could be translated into important losses of capital for the solar project. Besides, we could have the situation that only once the solar project is built, and the developer realizes that the project has a financial result significantly different than the one initially calculated for the specific design of the plant. This could make the project turns into a non-profitable and non-viable plan with significant losses for the estimated lifetime of the plant, which usually is expected to be around 30 years.

It is important to remark that the portion of the incoming solar radiation that reaches the Earth's surface exhibits large geographical and temporal variability due to its strong dependence on the atmospheric conditions. The most accurate way to analyse solar resource is by using ground measurements taken from radiometric stations; however, this information is often scarce and temporally limited in most parts of the world, and it is only specific to the location at which it was measured. As an alternative to ground-collected data, the solar resource can be modelled by satellite-derived data or numerical models as we have seen in the previous chapters. Radiometric stations are good at providing high frequency and accurate data (given well-maintained, high accuracy measuring equipment) for a given site. On the other hand, models provide data with a lower temporal frequency, but they can characterize a long history over wide territories. Both sources of data are needed since the solar renewable energy community depends on radiometric measurements to develop and validate solar energy models and the ground measurements can be used to correct the systematic biases of the long-term estimates from the models (Gueymard and Myers 2009).

The main aim of setting up a radiometric monitoring network in a country is to measure solar radiation (for a minimum period of one year to characterize the seasonal variability of solar radiation) to site adapt locally and validate the satellite estimations. The characterization of the complex spatial distribution of irradiance at the surface level would ideally be done by deploying as many sampling locations as possible. However, the cost would be costly unacceptable on a national level (Perez et al. 1997; Zelenka et al. 1999).

The diversity of climate zones present in a country in conjunction with its varied topography requires a detailed analysis for the determination of the optimal number of areas and best locations within each area for a solar monitoring network. To minimize costs, optimizing the number and placement of monitoring equipment is critical (Yang and Reindl 2015). In particular, we seek to establish a radiometric network that represents:

- Dispersed high solar irradiation zones with potentially high commercial exploitable solar resources.
- Areas with a similar dynamic in terms of solar radiation at ground surface level.
- Represent the variety of different local climates and/or topographical conditions in an area.

The methodology defined for such a site selection consists of three phases:

1. **Area selection.** First, zones with similar solar irradiance characteristics are grouped together in areas, so that a minimum number of stations can represent the whole region.
2. **Long-list candidate identification.** Within each previous area selected, several locations (size of the pixel of the satellite images or the grid of the numerical model) are identified. The selection is based on GIS techniques, a set of technical and logistic constraints and criteria to maximize the levelized cost of energy (LCOE) for PV or/and CSP power plants like proximity to grid networks, distance to cities or places of consumption, access to fresh water and others. The result of the interim site selection is the production of a long list of sites which are a candidate to place the radiometric stations.
3. **Specific site selection.** The final locations will be identified in a third phase based on a multi-criteria methodology and site visits. Once the long list of locations of interest to measure solar radiation has been defined, several ground-based limitations, constraints and criteria based on BSRN technical manual and procedures (Hegner et al. 1998; König-Langlo et al. 2013) are considered to select the topmost appropriate locations in each clustered area.

In this chapter, Sect. 2 contains the methodology for area selection according to the modelled solar resource based on ensemble clustering technique, as well as the methodology for the selection of the optimum number of areas. In Sect. 3, the procedure to select the long list of candidate sites for a solar monitoring network within each broad areas/cluster is presented. Section 4 presents the methodology to identify the final shortlist of sites following BSRN standards. Section 5 details two types of radiometric stations proposed to be installed in each site from the final shortlist. Finally, the conclusion section outlines the core points of the chapter. It is important to remark that the data examples used in this chapter are provided from the Energy Sector Management Assistance Programme (ESMAP[1]), more specifically, from the project executed over the African country of Tanzania (Bernardos et al. 2015; Bank 2015).

2 Clustering Method for Region Classification and Selection of the Optimal Number of Clusters

The simplest (and also the less efficient) monitoring network might consist of a regular grid, determined by only one design parameter: the inter-station spacing. In the case of solar energy mapping, the design parameter is given by the size of the pixels of the satellite or the grid of the numerical model. This distribution is also known as even sampling distribution or conventional. This arrangement is inefficient since a large number of stations would provide redundant information. To maximize efficiency, it is common to group spatial regions (with similar solar irradiance characteristics) into the same solar cluster.

In this section, a methodology is presented to automatically identify and detect a collection of solar irradiance patterns into clusters based on similarity criteria of the monthly solar resource. Such criteria would group satellite pixels into the same cluster where characteristics of the solar resource are similar to each other (Kaufman and Rousseeuw 1990; Vrahatis et al. 2002; Wu et al. 2014; Polo et al. 2015b).

2.1 Selection of the Optimal Number of Clusters

Clustering methods presented here are powerful and simple methods of pattern analysis and grouping. They are used commonly for applications such as explanatory data mining, machine learning and pattern classification.

The *k-means* algorithm (Hartigan and Wong 1979; Vrahatis et al. 2002; Ding and He 2004; Finley and Joachims 2008; Jain 2010; Kumar et al. 2011) is presented here to classify the solar irradiance data set into groups and to choose the optimum number of areas. The *"elbow criterion"* based on *k-means* variance is used to determine the optimal number of clusters (Zelnik-manor et al. 2004);

---

[1]http://www.esmap.org/

(Goutte et al. 1999). The *elbow* method looks at the percentage of decrease of a cost function previously defined. It exists with the idea to choose a number of clusters so that adding another cluster does not give significantly improved modelling of the data. The variance explained by the different number of clusters is analysed to select the optimum number. The first clusters will add much information, but at some point, the marginal gain when increasing the number of clusters will drop dramatically and will show an angle in the evolution of the variance as we will see in the next figure.

K-means clustering can be used in analysing the spatial variance to create a minimum and a maximum number of clusters using, for example, the following clustering evaluation criterion:

- Calinski-Harabasz (CH) (David and Averbuch 2012).
- Davies–Bouldin (DB) (Yang et al. 2006).
- Lihi Zelnik-Manor (LZM) (Zelnik-manor et al. 2004).

There are other evaluation criteria and methods in the literature that can be applied.

The optimal number of clusters (or broad regions) can be identified as the one which presents the less variance with the minimum number of clusters since we need areas which are spatially independent between them in terms of solar radiation. The number of clusters can be created from a minimum of 2 to a maximum defined previously (C_n) based on our experience and knowledge of the region under study.

Then, the clustering evaluation criteria—CH, DB and LZM methods—are applied to calculate the variance of each number of clusters. Finally, the optimum number of clusters can be chosen by applying the *Elbow* criteria. We will see some examples next to illustrate this methodology.

Next, we will see the *k-means* algorithm used on times series of cloud modification factor obtained from Meteosat Second Generation (MSG) imagery over Greece (Zagouras et al. 2013). The cloud modification factor is defined as the ratio of cloudy to clear sky irradiance. The aim is to identify regions of similar variability of clouds. Clustering regionalization of Greece is analysed to 90 different clusters, and 22 clusters were found as the optimal number to determine the optimal number of ground stations for solar radiation monitoring over the country.

Recently, additional works on regionalization of photosynthetically active radiation (PAR) have been published using also *k-means* (Vindel et al. 2018). In this case, the variable of interest to be used in *k-means* was the transparency index applied to PAR radiation, i.e. the ratio between measured PAR irradiation and the extraterrestrial PAR. The satellite-estimated gridded values of PAR over Spain were taken from CM SAF products (Posselt et al. 2012). Vindel et al. explore also three methods for determining the optimum number of clusters (*Elbow*, Silhouette and Davies–Bouldin methods). Finally, they present a novel methodology for determining the optimum number of ground stations for measuring PAR in Spain and their geographical positions.

Clustering techniques can be also used in regionalization of the variability of solar radiation for the development or improvement of the models (Polo et al. 2015a, b). This is the case of the study performed for Vietnam where Polo et al. used long-term data of sunshine duration measurements from 171 monitoring points in Vietnam. The *k-means* algorithm applied to the inter-quartile range of sunshine duration measurements resulted in a regionalization of the country coherent in relation to the *Köppen–Geiger* climatic regions (Kottek et al. 2006). This regionalization was used for the development of specific models for each region. The different regionalized models were applied to satellite-derived solar irradiation data that were finally used in the solar radiation mapping of Vietnam (Polo et al. 2015a).

Next, a detailed example of regionalization based on cluster analysis is presented. The methodology was applied to a data set of monthly GHI values from the years 2000–2015 (204 monthly values for each pixel). The data set was extracted for a terrestrial domain from the Meteosat First Generation over Indian Ocean satellite images field of view with high spatial resolution ($0.05° \times 0.05°$, which approximately corresponds to 5 km $\times$ 5 km).

Figure 1 presents the results of applying *k-means* clustering and variance spatial evaluation criteria to the GHI monthly data set. Table 1 shows the results of applying the method to a number of clusters from 2 to 16. In bold, it is shown the optimum number of clusters for each clustering variance evaluation criterion based on *Elbow* selection.

In this example, CH and DB fail to provide a physically expected number of cluster groups, since CH and DB divide the geographical area into only two broad areas. We expected a higher number of clusters since the region selected corresponds to Tanzania country with different topographical accidents and climates. Indeed, it contains from mountainous areas with local microclimates, to Selvatica

Table 1 Results of calculating the *k-means* clustering evaluation criteria—CH, DB and LZM—for a group of clusters from 2 to 16

Number of clusters	CH	DB	LZM
2	**1.2775**	**0.9468**	0.4743
3	0.9563	1.0594	0.3001
4	0.9037	1.1458	0.1953
5	0.8913	0.9896	0.1500
6	0.8459	1.0432	0.1428
7	0.7734	1.0998	**0.0918**
8	0.7216	1.1234	0.0783
9	0.6970	1.1686	0.0683
10	0.6683	1.1539	0.0624
11	0.6459	1.1559	0.0538
12	0.6297	1.1448	0.0487
13	0.6134	1.1181	0.0457
14	0.6116	1.0588	0.0432
15	0.5964	1.0678	0.0383
16	0.5875	0.9978	0.0289

Fig. 1 Differences in the variance within the cluster obtained from *k-means* and the number of clusters selected with *Elbow* criterion

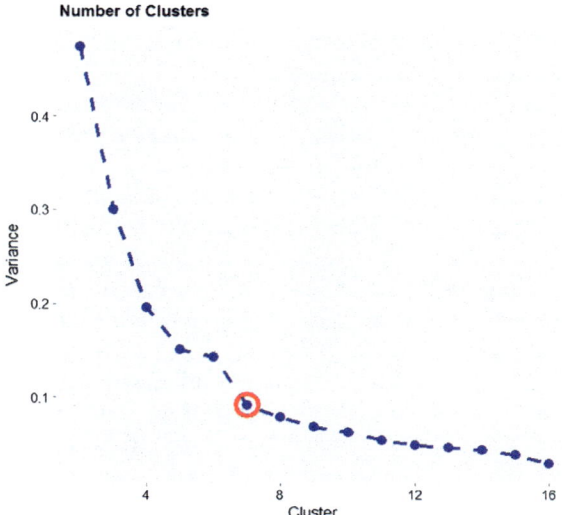

and tropical climates, coastal zones and very arid regions. LZM identify seven groups of clusters with *Elbow* criteria. LZM criteria provide more rational results regarding the number of clusters to identify the spatial monthly solar radiation variability in our data set.

LZM works better than CH and DB for automatically estimating the number of clusters. It has the property to work with multispectral and noise data and provides better clustering especially when the data includes multiple scales and when the clusters are placed within a cluttered background (Zelnik-manor et al. 2004). The variance within each cluster calculated from LZM is shown in Fig. 1. The variance presents a balance between the distinctiveness of the areas and their spatial dis-aggregation. The optimum number of clusters selected is highlighted with a red circle in the figure.

In this example, we can see that there is an incongruity observed in the results of the three different variance tests applied. To contrast these results, we will present next an additional methodology to identify the optimum number of clusters based on an ensemble analysis (Hornik 2005).

2.2   Ensemble Clustering

Cluster ensembles have emerged as a powerful meta-learning paradigm that provides improved accuracy and robustness by aggregating several input data according to statistical variability. Cluster ensemble methods combine multiple clustering of the same data set to yield a single overall clustering. It has been found that such a practice can improve robustness, as well as the quality of clustering results. The main objective of cluster ensembles is to combine different decisions of various clustering algorithms in such a way that achieves a superior accuracy compared with individual clustering.

LinkClueE (Iam-on and Garrett 2010) performs cluster ensemble link-based similarity methods which provide superior performance to the conventional and simpler co-association approach of the *k-means* method. The pairwise similarity approach is applied, in which the final data partition is derived based on relations among data points represented inside the similarity matrix. It is widely known as the *co-association* (CO) matrix (Fred and Jain 2005). The CO matrix denotes co-occurrence statistics between each pair of data points. The CO matrix can be regarded as a new similarity matrix, which is superior to the original distance-based counterpart which was presented previously (Jain and Law 2005).

LinkClueE is based on three link-based pairwise similarity matrices, named *connected-triple-based similarity (CTS)*, *SimRank-based similarity* (SRS) and *approximate SRS* (ASRS) matrices, respectively (Iam-On et al. 2008). Both methods work on the same conjecture of taking into consideration as much information, embedded in an ensemble cluster, as possible when finding the similarity between data points. To discover similarity values, they consider both the associations among data points as well as those among clusters in the ensemble using link-based similarity measures (Jeh and Widom 2001; Calado et al. 2006; Klink et al. 2006; Sun et al. 2011; Iam-On and Boongoen 2013). The techniques used to create the cluster are *single linkage* (SL), *complete linkage* (CL) and *average linkage* (AL). The combination of the techniques to create the cluster and the link-based methods provide cluster results for the following nine groups:

- CTS-SL: connected-triple-based similarity and single linkage.
- CTS-CL: connected-triple-based similarity and complete linkage.
- CTS-AL: connected-triple-based similarity and average linkage.
- SRS-SL: SimRank-based similarity and single linkage.
- SRS-CL: SimRank-based similarity and complete linkage.
- SRS-AL: SimRank-based similarity and average linkage.
- ASRS-SL: approximate SimRank-based similarity and single linkage.
- ASRS-CL: approximate SimRank-based similarity and complete linkage.
- ASRS-AL: approximate SimRank-based similarity and average linkage.

We provide this second methodology to the same data set of monthly solar radiation data presented before. Twenty (20) base clustering ensembles are produced for each group of clustering ensembles and link-based methods. After obtaining the cluster ensemble, a *consensus matrix* is generated to acquire the ultimate data partition. The evaluation of the quality of the data partitioning is performed with Compactness (Nguyen and Caruana 2007); Davies–Bouldin (Davies and Bouldin 1979) and Dunn (Dunn 1974) measures. For the monthly GHI data set of Tanzania, we present next the best clustering result which is obtained with CTS-AL criteria. Figure 2 shows the partitioning results based on seven group of clusters for CTS-AL criteria applied on Tanzania monthly solar radiation data.

Fig. 2 Optimal clustering of solar regimes into seven broader areas with CTS-AL criterion

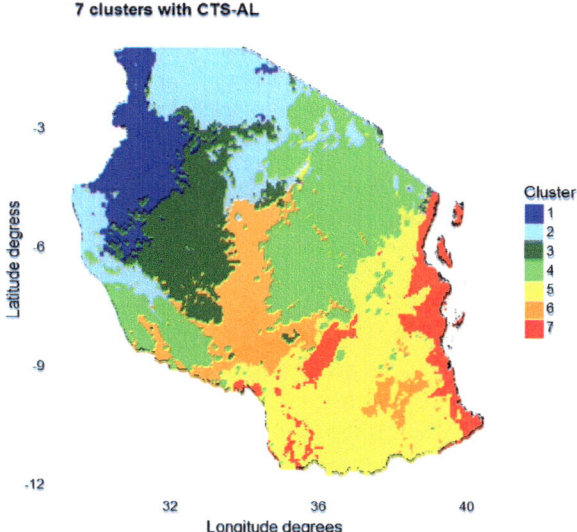

The results from the ensemble clustering methodology are similar to the selection based on LZM evaluation criteria and *k-means* presented in Sect. 2.1. A greater number of clusters provide regions with higher levels of disaggregation but less distinctiveness for the limits of our data set.

3   Methodology for Long-List Site Selection: GIS Limitation/Constraints Site Selection

The next step is the selection of a long list of candidate sites within each broad area for a solar monitoring network. In this section, we will apply the methodology to the results of Tanzania presented previously in Fig. 2.

The methodology presented here is based on GIS-MCDA (Multi-Criteria Decision Analysis) (Dodgson et al. 2009; Durbach and Stewart 2012) to obtain the long list of candidate sites for each broad area (Mendoza and Martins 2006; Wang et al. 2009; Estoque 2011; Huang et al. 2011; Velasquez and Hester 2013; Cinelli et al. 2014; Rikalovic et al. 2014). GIS-MCDA can be thought of as a process that transforms and combines geographical data and value judgments (the decision-maker's preferences) to obtain information for decision-making. The steps in the multi-criteria evaluation are the following:

1. Set the goal/define the problem.
2. Determine the criteria (factors/constraints).
3. Standardize the factors/criterion scores.

4. Determine the weight of each factor.
5. Aggregate the criteria to obtain the long list of site selection based on the best locations in each broad area.
6. Validate/verify the result to select the final shortlist of locations through ground truth verification (Sect. 4).

Using GIS-MCDA, each pixel from the satellite image or grid point from the numerical model, within each broad area, is ranked according to technical and logistic constraints and criteria. The objective is to maximize the levelized cost of energy (LCOE) for PV or/and CSP power plants like proximity to grid networks, distance to cities or places of consumption, access to fresh water and others. The main goal is to look up for suitable land for housing solar radiometric stations to develop CSP or/and PV solar power plants.

The *decision* is defined in GIS-MCDA as the choice between alternatives. In our case, the decision is the best locations among different site alternatives. *Criteria* are a set of guidelines or requirements used as the basis for a *decision*. There are two types of criteria: *factors* and *constraints*. A *factor* is a criterion that enhances or detracts from the suitability of a specific alternative for the application for solar energy. For example, distance to the road to maintain the radiometric station or develop the solar power plant where near is the most suitable and far is the least suitable. A constraint serves to limit the alternatives under consideration, elements that represent restrictions, land which is not preferred or considered unsuitable. The constraints are represented by a Boolean mask where 0 represents unsuitable areas and 1 suitable.

The combination of all the constraints is known as a decision rule which is a procedure by which the criteria are combined to arrive at a particular evaluation. The decision rule can be made through a function which provides a mathematical means for comparing alternatives with numerical exact decision rules or heuristic which specifies a procedure to be followed.

The constraints to be considered to make a GIS-MCDA analysis for the development of solar power plants include the following:

- Protected and environmentally sensitive areas.
- Rivers (or water lines).
- Lakes (or flooded land).
- Terrain slope higher than 2%.

These constraints can be described as either *avoidance* or *proximity* criteria. Other solar energy applications would need a different constraint criterion depending on the specific parameters which are needed to be considered for the development of the specific technology. We will see in the next chapters more solar energy applications where specific details on the constraints of each technology will be presented.

Figure 3 shows maps of avoidance criteria for Tanzania, where optimal areas are coloured in green and unfeasible areas in grey. Here, it should be noted that

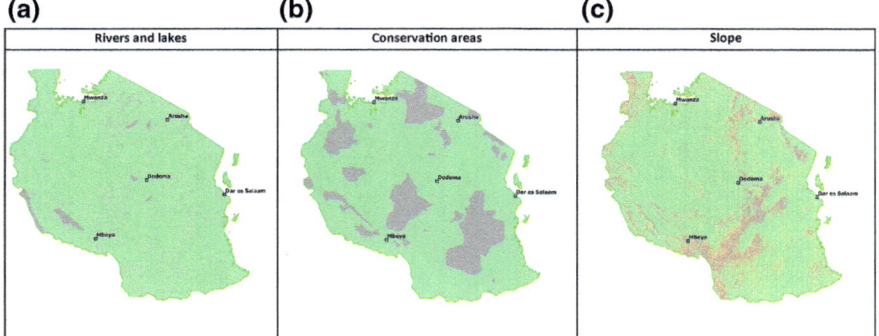

Fig. 3 Mapping of avoidance criteria. **a** rivers and lakes, **b** conservation areas and **c** slope

Fig. 4 Mapping of avoidance with combined criteria for Tanzania

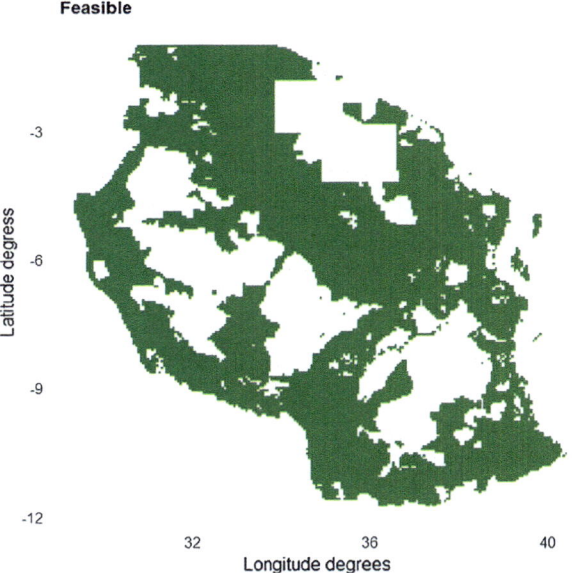

restrictions can vary from different country politics related to environment or safety policies. Figure 4 contains the final map with allowance areas in green.

The same procedure has been applied to the country of Qatar. Figure 5 shows the map of avoidance with the combination of multiple criteria, and Fig. 6 shows the final combined criteria, where optimal areas are coloured in green and unfeasible areas in red.

With regard to the factors or proximity criteria, we need to determine first the criteria, how much details are needed in the analysis (i.e. main roads vs. including minor roads) and standardize the factors to set the suitability values of the factors to

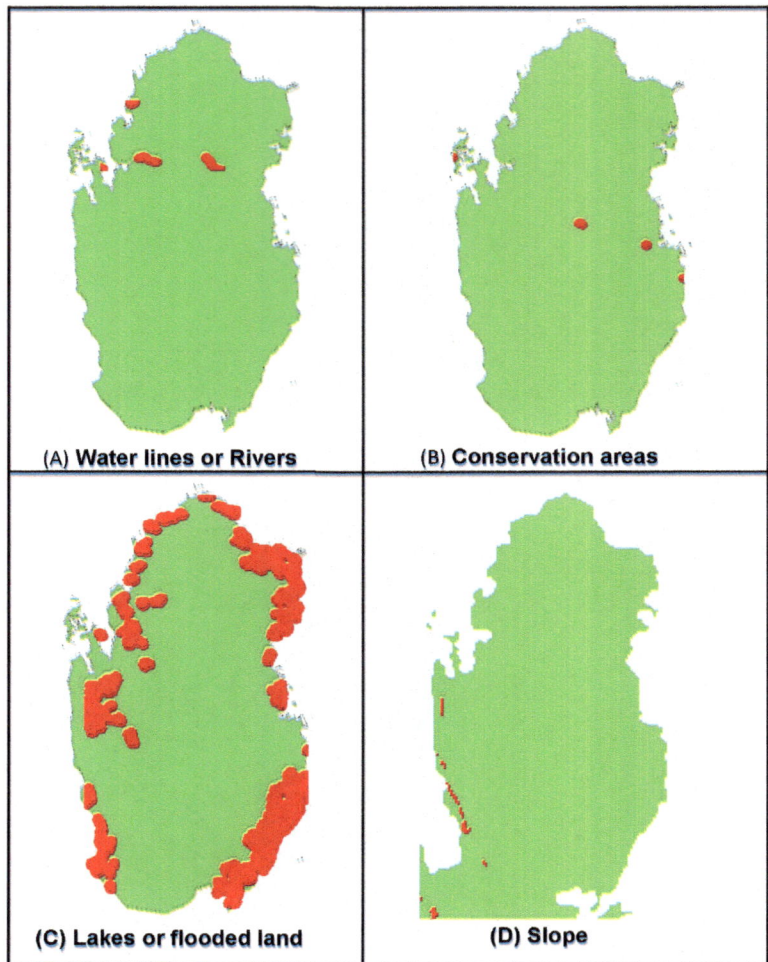

Fig. 5 Mapping of avoidance criteria. **a** rivers, **b** conservation areas, **c** lakes and **d** slope

a common scale. The main purpose is to make possible comparisons since, for example, elevation and slope are attributes which cannot be compared directly. All the factors need to be converted to a common range, i.e. 0–255 where 0 means least suitable and 255 most suitable; see Fig. 7. Other ways to standardize can be through linear, fuzzy or sigmoidal membership functions (Estoque 2011). The following factors can be considered for ease of access for commissioning and maintenance purposes and development of solar power plants:

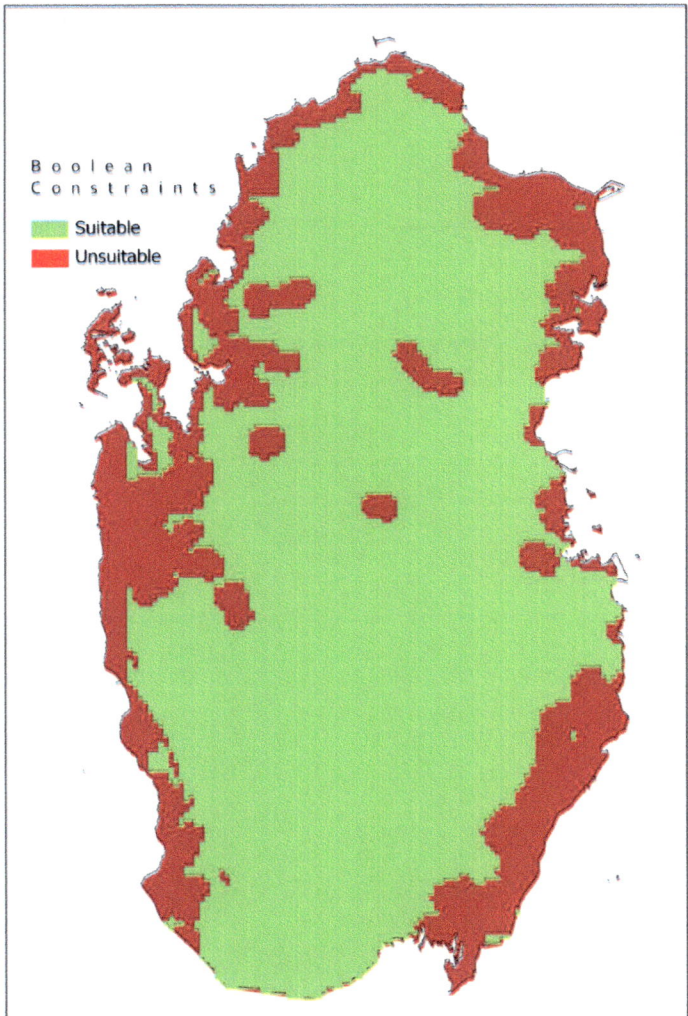

Fig. 6 Mapping of avoidance with combined criteria

- Transmission system.
- Cities/towns.
- Road networks.
- Railways network.

The factors and constraints are aggregated with the following weighted linear combination:

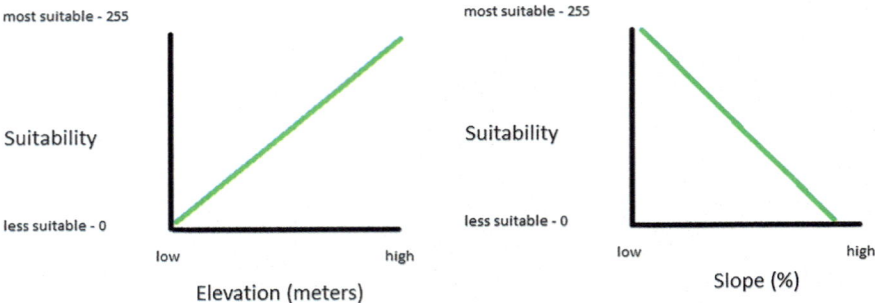

Fig. 7 Standardization of factor scores for elevation (m) and slope (%) (Estoque 2011)

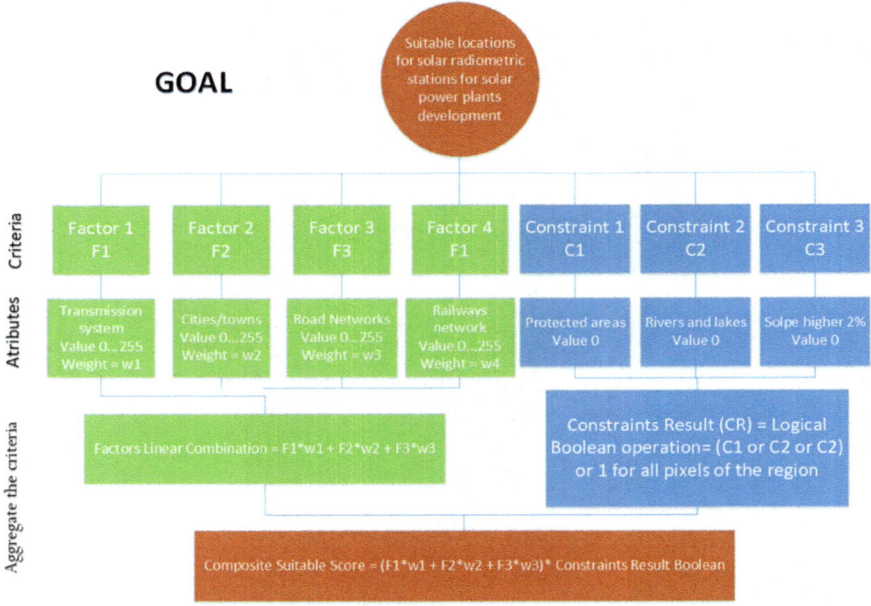

Fig. 8 Multi-criteria evaluation for the identification of best locations for the installation of radiometric stations

$$S = \sum w_i x_i x \qquad (10.1)$$

where S is the composite suitability score or final results which combines the factors and constraints, x_i is the factor scores (for each cell or pixel of the satellite), w_i is the weights assigned to each factor, c_j is the constraints expressed in Boolean form (as 0 or 1), $\sum$ is the sum of weighted factors, and $\prod$ is the Boolean logic product of constraints (1 suitable, 0 unsuitable) (Fig. 8).

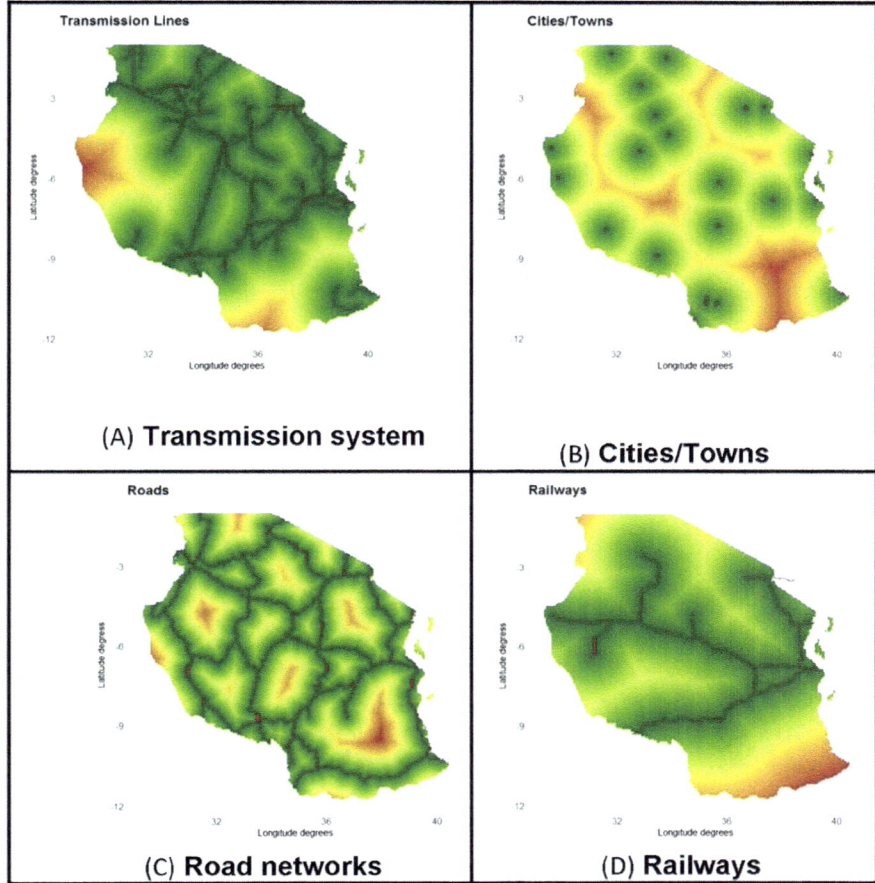

Fig. 9 Mapping of proximity criteria. **a** transmission system, **b** cities/towns, **c** road networks and **d** railways

Next figure summarizes the multi-criteria evaluation applied for the selection of radiometric station for the development of solar power plants:

Figure 9 shows maps proximity criteria for Tanzania, where optimal areas are coloured in green, unfeasibly areas in brown and intermediate suitable areas in yellow.

Figure 10 shows the map for Tanzania of proximity with the combination of MCDA constraints and factors, where optimal areas are coloured in green, unfeasible areas in red and intermediate areas in yellow. The weight of each factor is combined with the equal rating on a scale from 0 to 1 with an overall summation of 1 to have a normalized result.

Once the cluster analysis and the multi-criteria ranking methodology are completed, we are able to prepare a long list of feasible and candidate emplacements.

Fig. 10 Mapping of
proximity with combined
criteria

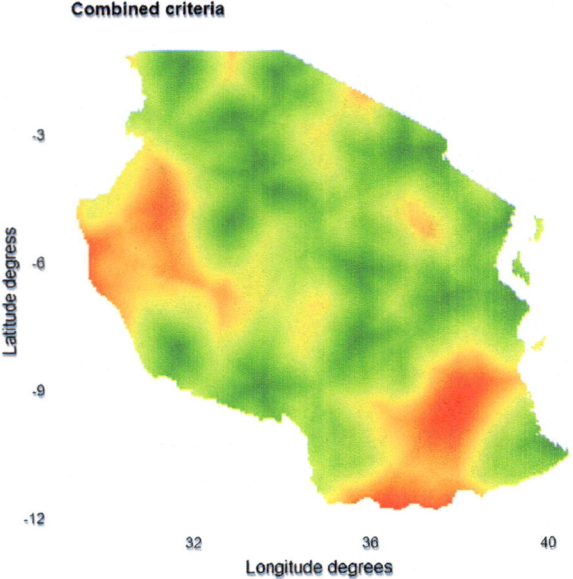

Each pixel of each cluster is ranked by means of the geographical information
methodology. It is imposed that the most ranked pixel for each cluster is included in
the long list. After that, the number of potential sites for each cluster is increased
proportionally to the final feasible area. Other criteria can be included in the long
list selection, such as economic prospection of the area, long-term solar resource,
future industrial investments and the population living in the area.

In the case of Tanzania, the seven (7) clusters identified and an approximation of
their sense is presented in Table 2—Characteristics of solar regimes found.

The long list obtained for Tanzania together with the clustered map obtained by
means of previously explained methodologies and criteria is shown in Fig. 11.

Table 2 Characteristics of solar regimes found

Cluster#	Colour code	Location
1	Dark blue	North West of the country, influenced by mountainous areas
2	Sky blue	North and West of the country, influenced by big lakes
3	Dark green	Center West of the country, wet savannah area
4	Orange	Center of the country, arid area
5	Light green	Center East of the country, arid savannah area and Kilimanjaro influence
6	Yellow	South of the country, jungle area
7	Red	East of the country, the Indian coast and island

Fig. 11 Locations of the selected areas and clusters

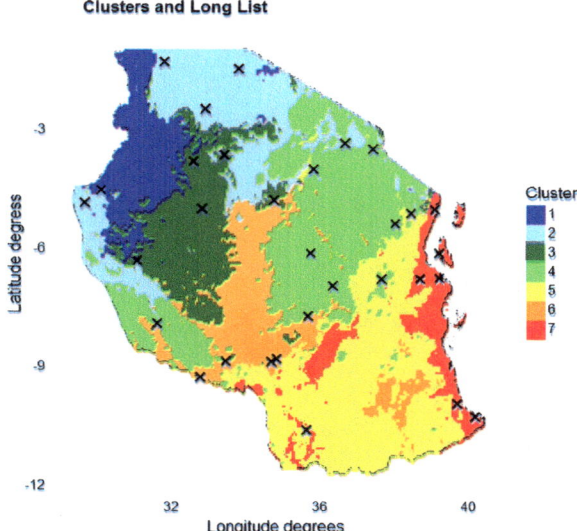

The ranking methodology applied categorizes the most representative solar regimes according to the monthly solar irradiation series (2000–2015). It has provided a sufficient understanding of the monitoring network required to characterize best the solar resource in Tanzania. Therefore, average values, inter-annual and intra-annual variations in solar irradiance have been considered up to this point besides economic and logistic criteria. Table 3 Long list of possible site locations contains the final long list identified for each cluster. It should be noted that the clusters 1 and 5 correspond to areas with largely protected zones and low levels of population, being proposed only one and two emplacements, respectively, in the long list. Finally, Fig. 12 shows the long list with a different colour by cluster, and the size of each point reflects the ranking obtained in the respective cluster.

4 Shortlist Solar Radiation Selection

The final aim of the long list is to dispose of enough alternatives that ensure that the measurement campaign reduces the uncertainty of solar resource knowledge and its characterization. As it has been seen in the previous section, at this point there is available a long list of interesting sites to install the measurements stations to collect the solar resource variability, both spatial and temporal besides accomplishing economic and logistic criteria. However, the normal situation is that, for different reasons, the long list is still too long to install one station for each location. So, a short list is required to propose the final network. In much more cases, the available budget will be the criteria that decide the number of final stations but as general one

Table 3 Longlist of possible site locations

Nearest city	Latitude (°)	Longitude (°)	Cluster	Ranking
Kasulu	-4.55	30.10	1	1
Mwanza	-2.50	32.90	2	1
Kigoma	-4.90	29.65	2	3
Musoma	-1.50	33.80	2	2
Bukoba	-1.30	31.80	2	4
Kahama	-3.85	32.60	3	4
Tabora	-5.05	32.80	3	1
Shinyanga	-3.70	33.40	3	3
Singida	-4.80	34.75	3	2
Mpanda	-6.35	31.10	3	5
Arusha	-3.40	36.70	4	2
Dodoma	-6.15	35.75	4	1
Morogoro	-6.80	37.65	4	4
Moshi	-3.50	37.40	4	3
Babati	-4.05	35.80	4	5
Handeni	-5.40	38.00	4	6
Mahinga	-7.00	36.35	4	7
Songea	-10.65	35.65	5	1
Korogwe	-5.15	38.45	5	2
Mbeya	-8.90	33.45	6	2
Sumbawanga	-7.95	31.60	6	6
Iringa	-7.75	35.70	6	1
Njombe	-8.90	34.70	6	4
Tunduma	-9.30	32.75	6	3
Makambako	-8.85	34.85	6	5
Dar es Salaam	-6.80	39.20	7	1
Tanga	-5.05	39.10	7	5
Zanzibar	-6.15	39.20	7	2
Kibaha	-6.80	38.70	7	4
Mtwara	-10.30	40.20	7	3
Lindi	-10.00	39.70	7	6

station by cluster should be installed. Furthermore, the first ranked emplacement for each cluster will be included in the final shortlist, and therefore, it will host a solar radiation station.

The different clusters to install the radiometric stations will have, at the same time, scale of priority. This scale should be based on the solar resource information and the local stakeholder's knowledge among other criteria. As a general recommendation, the next list presents the criteria to increase the final list of monitoring stations to be installed in each cluster:

Long list by cluster and rank

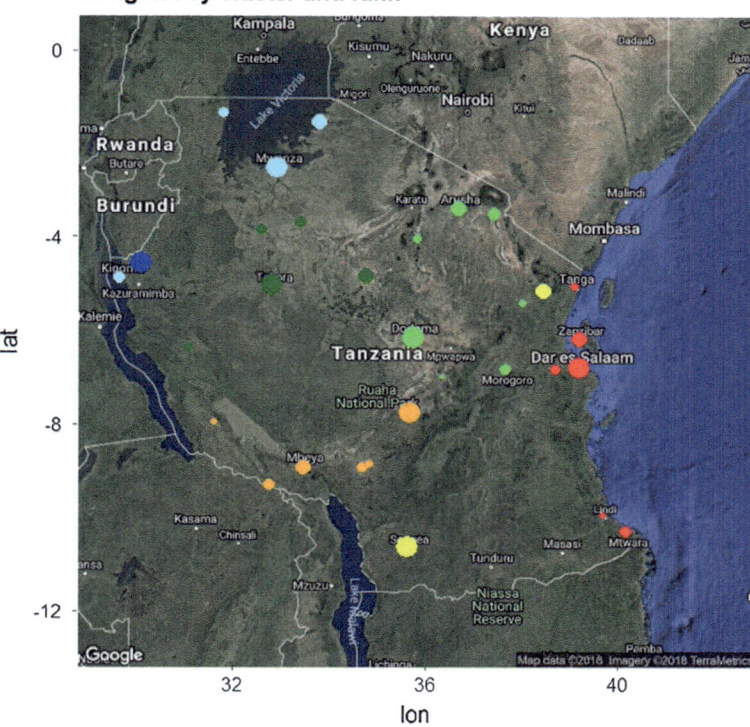

Fig. 12 Locations of the selected areas and clusters

1. Location with the largest level of solar resource.
2. Location with the largest rank of multi-criteria.
3. Cluster with the larger geographical area.
4. Cluster with the major population.

4.1 Selection of Network Host

For each one of the candidate sites in the shortlist, the potential feasibility of each location to host a radiometric station must be identified. We propose a ranking criterion based on the evaluation of the concrete emplacements after visiting them. The next table presents the criteria to be ranked. The total value of all the criteria will be used to rank each station and compared their preference within each cluster (Table 4).

Table 4 Criteria to rank the long list of possible site locations

Accessibility and safety	
Availability of maintenance personnel	
Mobile data connection availability	
Area of visibility surface, km^2	
Solar resource in the area, GHI component	

5 Type of the Measurement Stations to be Installed in Each Cluster

After selecting the emplacements, we have to choose the type of measurement stations to be deployed, in order to maximize the cost-benefit ratio of the monitoring network. A Tier 1 station is recommended for installation on the main sites, and for the remaining more isolated sites or with less potential for the development of solar energy, Tier 2 stations are recommended. A brief description of the characteristics of these stations is presented below:

5.1 Tier 1 Station

Tier 1 is a quality research station that provides the highest accuracy measurements, collecting data according to the protocols of the Baseline Surface Radiation Network (http://www.bsrn.awi.de/). A Tier 1 station's design may be characterized by:

- Solar tracker and shadow accessories for pyrheliometers and pyranometers. Sun tracker and shading disc kit.
- Two thermoelectric pyranometers Secondary Standard for measures of global (GHI) and diffuse (DHI) components.
- A thermoelectric pyrheliometer first class for DNI measurements.
- Titled thermopile or/and PV sensor.

The Tier 1 station also may include a weather station with temperature, humidity, barometric pressure, wind speed and direction sensors. The quality and variety of types of data collected by this type of station make it a valuable tool for many different kinds of scientific researches. Such equipment requires trained personnel readily available to operate and maintain properly the station daily.

5.2 Tier 2 Station

Tier 2 is a lower quality station to measure GHI and GTI. The maintenance of this station is done every 1 or 2 weeks. The equipment included in these stations should be the following:

- One thermoelectric pyranometer first class.
- One PV cell to measure GHI.
- Titled thermopile or/and PV sensor.

There are should be several common elements for both Tier 1 and Tier 2 stations, principally data acquisition systems, remote communication, and autonomy regarding solar panels and batteries for electricity supply and meteorological sensors. The specifications for such common elements are as follows:

- Remote Automatic Data Acquisition and Transmission Unit. Communications Add-ins (GPRS, Ethernet, satellite Internet).
- Lightning protection.
- Solar PV panel PV peak power.
- Meteorological tower at 10 m of height (3 m for Tier 2).
- Wind speed and direction sensors.
- Relative humidity and air temperature sensors.
- Barometric pressure sensor.

6 Example of a Radiometric Network to Map Solar Radiation and for Different Applications

In this section, we present examples of different examples around the world and the radiometric network of India used to map the resources of the country ((C-WET) 2012; Kumar et al. 2013; Schwandt et al. 2014).

Geónica S.A. (Spain) and Indian Distributor, SGS Weather, supplied, installed and maintained a network of measuring instruments consisting of >124 (Phase-1, Phase-2 and MEDA) remote measurement stations for solar radiation resource assessment in India. The network is used by National Institute of Wind Energy Technology (NIWE) (formerly C-WET), an autonomous R&D institution under the India Ministry of New and Renewable Energy. The purpose of the network is to generate the "solar map" of India. Solar maps, showing real data of yearly solar radiation levels, are used for the design of solar power plants. Detailed historical data series are made available by NIWE as a commercial service.

The network continuously monitors and stores nationwide solar radiation and weather parameters and communicates these to a Data Receiving Center (DRC) with redundancy located at C-WET Headquarters in Chennai. Each measurement station is equipped with high accuracy meteorological sensors. More specifically, every station includes a METEODATA Data Logger/Controller, a SunTracker-3000 and several solar radiation sensors, such as a pyrheliometer mounted on the solar tracker, and two pyranometers (one shaded for the measurement of the diffuse radiation). Data is transferred via GPRS cellular network to the Data Receiving Center for analysis and final archiving. In Phase-2, there are also four "advanced" stations measuring albedo irradiance, far infrared irradiance (pyrgeometer) and AOD. Real-time data is also available on the Internet by means of GEONICA WEBTRANS Ubiquitas Internet Platform.

Installation and commissioning of the Phase-2 and MEDA were carried out on September 2013–February 2014.

Phase-1 was completed on November 2011; see Figs. 13 and 14.

Fig. 13 National Institute of Wind Energy Technology (NIWE) ground-based stations network (from Geonica) (Phase-1, Phase-2 and MEDA). Courtesy of Geonica

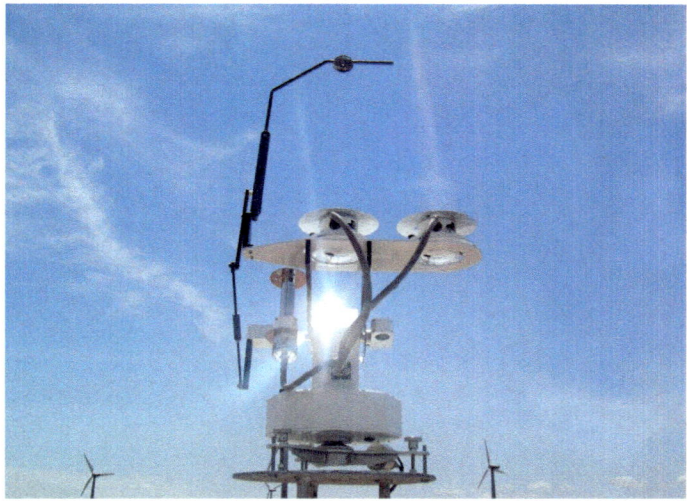

Fig. 14 Ministry of New Renewable Energy (MNRE), India, C-WET project solar national. Network-4LR. Courtesy of Geonica ©

Each one of the stations of Phase-1 and Phase-2 (red points in the map) is measuring:

- Global Horizontal Irradiance (GHI).
- Direct Normal Irradiance (DNI).
- Diffuse Horizontal Irradiance (DHI).
- Wind speed and direction.
- Ambient temperature and relative humidity.
- Atmospheric pressure.

Each one of the four "advanced" radiation stations (blue points in the map) is measuring:

- Albedo irradiance (albedometer).
- Far infrared radiation (pyrgeometer).
- AOD (sunphotometer).

The next figures present examples of stations around the world in different applications: (Figs. 15, 16, 17, 18, 19 and 20).

Fig. 15 Solar meteo station with sun tracker. Courtesy of Geonica ©

Fig. 16 Radiometric station developed in Armenia for the project utility-scale solar power project from the world bank. Courtesy of Geonica ©

Fig. 17 General view of a radiometric (from) and meteorological station (back) in a remote location working autonomously with power supplied from two PV panels. Courtesy of Geonica ©

Fig. 18 Detailed view of a radiometric (from) and meteorological station (back) in a remote location working autonomously with power supplied from two PV panels. Courtesy of Geonica ©

Fig. 19 Sunpower system real-time monitoring at CPV plant. Courtesy of Geonica ©

Fig. 20 SunTracker-3000 at concentrated solar power (CSP) plant. Courtesy of Geonica ©

7   Conclusions

This chapter has presented a complete methodology to design a network of solar radiation stations for a geographical area. The solar resource variability both in spatial and temporal point of view is considered to identify the solar regimes presented in the country. The geographical information is used to evaluate the feasibility of each pixel producing a ranking among them that allows the identification of a long list of potential candidates to install the stations. Besides, it has presented the criteria to reduce the initial long list of candidate stations to a feasible short list of stations.

The procedure has been shown by using data coming from the ESMAP project of Tanzania where authors participated but it can be applied to any geographical area of interest.

References

(C-WET) (2012) Summary of monthly values of solar radiation and meteorological parameters. 1–2

Bank TW (2015) Solar resource mapping in Tanzania : site identification report. The World Bank

Bernardos A, Gaston M, Fernandez-Peruchena C, et al (2015) Solar resource mapping in Tanzania : solar modelling report. The World Bank

Calado P, Cristo M, Gonçalves MA et al (2006) Link-based similarity measures for the classification of web documents. J Am Soc Inf Sci Technol 57:208–221. https://doi.org/10.1002/asi.20266

Cinelli M, Coles SR, Kirwan K (2014) Analysis of the potentials of multi criteria decision analysis methods to conduct sustainability assessment. Ecol Indic

David G, Averbuch A (2012) SpectralCAT: Categorical spectral clustering of numerical and nominal data. Pattern Recognit 45:416–433. https://doi.org/10.1016/j.patcog.2011.07.006

Davies DL, Bouldin DW (1979) A cluster separation measure. IEEE Trans Pattern Anal Mach Intell 1:224–227. https://doi.org/10.1109/TPAMI.1979.4766909

Ding C, He X (2004) K -means clustering via principal component analysis. In: Proceedings of the twentyfirst international conference on machine learning ICML 04 Cl:29. https://doi.org/10.1145/1015330.1015408

Dodgson JS, Spackman M, Pearman A, Phillips LD (2009) Multi-criteria analysis : a manual

Dunn JC (1974) Well-separated clusters and optimal fuzzy partitions. J Cybern 4:95–104. https://doi.org/10.1080/01969727408546059

Durbach IN, Stewart TJ (2012) Modeling uncertainty in multi-criteria decision analysis. Eur. J. Oper. Res

Estoque RC (2011) GIS-based multi-criteria decision analysis. Geogr Compass. https://doi.org/10.1111/j.1749-8198.2011.00431.x

Finley T, Joachims T (2008) Supervised k-means clustering. Learning

Fred ALN, Jain AK (2005) Combining multiple clusterings using evidence accumulation. IEEE Trans Pattern Anal Mach Intell 27:835–850. https://doi.org/10.1109/TPAMI.2005.113

Goutte C, Toft P, Rostrup E et al (1999) On clustering fMRI time series. Neuroimage 9:298–310. https://doi.org/10.1006/nimg.1998.0391

Gueymard CA, Myers DR (2009) Evaluation of conventional and high-performance routine solar radiation measurements for improved solar resource, climatological trends, and radiative modeling. Sol Energy 83:171–185. https://doi.org/10.1016/j.solener.2008.07.015

Hartigan JA, Wong MA (1979) A k-means clustering algorithm. Appl Stat 28:100–108. https://doi.org/10.2307/2346830

Hegner H, Müller G, Nespor V, et al (1998) Update of the technical plan for BSRN data. 3:38

Hornik K (2005) Cluster ensembles. In: Studies in classification, data analysis, and knowledge organization, pp 65–72

Huang IB, Keisler J, Linkov I (2011) Multi-criteria decision analysis in environmental sciences: ten years of applications and trends. Sci. Total Environ

Iam-On N, Boongoen T (2013) Revisiting link-based cluster ensembles for microarray data classification. In: Proceedings—2013 IEEE international conference on systems, man, and cybernetics, SMC 2013. pp 4543–4548

Iam-On N, Boongoen T, Garrett S (2008) Refining pairwise similarity matrix for cluster ensemble problem with cluster relations. In: Lecture notes in computer science (including subseries Lecture Notes in artificial intelligence and lecture notes in bioinformatics). pp 222–233

Iam-on N, Garrett S (2010) LinkCluE: a matlab package for link-based. J Stat Softw 36:1–36

Jain A, Law M (2005) Data Clustering: A User's Dilemma. Pattern Recognit. Mach. Intell. 1–10

Jain AK (2010) Data clustering: 50 years beyond K-means. Pattern Recognit Lett 31:651–666. https://doi.org/10.1016/j.patrec.2009.09.011

Jeh G, Widom J (2001) SimRank : a measure of structural-context similarity. In: Proceedings of the eighth ACM SIGKDD international conference on knowledge discovery and data mining. 1–11. https://doi.org/10.1145/775047.775126

Kaufman L, Rousseeuw P (1990) Finding groups in data: an introduction to cluster analysis, Wiley-Interscience

Klink S, Reuther P, Weber A et al (2006) Analysing social networks within bibliographical data. Database Expert Syst Appl 234–243. https://doi.org/10.1007/11827405_23

König-Langlo G, Sieger R, Schmithüsen H et al (2013) Baseline surface radiation network (BSRN) update of the technical plan for BSRN data management October 2013. 30

Kottek M, Grieser J, Beck C et al (2006) World map of the Köppen-Geiger climate classification updated. Meteorol Zeitschrift 15:259–263. https://doi.org/10.1127/0941-2948/2006/0130

Kumar A, Gomathinayagam S, Giridhar G et al (2013) Field experiences with the operation of solar radiation resource assessment stations in India. In: Energy Procedia

Kumar J, Mills RT, Hoffman FM, Hargrove WW (2011) Parallel k-means clustering for quantitative ecoregion delineation using large data sets. Procedia Comput Sci 4:1602–1611. https://doi.org/10.1016/j.procs.2011.04.173

Mendoza GA, Martins H (2006) Multi-criteria decision analysis in natural resource management: a critical review of methods and new modelling paradigms. For Ecol Manage 230:1–22. https://doi.org/10.1016/j.foreco.2006.03.023

Nguyen N, Caruana R (2007) Consensus clusterings. In: Proceedings - IEEE international conference on data mining, ICDM. pp 607–612

Perez R, Seals R, Zelenka A (1997) Comparing satellite remote sensing and ground network measurements for the production of site/time specific irradiance data. Sol Energy 60:89–96. https://doi.org/10.1016/S0038-092X(96)00162-4

Polo J, Bernardos A, Navarro AA et al (2015a) Solar resources and power potential mapping in Vietnam using satellite-derived and GIS-based information. Energy Convers Manag 98:348–358. https://doi.org/10.1016/j.enconman.2015.04.016

Polo J, Gastón M, Vindel JM, Pagola I (2015b) Spatial variability and clustering of global solar irradiation in Vietnam from sunshine duration measurements. Renew Sustain Energy Rev 42:1326–1334. https://doi.org/10.1016/j.rser.2014.11.014

Posselt R, Mueller RW, Stöckli R, Trentmann J (2012) Remote sensing of solar surface radiation for climate monitoring—the CM-SAF retrieval in international comparison. Remote Sens Environ 118:186–198. https://doi.org/10.1016/j.rse.2011.11.016

Rikalovic A, Cosic I, Lazarevic D (2014) GIS based multi-criteria analysis for industrial site selection. In: Procedia engineering

Schwandt M, Chhatbar K, Meyer R, et al (2014) Quality check procedures and statistics for the Indian SRRA solar radiation measurement network. In: Energy procedia

Sun L, Cheng R, Cheung DW, Han J (2011) On Link-based similarity join. Vldb 714–725

Velasquez M, Hester PT (2013) An analysis of multi-criteria decision making methods. Int J Oper Res. https://doi.org/10.1007/978-3-319-12586-2

Vindel JM, Valenzuela RX, Navarro AA, Zarzalejo LF (2018) Methodology for optimizing a photosynthetically active radiation monitoring network from satellite-derived estimations: a case study over mainland Spain. Atmos Res 212:227–239. https://doi.org/10.1016/j.atmosres.2018.05.010

Vrahatis MN, Boutsinas B, Alevizos P, Pavlides G (2002) The new k-windows algorithm for improving thek -means clustering algorithm. J Complex 18:375–391. http://dx.doi.org/10.1006/jcom.2001.0633

Wang JJ, Jing YY, Zhang CF, Zhao JH (2009) Review on multi-criteria decision analysis aid in sustainable energy decision-making. Renew Sustain Energy Rev 13:2263–2278. https://doi.org/10.1016/j.rser.2009.06.021

Wu X, Zhu X, Wu G-Q, Ding W (2014) Data mining with big data. IEEE Trans Knowl Data Eng 26:97–107. https://doi.org/10.1109/TKDE.2013.109

Yang C, Wan B, Gao X (2006) Effectivity of internal validation techniques for gene clustering. Lect Notes Comput Sci 4345:49

Yang D, Reindl T (2015) Solar irradiance monitoring network design using the variance quadtree algorithm. Renewables Wind Water, Sol 2:1–8. https://doi.org/10.1186/s40807-014-0001-x

Zagouras A, Kazantzidis A, Nikitidou E, Argiriou AA (2013) Determination of measuring sites for solar irradiance, based on cluster analysis of satellite-derived cloud estimations. Sol Energy 97:1–11. https://doi.org/10.1016/j.solener.2013.08.005

Zelenka A, Perez R, Seals R, Renne D (1999) Effective accuracy of satellite-derived hourly irradiances. Theor Appl Climatol 62:199–207

Zelnik-manor L, Zelnik-manor L, Perona P, Perona P (2004) Self-tuning spectral clustering. Adv Neural Inf Process Syst 17(2):1601–1608. https://doi.org/10.1184/7940

Chapter 11
Solar Power Plant Performance

Jesús Polo

Abstract A key part in the development of any project for deployment a solar power plant is the analysis of the expected energy yield production. The system energy production depends on the plant design, the technology used for power conversion, the solar resource, and the characteristics of the site. Due to the intrinsic variability of the solar resource, the prediction of long-term electricity production is also crucial for the financial evaluation of solar power plants. The energy yield performance is thus the process of predicting the annual average energy output for the lifetime of the solar power plant. For that purpose, a number of system performance models and tools have been developed; many of them are updated regularly. In addition, several international programs deliver recommendations and guidelines for yield performance analysis. Thus, in the case of photovoltaic (PV) plants the PVPS program from the International Energy Agency (IEA) publishes regularly updated reports on many aspects of PV generation (http://www.iea-pvps.org/). In addition, the Sandia National Laboratories is facilitating a collaborative group called PV Performance Modeling Collaborative (PVPMC) with regular activities focused on improving the accuracy of PV performance analysis (https://pvpmc.sandia.gov/). On the other hand, in the case of Concentrated Solar Power (CSP), the SolarPACES program of the IEA (http://www.solarpaces.org/) is developing guidelines for solar thermal energy (STE) yield assessment. This chapter summarizes the main aspects included in the tools and software for estimating yield performance of PV and CSP power plants and the long-term characterization of yield energy, risk analysis, and uncertainty quantification.

J. Polo (✉)
Photovoltaic Solar Energy Unit, Renewable Energy Division (Energy Department),
CIEMAT, Avda Computense 40, 28040 Madrid, Spain
e-mail: jesus.polo@ciemat.es

© Springer Nature Switzerland AG 2019
J. Polo et al. (eds.), *Solar Resources Mapping*, Green Energy and Technology,
https://doi.org/10.1007/978-3-319-97484-2_11

1 Tools and Models to Simulate Energy Yield

1.1 *Photovoltaic Performance*

1.1.1 Modeling

There are several solar PV modeling software packages available on the market. PVsyst has become probably the most widely used model for PV performance analysis (Mermoud and Wittmer 2014). However, there are additional open-source software and no cost models that are becoming also standards in PV performance modeling. The two most well known and used are the Sandia Array performance Model (SAPM) which is included in the PV-Lib package and the System Advisor Model (SAM). According to PVPMC, the estimation of the performance of any PV system or plant involves several steps that are illustrated in Fig. 1 (Stein and Farnung 2017).

The PV system design parameters include the site characteristics (latitude, longitude, and elevation above sea level) and the system parameters. The latter basically refers to module and inverter technical data, array configuration (number of strings and number of modules per string), tilt and azimuth angles or tracking details, and albedo and shading information.

Fig. 1 Sequence of standard PV modeling steps

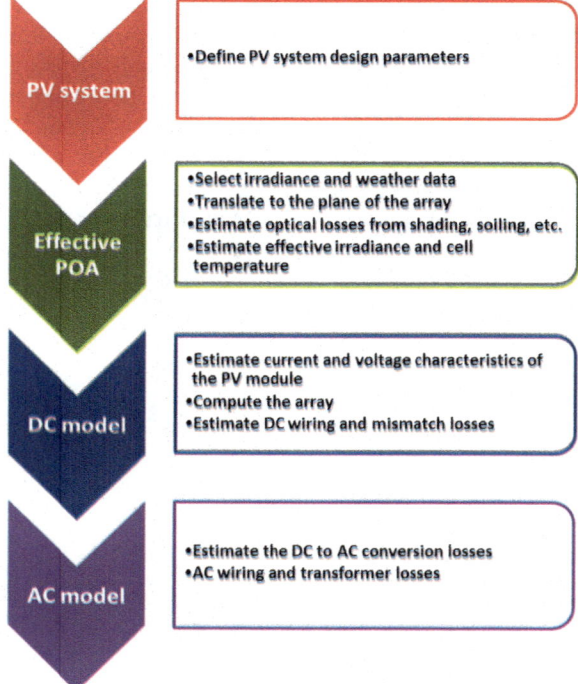

PV system

• Define PV system design parameters

Effective POA

• Select irradiance and weather data
• Translate to the plane of the array
• Estimate optical losses from shading, soiling, etc.
• Estimate effective irradiance and cell temperature

DC model

• Estimate current and voltage characteristics of the PV module
• Compute the array
• Estimate DC wiring and mismatch losses

AC model

• Estimate the DC to AC conversion losses
• AC wiring and transformer losses

The effective irradiance at the plane of the array (POA) is the incident radiation on the PV cells that can be directly converted into electrical current and can be estimated from the incident solar irradiance taking into account the different optical losses. POA irradiance can be measured by a thermopile pyranometer or a calibrated cell placed at the same plane of the PV array. In many situations, POA irradiance is not measured and must be estimated from the other components of the solar radiation: direct normal irradiance (DNI), global horizontal irradiance (GHI), and diffuse horizontal irradiance (DHI). These components are generally available from a complete measuring station or, in most situations, must be derived from modeling (mainly from satellite imagery). There are several models for estimating POA irradiance from the other components, generically known by the name of "Transposition Models" (Padovan and Del Col 2010). A thorough classical review can be found elsewhere (Muneer and Saluja 1985). The incident global irradiance on a tilted surface is generally expressed in all models as the sum of three contributions: the projection of the direct solar irradiance onto the surface, the sky diffuse component, and the ground-reflected component. The direct can be computed straightforwardly from direct normal irradiance (DNI) and the angle of incidence of the sun. Therefore, the main differences among the models are mostly based on the isotropic or anisotropic approaches for computing the diffuse and reflected components (Demain et al. 2013). Furthermore, nearly most of the models use the isotropic approach for the ground-reflected component, which depends on the global horizontal irradiance, the ground albedo, and the surface tilt angle (Loutzenhiser et al. 2007), so transposition models vary primarily in the estimation of the sky diffuse irradiance. The accuracy of transposition models has been extensively tested in many different climatic conditions so far (Kambezidis et al. 1994; Mefti et al. 2003; Cucumo et al. 2007; Pandey and Katiyar 2011; Demain et al. 2013; Lee et al. 2013; Khalil and Shaffie 2013; Khorasanizadeh et al. 2014; Mohammadi and Khorasanizadeh 2015; Wattan and Janjai 2016).

The models for calculating the DC current and voltage of a PV system can be classified as equivalent circuit diode models, semi-empirical models, and simple efficiency models (Cameron et al. 2011).

The diode equivalent circuit model represents the solar cell or PV module by an electrical circuit Fig. 2. The single-diode circuit is one of the most widely used analogy to model the performance of a PV module, and the five parameter equation for solving the circuit is described by (De Soto et al. 2006),

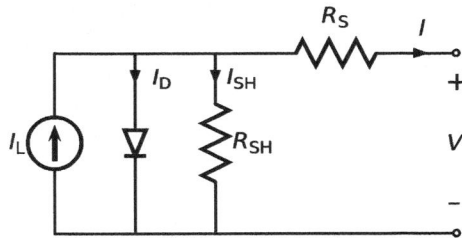

Fig. 2 Single-diode equivalent circuit of a solar cell

$$I = I_L - I_D \left[\exp\left(\frac{V + IR_S}{a}\right) - 1 \right] - \frac{V + IR_S}{R_{SH}} \tag{1}$$

This model solves the current I and voltage V of the circuit by using five module parameters $(I_L, I_D, a, R_S, \text{and} R_{SH})$. However, it is also common to find more sophisticated versions that include additional diodes and thus more parameters (Elbaset et al. 2014). Nevertheless, the single-diode five parameter is likely the most widely used model by performance tools like PVSyst or SAM (Mermoud and Wittmer 2014; Gilman 2015). Since Eq. (1) is a nonlinear function of the current and voltage, the solution is not always easy and there are several approximations proposed to reduce the parameters or solving algorithms based on the Lambert W-function (Jain 2004; Ghani et al. 2014; Mares et al. 2015; Ayodele et al. 2016; Sudhakar Babu et al. 2016; Et-Torabi et al. 2017).

One of the most well-known semi-empirical models for PV performance is the Sandia Photovoltaic Array performance (King et al. 2004). Sandia PV Array Performance Model (SAPM) is versatile and accurate for almost all PV technologies (Peng et al. 2015). The model is available from PV Performance Modeling Collaborative (PVPMC, https://pvpmc.sandia.gov/) initiative facilitated by Sandia National Laboratories (King et al. 2004, 2007), and it is also implemented in NREL's SAM software. The SAPM model consists basically of modeling five points of the I-V curve, and it requires also many additional empirical parameters for modeling the different losses and temperature effects. The model is basically formed by the following equations (King et al. 2004):

$$I_{mp} = I_{mp0}\left(C_0 E_e + C_1 E_e^2\right)\left(1 + \alpha_{mp}(T_c - T_0)\right) \tag{2}$$

$$I_{sc} = I_{sc0} f(AM_a) E_e (1 + \alpha_{sc}(T_c - T_0)) \tag{3}$$

$$V_{mp} = V_{mp0} + C_2 N_s \delta(T_c) \ln(E_e) + C_3 N_s \delta(T_c) \ln(E_e)^2 + \beta_{mp}(T_c - T_0) \tag{4}$$

$$V_{oc} = V_{oc0} + N_s \delta(T_c) \ln(E_e) + \beta_{oc}(T_c - T_0) \tag{5}$$

$$\delta(T_c) = nk\frac{(T_c + 273.15)}{q} \tag{6}$$

where I_{mp} is the current at the maximum power point, I_{sc} is the short-circuit current, V_{mp} is the voltage at the maximum power point, V_{oc} is the open circuit voltage, E_e is the effective irradiance defined as the fraction of total irradiance incident on the module to which the cells inside actually respond, T_c is the PV cell temperature inside the module, α is the temperature coefficient of current, β the temperature coefficient of voltage, $f(AM_a)$ is the empirical function of absolute air mass, N_s is the number of solar cells in series, k is the Boltzmann's constant, n the empirical diode factor, and q is the elementary charge. C_i are empirical parameters and subscript 0 indicates STC. AC power is estimated from DC power by an empirical function (King et al. 2007). The main limitation of the SAPM model is the need of many empirical parameters as input to the model. Nevertheless, several procedures have been proposed in the literature to obtain all the empirical coefficients that SAPM model requires (Peng et al. 2015; King et al. 2016).

DC wiring and mismatch losses are usually modeled on performance tools by derating factors. Thus, for instance, in SAM the DC losses are computed by the product of several loss factors that account for wiring, diode connections, wiring, tracking error, and so on that result in a net DC loss factor that multiplies the DC power.

Finally, an inverter sub-model is normally used to convert the array's net DC power output to AC power output. Power losses usually occur in this process and should be taken into account for modeling the PV performance. The Sandia inverter model proposes an empirical expression to estimate the AC power output as a function of DC input power and voltage and of DC power and the voltage at reference conditions (King et al. 2007). The Sandia model has several empirical parameters that depend on the technical characteristics of the inverter and conse-quently, PV-Lib, SAM, and other tools usually have a database of inverters that includes all technical and empirical parameters. It is very common implementation of maximum power point tracking (MPPT) algorithms in the inverters. These methods usually control the operating voltage to reach the maximum power, which varies with irradiance and temperature. Similarly to other losses, most PV perfor-mance models assume MPPT losses as a derate factor. For instance, SAM models a power clipping that happens when the array operating voltage exceeds the inverter rated MPPT limits. Moreover, in future versions of SAM, it is planned to incor-porate multi-MPPT inverters modeling.

In PV, performance analysis is very frequent to estimate the energy production or system yield together with the performance ratio (PR). The PR estimates the overall efficiency of energy conversion (Richter et al. 2015). The final energy yield (Y_f) is defined as the ratio between the produced energy and the nominal system power, and the PR is then defined as the ratio between the final system yield and the reference yield (Y_{ref}). The reference yield is derived as the quotient between the incoming solar irradiation over the period and the irradiation at Standard Test Conditions (STCs) that usually is 1000 W m^{-2}.

$$\mathrm{PR} = \frac{Y_f}{Y_{ref}} \qquad (7)$$

1.2 CSP Performance Modeling

The main deployed CSP technologies are parabolic trough, central receiver systems also known as solar towers, and linear Fresnel collectors. Many of these plants integrate a thermal storage system (TES) based on molten salts whose capacity is usually expressed in terms of the time (hours) of the plant working without solar radiation incoming to the collectors. Notwithstanding parabolic trough systems are currently the dominating technology in CSP plants, the interest in solar power plants is increasing due to the capacity to achieve higher temperatures and thus higher efficiencies. Two main schemes have emerged in solar tower technologies: solar towers that use molten salt as heat transfer fluid having thus a TES integrated, and solar towers with direct steam generation without any TES capacity.

The typical approach to model the energy production of a CSP plant includes four main components: solar collectors that concentrate solar radiation, receivers to convert solar energy into thermal energy, the TES to store thermal energy, and the power block to convert thermal energy into electricity. The first two components are normally referred to as the solar field. The most widely used tools for performance modeling and pre-feasibility studies of CSP plants are SAM (NREL, https://sam. nrel.gov/) and Greenius (DLR, http://freegreenius.dlr.de/). Two kinds of input are required to model a CSP plant performance: the solar resource and meteorological input, and the plant technical specification input. The latter basically consists of all the parameters for determining the solar collector and receiver data that will determine the solar field, heat transfer fluid type, and working temperatures and pressure, and power block. SAM, for instance, includes specific models for parabolic trough, molten salts, direct steam solar towers, and linear Fresnel plants with component libraries providing default values for most main technical parameters (Wagner and Gilman 2011; Blair et al. 2014; Dobos et al. 2014). In SAM, the solar plant is represented by interconnected components (solar collectors, receivers, heat exchangers, piping, TES, and power cycles) that are modeled iteratively for matching pressures, mass flow rates, and temperatures at the interfaces. The first versions of SAM were based on the TRNSYS kernel, but it was changed to a specific Transient Component Simulation (TCS) kernel which offers improvements and higher performance simulations (Dobos et al. 2014). The regular updates and improvements made by NREL in SAM software are making this tool a standard for CSP yield analysis (Wagner 2008; Wagner and Gilman 2011; Wagner and Zhu 2012).

The complexity of modeling CSP plants and the detailed information needed to vary through the different phases in a CSP project (pre-feasibility, feasibility studies, project development, due diligence, etc.). In order to standardize the methodology for CSP plant modeling and performance studies, the SolarPACES program Task I started in 2009 with the project guiSmo as an ongoing work for creating guidelines for CSP yield assessment studies, and a first document was delivered in January 2017 (Hirsch et al. 2017).

2 Typical Meteorological Year and Typical Yield Year

The tools for modeling the performance of any solar system require solar irradiance and meteorological data as input. The most common procedures so far consist of using time series of the main variables (solar irradiance components and other weather variables) for one year in hourly basis that represents the long-term meteorological conditions expected for the target site. A year of hourly values of solar irradiance components and some other meteorological variables are often referred to as Typical Meteorological Year (TMY), and performance tools like PVSyst or SAM include the option to read common formats of this input files. Therefore, yield analysis of PV and CSP plants assume that modeling a TMY will represent the long-term performance of the plant.

The first TMY methodology was developed by Sandia National Laboratories to provide representative data for solar heating and cooling applications for over 200 sites in the USA (Hall et al. 1978). Later versions, TMY2, used data from 30 years and extended the number of sites (Marion and Urban 1985). The latest version, TMY3, was designed to maximize both the number of stations and the number of years (Wilcox and Marion 2008). The main variables involved in TMY3 include three components of solar irradiance (GHI, DNI, and DHI), dry-bulb and dew-point temperatures, relative humidity, pressure, wind speed and direction, and albedo.

In general, a TMY is constructed by the concatenation of 12 months selected from the individual years and sorted to form a complete year of data. Figure 3 illustrates, with intensity plots, the TMY concept with an example of a TMY generated from over 30 years of hourly values of solar irradiance. The Sandia methodology uses the Finkelstein–Schafer (FS) statistic to select a representative month (Finkelstein and Schafer 1971). The FS statistic for a specific meteorological variable is given by,

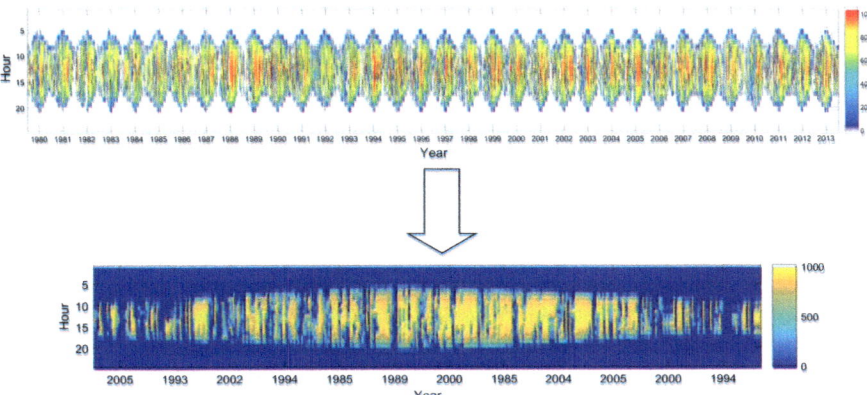

Fig. 3 Intensity plots illustrating the TMY generation procedure

$$FS = \frac{1}{N} \sum_{i=1}^{N} \left| CDF_m(d_i) - CDF_{y,m}(d_i) \right| \tag{8}$$

where CDF_m is the long-term cumulative distribution function of daily values, d_i, for month m, $CDF_{y,m}$ is the short-term (corresponding to year y) cumulative distribution function of daily values, and N is the number of bins or days, usually considered as 31. There are other proposals elsewhere which use different methodologies for the statistical characterization of the long-term (Festa and Ratto 1993; Lund 1995).

For each parameter, the representative or candidate month is the month with minimum FS value and it represents the long-term. Since there are several variables involved, the final long-term month is obtained from a weighted sum of the FS of all the variables,

$$WS = \sum w_i FS_i \tag{9}$$

where w_i is the weight assigned to the variable i. In Sandia procedure, the individual months are ranked in ascending order of the WS values. Among the five months with the lowest WS value, the one with the smallest deviation from the long-term CDF is selected as the typical month. Finally, the 12 selected typical months are concatenated to generate a complete year. The Sandia procedure is considered by many researchers the best methodology to generate hourly data of TMY (Skeiker 2004; Janjai and Deeyai 2009; Jiang 2010; Pusat et al. 2015).

There are different proposals regarding the weights to be applied to the different meteorological variables. Besides, TMY was conceived to typify the meteorological conditions of a site, and this might not be necessarily the same as the typical solar year. Thus, typical solar radiation years have been studied for solar system applications where the weight is 100% to the GHI or DNI (Habte et al. 2014), increased for solar radiation parameters (Cebecauer and Suri 2015), or can be shared between both (Ramírez et al. 2012). A revision and discussion on practices and uses of meteorological datasets for CSP performance modeling are being conducted within the SolarPACES program (Hirsch et al. 2017; Pagh Nielsen et al. 2017; Ramírez et al. 2017).

There are several sources of TMY and weather data for modeling the performance of a solar system. The National Renewable Energy Laboratory (NREL) has available NSRDB (National Solar Radiation Database, https://nsrdb.nrel.gov/) that includes 4 × 4 km gridded data for the USA, Central America, and part of South America and Asia. There are also commercial software like Meteonorm (http://www.meteonorm.com/) that supply synthetic TMY for every part of the world by interpolation and stochastic techniques from the Global Energy Balance Archive Data (GEBA). In addition, there are several suppliers of solar radiation data, mostly based on satellite-derived data, which provide also TMY formats. A thorough list of them can be found in the deliverables from the Tasks 36 and 46 of IEA Solar Heating and Cooling program (Sengupta et al. 2017).

The universality of TMY for solar plant performance analysis is being questioned and reviewed by several researchers and collaborative programs (Fernández-Peruchena et al. 2015; Pelland et al. 2016; Polo et al. 2017; Ramírez et al. 2017). On the one hand, the methodology to generate the TMY might be different according to the solar technology for the application (e.g., PV, CSP, CPV) or even the meteorological variables that should be included are also technology dependent. Accordingly, the ENDORSE project proposed the generation of a driver long-term time series of relevant meteorological parameters that are combined to create a composite of the time series as much as possible linearly dependent to the energy production of the system (Espinar et al. 2012). The validation results confirmed that the nonlinear effects of the system with respect to the row meteorological main values might have an influence. Recent studies in modeling different CSP technologies (solar towers, parabolic trough, and linear Fresnel) showed also these differences related to the nonlinear relationship between the DNI and plat power output (Polo et al. 2017). Figure 4 compares the daily energy estimated with SAM for a parabolic trough plant using a TMY as input with the daily yield resulted from modeling over 30 years of meteorological input where significant differences can be appreciated. This work proposes to perform the long-term characterization by the statistical analysis of multi-year time series of energy yield production instead of on solar irradiance data and distinguishes between Typical Yield Year (TYY) and TMY, being the former the year referring to long-term energy production estimated by multi-year modeling of a solar power plant. The possibility of simulating multiple years is also recognized in guiSmo guidelines, but in the 2017 version of the document, it still recommends finally to calculate the yield performance using a TMY (Hirsch et al. 2017).

In addition to the methodologies, the importance to include uncertainty analysis with the long-term yield calculation is also recognized. CSP yields with

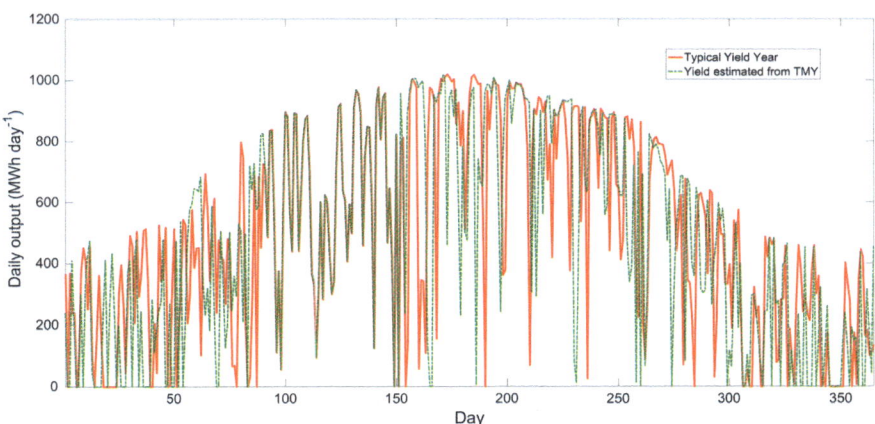

Fig. 4 Daily energy output calculated for a parabolic trough plant of 50 MWe using multi-year (TYY) and single TMY modeling

probabilistic meteorological data have been studied for Brazil in order to evaluate the impact of all independent uncertainties of DNI in the energy yield (Röttinger et al. 2015). On the other hand, a clustering technique is proposed to select a number of individual days that represent the long-term performance of a solar power plant (Peruchena et al. 2016). The uncertainty estimation of the long-term characterization according to the GUM (Guide to Expression of Uncertainty in Measurement) is being recommended and used in most of the recent studies (Cebecauer and Suri 2015; Polo et al. 2016; Hirsch et al. 2017).

3 Financial Feasibility of Solar Power Plants

The economic value of a solar plant largely depends on the availability of the solar resource. Therefore, uncertainties and inter-annual variability of the solar resource are key factors in determining the economic risk and the feasibility of a solar power plant. Solar radiation is variable in time and space as any other weather variable. Consequently, long time series of data are needed to evaluate properly the inter-annual variability of the solar resource. Figure 5 illustrates the inter-annual variability of global horizontal irradiance from 20 years of satellite estimations for a site in South Spain. Therefore, as a result of the inter-annual variability of solar radiation, TMY for a given site is not enough information for a feasible and bankable characterization of the long-term yield. Indeed, TMY does not include information on the expected variability of the solar resource from year to year (Vignola et al. 2012). Thus, it is then very frequent to require the probabilities of exceedance in addition to the average yield value for long-term performance

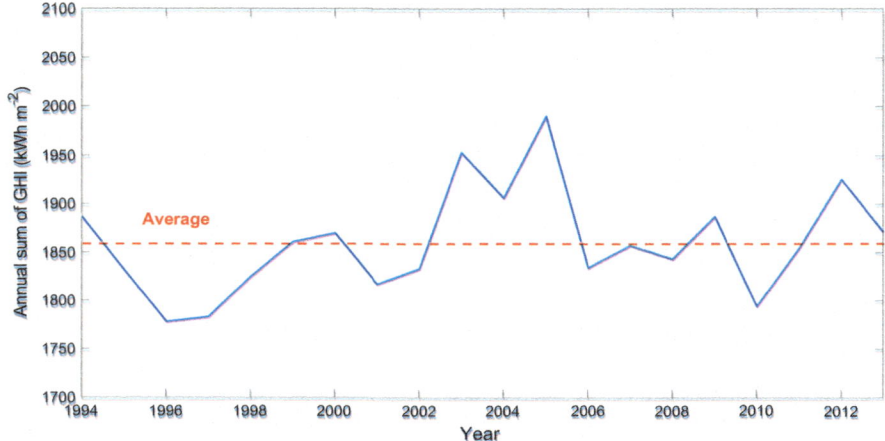

Fig. 5 Example of inter-annual variability from satellite-derived GHI over 20 years

analysis in a solar power project. In this context, the probability of exceedance describes the likelihood of annual solar irradiation or annual energy production being exceeded in a given period. In other words, the probability of exceedance at confidence X% refers to the annual energy produced that will be exceeded with a probability of X%. In order to quantify properly the risk, it is usually required the P90 or P95 as parameters to represent statistically the worse years and their relative comparison with the P50 (assumed as representative of the long-term or TMY). This information is used to assess the financial risk of a solar power project over the plant's life (20–30 years).

The probability of exceedance is complementary to the percentile of the distribution function, and its quantification depends on the distribution function of the data. Assuming that the annual solar irradiation or annual energy follows a normal distribution with mean μ and standard deviation σ, the probability of exceedance can be determined straightforward from the tables of the error function and its inverse (Petrucelli et al. 1999).

$$P50 = \mu$$
$$P90 = \mu - 1.282\sigma \tag{10}$$

Figure 6 illustrates the P50, which is the same as the median and the mean in a normal distribution function, and P90 in the case of a time series of annual GHI that can be assumed to be normally distributed.

In the case of no analytical statistical probability distribution that fit the data, an empirical cumulative distribution function (CDF) can be used to estimate the probabilities of exceedance. The empirical method requires to have a long time series of annual data in order to get more accurate estimates. The empirical method basically consists of sorting the data in ascending order and assigning each data point an equal fraction of the total probability. SAM includes both methods for estimating P50 and P90 in yield analysis and a description and comparison of both methods can be found in the NREL reports (Dobos et al. 2012).

Departure from normal distribution fits can occur when exceptional years of low solar irradiation are included corresponding to extreme meteorological conditions such as large volcano eruptions. The volcano eruptions of El Chicón, in 1982, and Pinatubo, in 1992, are examples of these events that resulted in extremely low DNI annual values. These events have been observed in the long time series of DNI measurements in the Solar Radiation Monitoring Laboratory network of the University of Oregon (Riihimaki and Vignola 2005; Riihimaki et al. 2005; Lohmann et al. 2006, 2007). The inclusion of extremely low solar irradiation years would result in tailed distribution functions different from Gaussian distributions (symmetrical). The probability distribution function of DNI for a long period in Burns station (University of Oregon) was fitted to a Weibull distribution function, and specific procedures for estimating the probability of exceedance have been recently proposed in the literature (Fernández Peruchena et al. 2016a). The need to include or not those extreme years, such as large volcano eruptions, in the long-term yield analysis has no general consensus among the scientific community so far.

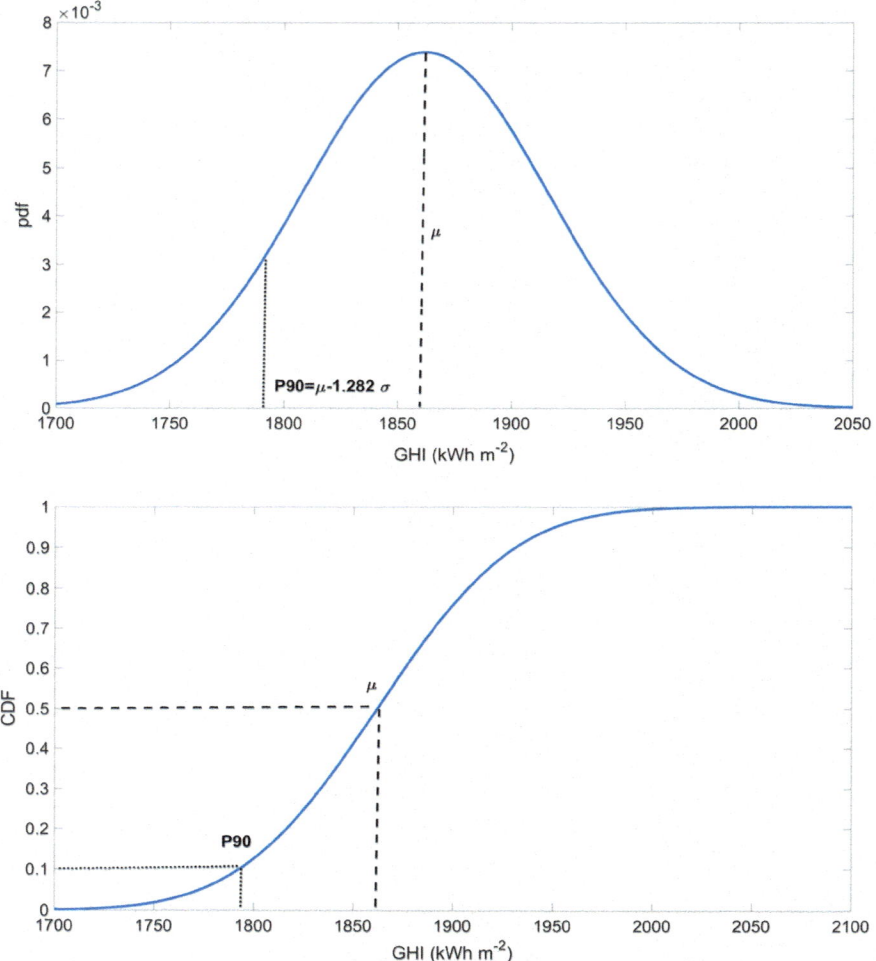

Fig. 6 Examples of normal probability and cumulative distribution functions showing the P90 and P50 values

The bankability and feasibility requirements can be strongly affected by the consideration of those extreme meteorological years of low probability. In this sense, for instance, the SolarPACES guiSmo project recommends removing those exceptional years from the long-term analysis for CSP plants (Hirsch et al. 2017).

In the determination of the probability of exceedance for yield performance, there are also two main approaches. One is based on the construction of a synthetic meteorological year that represents the probability of exceedance in terms of solar irradiation that could be used to simulate the plant and to determine consequently the probability of exceedance of the energy yield (Fernández Peruchena et al.

2016b; Fanego et al. 2017). On the other hand, other authors propose to perform multi-year modeling estimation of the solar plant in order to estimate the probability of exceedance directly from the annual energy yield time series instead of on the solar resource (Dobos et al. 2012; Polo et al. 2016). In the solar bankability project under European Union H2020 program, a revision was performed on uncertainty and risk in yield estimation during PV planning phase observing an overall impact as high as 10% (mostly due to solar resource uncertainties) on the estimated energy yield (Tjengdrawira et al. 2017). The recommendation is to calculate exceedance probabilities using empirical probability distribution functions.

4   Uncertainty in Yield Performance Analysis

The proper long-term yield characterization and also the probabilities of exceedance must include the associated uncertainties. The sources of uncertainties are variable and depend on the technology (e.g., PV or CSP for the same emplacement plants might have different combined uncertainties). However, it is essential to collect and estimate all the uncertainties sources in order to elaborate a reliable total combined uncertainty.

For instance, as reported by a Natural Resources Canada project, in PV yield analysis uncertainties were estimated in the best cases as 4% for inter-annual variability, 5% for solar resource estimation on horizontal plane, 3% for conversion of solar irradiance to the plane of the array, 3% for power estimation of the arrays, and up to 5% due to losses and other sources of error (Thevenard et al. 2010). This study reported a total combined uncertainty of near 8% for the average yield over the PV system lifetime. It is generally recognized that the most contributor to the total combined uncertainty in PV yield assessment is the measured or estimated solar resource (Richter et al. 2015). The total combined uncertainty must be usually estimated by the square root of the sum of squares of individual uncertainties (solar resource uncertainty, yield modeling uncertainty, uncertainties due to losses, etc.). The expanded uncertainty is obtained by multiplying the total combined uncertainty by a coverage factor that depends on the level of confidence required, 2 at 95% and 3 at 99% of confidence (JCGM 2008).

In the guidelines for CSP yield analysis delivered by SolarPACES program, it is stated that the probabilistic uncertainty evaluation is the most effective method for CSP plant performance uncertainty analysis (Hirsch et al. 2017). This methodology includes the aspects of uncertainty and variability in solar resource and performance modeling, considering all the independent uncertainties to produce a distribution function of energy yield (Röttinger et al. 2015). The main sources of uncertainty in CSP yield analysis are the modeling approach, the technical parameters, and the boundary conditions. The latter includes the uncertainty in the direct normal irradiance (DNI) which is considered the dominating source of uncertainty. The guiSmo guidelines proposed the following expression to compute the overall uncertainty (Hirsch et al. 2017).

$$\sigma = \sqrt{\sigma_{parameter}^2 + \sigma_{model}^2 + \sigma_{meteo}^2} \tag{11}$$

According to this guidelines, the uncertainty of model parameters, $\sigma_{parameter}$, can be best determined by means of probabilistic modeling; the model uncertainty, σ_{model}, depends on the degree of maturity and complexity of the model, and the uncertainty of the long-term mean associated to meteorological input is proposed to be determined as a function of the P50 and P90.

Statistical simulations of yearly yield for PV plants, using the Latin Hypercube Sampling method with SAM, were also used to estimate the long-term uncertainty, and the methodology was compared with the simplified methodology of combining individual uncertainties with rather an agreement between both methods (Thevenard and Pelland 2013).

5 Final Remarks

Modeling the yield performance of a PV or CSP power plant involves the estimation of the annual energy expected for the plant life (the long-term yield) accompanied by a proper determination of the associated uncertainties and financial risks. Long-term time series of solar irradiance data along with additional meteorological variables with high quality and accuracy and with a proper timestamp (hourly or even sub-hourly) are needed to accomplish this analysis. The two main approaches to determine the long-term yield energy are the generation of a representative meteorological or solar year to input the energy modeling or the multi-year modeling of a long number of meteorological input files and to estimate the long-term yield on the energy output time series. There are several free tools that are becoming standard in pre-feasibility and yield performance analysis for both PV and CSP systems. High-quality input data, particularly solar resource accurate data, are desirable in every yield performance study. Solar resource data should cover long periods (at least more than 10 years) in order to characterize the inter-annual variability. Finally, it is strongly recommended that any yield performance study is accompanied by a complete determination of the overall uncertainty of the yield energy.

Acknowledgements The author/editor, Jesús Polo, wishes to acknowledge the PVCastSOIL Proyect (ENE2017-83790-C3-1, 2 and 3), which is funded by the Spanish Ministerio de Economía y Competitividad and co-financed by the European Regional Development Fund.

References

Ayodele TR, Ogunjuyigbe ASO, Ekoh EE (2016) Evaluation of numerical algorithms used in extracting the parameters of a single-diode photovoltaic model. Sustain Energy Technol Assessments 13:51–59. https://doi.org/10.1016/j.seta.2015.11.003

Blair N, Dobos AP, Freeman J et al (2014) System advisor model, sam, 2014.1.14: General description. Tech Report, NREL/TP-6A20-61019, Golden, USA

Cameron CP, Stein JS, Tasca CA (2011) PV performance modeling workshop summary report. Rep SAND2011-3419, Alburquerque, USA 1–92

Cebecauer T, Suri M (2015) Typical meteorological year data: solarGIS approach. Energy Procedia 69:1958–1969. https://doi.org/10.1016/j.egypro.2015.03.195

Cucumo M, De Rosa A, Ferraro V et al (2007) Experimental testing of models for the estimation of hourly solar radiation on vertical surfaces at Arcavacata di Rende. Sol Energy 81:692–695. https://doi.org/10.1016/j.solener.2006.09.002

De Soto W, Klein SA, Beckman WA (2006) Improvement and validation of a model for photovoltaic array performance. Sol Energy 80:78–88. https://doi.org/10.1016/j.solener.2005.06.010

Demain C, Journée M, Bertrand C (2013) Evaluation of different models to estimate the global solar radiation on inclined surfaces. Renew Energy 50:710–721. https://doi.org/10.1016/j.renene.2012.07.031

Dobos A, Neises T, Wagner M (2014) Advances in CSP simulation technology in the system advisor model. Energy Procedia 49:2482–2489. https://doi.org/10.1016/j.egypro.2014.03.263

Dobos AP, Gilman P, Kasberg M (2012) P50/ P90 analysis for solar energy systems using the system advisor model. In: 2012 World renewable energy forum

Elbaset AA, Ali H, Abd-El Sattar M (2014) Novel seven-parameter model for photovoltaic modules. Sol Energy Mater Sol Cells 130:442–455. https://doi.org/10.1016/j.solmat.2014.07.016

Espinar B, Blanc P, Wald L (2012) Report on the production S4 "TMY FOR PRODUCTION." Proj ENDORSE 12

Et-Torabi K, Nassar-Eddine I, Obbadi A et al (2017) Parameters estimation of the single and double diode photovoltaic models using a gaussian seidel algorithm and analytical method: a comparative study. https://doi.org/10.1016/j.enconman.2017.06.064

Fanego VL, Rubio JP, Peruchena CMF, et al (2017) A novel procedure for generating solar irradiance TSYs. In: AIP conference proceedings 1850. https://doi.org/10.1063/1.4984523

Fernández-Peruchena CM, Gastón M, Sánchez M et al (2015) MUS: a multiscale stochastic model for generating plausible meteorological years designed for multiyear solar energy yield simulations. Sol Energy 120:244–256. https://doi.org/10.1016/j.solener.2015.07.037

Fernández Peruchena CM, Ramírez L, Silva-Pérez MA et al (2016a) A statistical characterization of the long-term solar resource: towards risk assessment for solar power projects. Sol Energy 123:29–39. https://doi.org/10.1016/j.solener.2015.10.051

Fernández Peruchena CM, Ramírez L, Silva M et al (2016b) A methodology for calculating percentile values of annual direct normal solar irradiation series. 101:120010–150005. https://doi.org/10.1063/1.4949234

Festa R, Ratto CF (1993) Proposal of a numerical procedure to select reference years. Sol Energy 50:9–17. https://doi.org/10.1016/0038-092X(93)90003-7

Finkelstein JM, Schafer RE (1971) Improved goodness-of-fit tests. Biometrika 58:641–645. https://doi.org/10.1093/biomet/58.3.641

Ghani F, Rosengarten G, Duke M, Carson JK (2014) The numerical calculation of single-diode solar-cell modelling parameters. Renew Energy 72:105–112. https://doi.org/10.1016/j.renene.2014.06.035

Gilman P (2015) SAM photovoltaic model technical reference. Technical Report NREL/TP-6A20-64102, Golden CO, USA

Habte A, Lopez A, Sengupta M, Wilcox S (2014) Temporal and spatial comparison of gridded TMY, TDY, and TGY data sets. NREL/TP-5D00-60886, National Renewable Energy lab Report, Golden Co

Hall IJ, Prairie RR, Anderson HE, Boes EC (1978) Generation of typical meteorological years for 26 SOLMET stations. Albuquerque (USA)

Hirsch T, Dernsch J, Fluri T et al (2017) SolarPACES guideline for bankable STE yield assessment. IEA technology collaboration programme solarPACES

Jain A (2004) Exact analytical solutions of the parameters of real solar cells using Lambert W-function. Sol Energy Mater Sol Cells 81:269–277. https://doi.org/10.1016/j.solmat.2003.11.018

Janjai S, Deeyai P (2009) Comparison of methods for generating typical meteorological year using meteorological data from a tropical environment. Appl Energy 86:528–537. https://doi.org/10.1016/j.apenergy.2008.08.008

JCGM (2008) Evaluation of measurement data—guide to the expression of uncertainty in measurement. Int Organ Stand Geneva ISBN 50:134. https://doi.org/10.1373/clinchem.2003.030528

Jiang Y (2010) Generation of typical meteorological year for different climates of China. Energy 35:1946–1953. https://doi.org/10.1016/j.energy.2010.01.009

Kambezidis HD, Psiloglou BE, Gueymard C (1994) Measurements and models for total solar irradiance on inclined surface in Athens, Greece. Sol Energy 53:177–185. https://doi.org/10.1016/0038-092X(94)90479-0

Khalil SA, Shaffie AM (2013) A comparative study of total, direct and diffuse solar irradiance by using different models on horizontal and inclined surfaces for Cairo Egypt. Renew Sustain Energy Rev 27:853–863. https://doi.org/10.1016/j.rser.2013.06.038

Khorasanizadeh H, Mohammadi K, Mostafaeipour A (2014) Establishing a diffuse solar radiation model for determining the optimum tilt angle of solar surfaces in Tabass. Iran. Energy Convers Manag 78:805–814. https://doi.org/10.1016/j.enconman.2013.11.048

King BH, Hansen CW, Riley D et al (2016) Procedure to determine coefficients for the Sandia Array Performance Model (SAPM). Sandia Report

King DL, Boyson WE, Kratochvill JA (2004) Photovoltaic array performance model. Sandia Rep. https://doi.org/10.2172/919131

King DL, Gonzalez S, Galbraith GM, Boyson WE (2007) Performance model for grid-connected photovoltaic inverters, SAND2007-5036. Contract 38:655–660

Lee K, Yoo H, Levermore GJ (2013) Quality control and estimation hourly solar irradiation on inclined surfaces in South Korea. Renew Energy 57:190–199. https://doi.org/10.1016/j.renene.2013.01.028

Lohmann S, Riihimaki L, Vignola F, Meyer R (2007) Trends in direct normal irradiance in Oregon: comparison of surface measurements and ISCCP derived irradiance. Geophys Res Lett 34:1–4

Lohmann S, Schillings C, Mayer B, Meyer R (2006) Long-term variability of solar direct and global radiation derived from ISCCP data and comparison with reanalysis data. Sol Energy 80:1390–1401. https://doi.org/10.1016/j.solener.2006.03.004

Loutzenhiser PG, Manz H, Felsmann C et al (2007) Empirical validation of models to compute solar irradiance on inclined surfaces for building energy simulation. Sol Energy 81:254–267. https://doi.org/10.1016/j.solener.2006.03.009

Lund H (1995) The design reference year. Users manual. IEA, Solar heating and Cooling Task 9 Report

Mares O, Paulescu M, Badescu V (2015) A simple but accurate procedure for solving the five-parameter model. Energy Convers Manag 105:139–148. https://doi.org/10.1016/j.enconman.2015.07.046

Marion W, Urban K (1985) User's Manual for TMY2 s. Typical Meteorological Years. NREL/TP-463-7668, National Renewable Enery Laboratory, Golden CO

Mefti A, Bouroubi MY, Adane A (2003) Generation of hourly solar radiation for inclined surfaces using monthly mean sunshine duration in Algeria. Energy Convers Manag 44:3125–3141. https://doi.org/10.1016/S0196-8904(03)00070-0

Mermoud AA, Wittmer B (2014) Pvsyst user's manual. PVSYST SA

Mohammadi K, Khorasanizadeh H (2015) A review of solar radiation on vertically mounted solar surfaces and proper azimuth angles in six Iranian major cities. Renew Sustain Energy Rev 47:504–518. https://doi.org/10.1016/j.rser.2015.03.037

Muneer T, Saluja GS (1985) A brief review of models for computing solar radiation on inclined surfaces. Energy Convers Manag 25:443–458. https://doi.org/10.1016/0196-8904(85)90009-3

Padovan A, Del Col D (2010) Measurement and modeling of solar irradiance components on horizontal and tilted planes. Sol Energy 84:2068–2084. https://doi.org/10.1016/j.solener.2010. 09.009

Pagh Nielsen K, Blanc P, Vignola F et al (2017) Discussion of current used practices for: creation of meteorological data sets for CSP/STE performance simulations. Technical report SolarPACES Task V

Pandey CK, Katiyar AK (2011) A comparative study of solar irradiation models on various inclined surfaces for India. Appl Energy 88:1455–1459. https://doi.org/10.1016/j.apenergy. 2010.10.028

Pelland S, Maalouf C, Kenny R et al (2016) Solar Energy Assessments: When Is a Typical Meteorological Year Good Enough? In: Proceedings of the American solar energy society national 2016 conference 1–7. https://doi.org/10.18086/solar.2016.01.17

Peng J, Lu L, Yang H, Ma T (2015) Validation of the Sandia model with indoor and outdoor measurements for semi-transparent amorphous silicon PV modules. Renew Energy 80:316–323. https://doi.org/10.1016/j.renene.2015.02.017

Peruchena CMF, García-Barberena J, Guisado MV, Gastón M (2016) A clustering approach for the analysis of solar energy yields: a case study for concentrating solar thermal power plants. p 070008

Petrucelli JD, Nandram B, Chen M (1999) Applied statistics for scientists and engineers, Prentice-Hall Inc

Polo J, Fernández-Peruchena C, Gastón M (2017) Analysis on the long-term relationship between DNI and CSP yield production for different technologies. Sol Energy 115:1121–1129. https://doi.org/10.1016/j.solener.2017.07.059

Polo J, Téllez FM, Tapia C (2016) Comparative analysis of long-term solar resource and CSP production for bankability. Renew Energy 90:38–45. https://doi.org/10.1016/j.renene.2015.12. 057

Pusat S, Ekmekçi İ, Akkoyunlu MT (2015) Generation of typical meteorological year for different climates of Turkey. Renew Energy 75:144–151. https://doi.org/10.1016/j.renene.2014.09.039

Ramírez L, Barnechea B, Bernardos A et al (2012) Towards the standardization of procedures for solar radiation data series generation. In: Proceedings of the solarPACES conference. p 5

Ramírez L, Pagh Nielsen K, Vignola F et al (2017) Road map for creation of advanced meteorological data sets for CSP performance simulations. Technical report SolarPACES Task V

Richter M, Kalisch J, Schmidt T et al (2015) Best Practice Guide On Uncertainty in PV Modelling

Riihimaki L, Vignola F (2005) Trends in direct normal solar irradiance in Oregon from 1979–2003. In: ISES Solar world congress 2005

Riihimaki L, Vignola F, Lohmann S, Meyer R (2005) Observing changes of surface solar irradiance in Oregon: a comparison of satellite and ground-based time-series. In: AGU fall meeting, 2005-12-05–2005-12-09, San Francisco, CO (USA). p

Röttinger N, Remann F, Meyer R, Telsnig T (2015) Calculation of CSP Yields with probabilistic meteorological data sets: a case Study in Brazil. Energy Procedia 69:2009–2018. https://doi. org/10.1016/j.egypro.2015.03.210

Sengupta M, Habte A, Gueymard C, Wilbert S, Renné D (2017) Best practices handbook for the collection and use of solar resource data for solar energy applications, second edn. https://doi. org/10.18777/ieashc-task46-2015-0001

Skeiker K (2004) Generation of a typical meteorological year for Damascus zone using the Filkenstein-Schafer statistical method. Energy Convers Manegement 45:99–112. https://doi.org/10.1016/S0196-8904(03)00106-7

Stein JS, Farnung B (2017) PV performance modeling methods and practices results from the 4th PV performance modeling collaborative workshop. IEA PVPS Task 13

Sudhakar Babu T, Prasanth Ram J, Sangeetha K et al (2016) Parameter extraction of two diode solar PV model using Fireworks algorithm. Sol Energy 140:265–276. https://doi.org/10.1016/j.solener.2016.10.044

Thevenard D, Driesse A, Turcotte D et al (2010) Uncertainty in long-term photovoltaic yield predictions

Thevenard D, Pelland S (2013) Estimating the uncertainty in long-term photovoltaic yield predictions. Sol Energy 91:432–445. https://doi.org/10.1016/j.solener.2011.05.006

Tjengdrawira C, Moser D, Jahn U et al (2017) PV investment technical risk management: best practice guidelines for risk identification, assessment and mitigation

Vignola F, Grover C, Lemon N, McMahan A (2012) Building a bankable solar radiation dataset. Sol Energy 86:2218–2229

Wagner MJ (2008) Simulation and predictive performance modeling of utility-scale central receiver system power plants. Thesis at the University of Wisconsin-Madison

Wagner MJ, Gilman P (2011) Technical manual for the SAM physical trough model. NREL Report No. NREL/TP-5500-51825

Wagner MJ, Zhu G (2012) A direct-steam linear fresnel performance model for NREL's system advisor model. In: Proceedings of the ASME 2012 6th international conference on energy sustainability and 10th fuel cell science, engineering and technology conference, San Diego, CA, USA, 23–26 July 2012

Wattan R, Janjai S (2016) An investigation of the performance of 14 models for estimating hourly diffuse irradiation on inclined surfaces at tropical sites. Renew Energy 93:667–674. https://doi.org/10.1016/j.renene.2016.02.076

Wilcox S, Marion W (2008) Users manual for TMY3 data sets. NREL/TP-581-43156, National Renewable Energy Laboratory, Golden CO

Chapter 12
Solar Radiation Spatio-Temporal Analysis and Its Implications for Power Grid Management

Jan Remund

Abstract Photovoltaic power has the temporal and spatial variability associated with the natural intermittency of solar resource. This variability results in potential threats to the grid since it must be able to accommodate safely the foreseeing variability. This chapter gives an overview of the variability scales and its quantification. The smoothing effect and other tools in power grid management are also described.

1 Introduction

Photovoltaic (PV) has become, in 2018, the cheapest way of production of electricity—leading to a global boom. Two main trends regarding PV are currently seen: on the one hand high penetration in densely populated regions and on the other hand the installation of very big PV plants mainly in sunny climates. Both trends raise the question of the variability of the solar resource.

PV is intermittent, depending mainly on clouds and solar position. The grid needs to accommodate this variability. This kind of variability is new to electricity systems. Therefore, it is not astonishing that transmission or distribution system operators are concerned about keeping their grid stable. Those concerns are real but can be reassured with two answers: the first variability is smoothed out in space and time significantly, and the second variability can be modelled. Both will be shown in this chapter.

Many papers are published lately on this subject. This chapter is mainly based on the work of Hoff and Perez (2012) and Remund et al. (2015)—including updates based on the work of Lohmann et al. (2016, 2017). A good overview of the topic is also given by Perez et al. (2016).

Most figures shown in this chapter were based on the high-resolution measurement site located in Oahu, Hawaii, USA, run by the National Renewable

J. Remund (✉)
Energy and Climate, Meteotest, Bern, Switzerland
e-mail: jan.remund@meteotest.ch

J. Polo et al. (eds.), *Solar Resources Mapping*, Green Energy and Technology,
https://doi.org/10.1007/978-3-319-97484-2_12

Energy Laboratory (Sengupta and Andreas 2011). The system consists of a 17 global horizontal irradiance sensor grids spread across approximately 0.76 km², with one reading per second. The data is available online. We analysed 15 randomly chosen days for all seasons from May 2010 until July 2011.

2 Temporal and Spatial Scales

PV systems' power output depends essentially on the global irradiance to which they are subjected. Therefore, it is necessary to understand irradiance variability and its consequences. Changes in solar irradiance that affect PV systems occur in a wide range of timescales, from few milliseconds to several decades. Each of these changes will cause a different kind of impact on the power system (Table 1).

The sun is a variable star at all observed timescales and at all wavelengths. However, the sun's variability of the time range between seconds and hours is by far lower than the one induced by clouds (Fig. 1).

Table 1 Potential power system impacts of solar irradiance variability

Timescale of changes in solar irradiance	Potential power system impact
Seconds	Power quality (e.g. voltage flicker)
Minutes	Regulation reserves
Minutes to hours	Load following
Hours to days	Unit commitment
Months to years	Missing storage and/or capacity

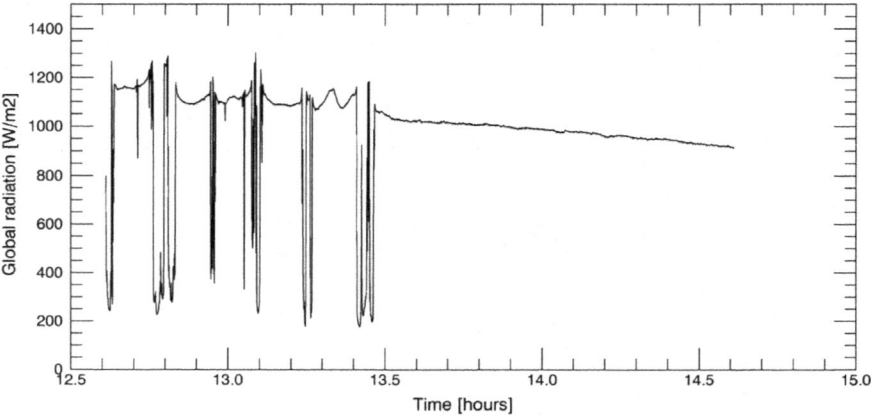

Fig. 1 Global radiation measured at Oahu on 30 March 2010, in 1 s time resolution. Until 13:30, effects of clouds are clearly visible (strong dips and ridges). After 13:30, the radiation is lowering slightly due to the sun's orbit but is otherwise almost stable

Solar variations in sub-seconds are scarcely investigated but can be assumed as small for global radiation and PV production (Lohmann et al. 2016). Possible sources are small objects (other than clouds) passing over the PV installations.

Spatial scales go from metres (small PV installations) over kilometres (big PV installations, quarters) to thousands of kilometres (transmission systems). Spatial and temporal scales are not independent as we will show in this chapter.

The influence of the sun's orbit is strong, but precisely predictable and therefore not handled here. Strong and regionally highly correlated gradients are induced by solar eclipses as seen in Europe in March 2015 (Fritz 2016). Those are, however, seldom events—the next one in Central Europe with a noticeable effect will be in 2048—and can be calculated well in advance and are therefore also not covered in this report.

3 Quantifying Variability

Many different ways of quantifying variability are possible and have been proposed lately. We suggest to use the relative ramp rate as a variability measure as described by Hoff and Perez (2012). In other publications, this measure is called increment changes (Lohmann et al. 2017) or step change (Widén 2015). As PV production is linearly depending on global radiation to a great extent, this value can be used in lieu of suitable PV power ramp data—which is by far more scarcely measured and available.

Relative ramp rates are the difference of two time steps—following each other— of global horizontal irradiance (GHI) divided by the clear sky radiation (clearness index) (Eqs. 1 and 2).

$$k_c(t) = \frac{\text{GHI(t)}}{\text{GHI}_{\text{cs}}(t)} \tag{1}$$

$$\Delta k_c(t) = k_c(t+1) - k_c(t) \tag{2}$$

Two ways of visualizations of variability are most popular and give a quick and good overview of the variability situation: the partial distribution function (PDF) of increment changes (Fig. 2) and the correlation factor of relative ramps versus distance (Fig. 2).

PDFs are not Gaussian but rather show a Laplacian distribution. This is true for high resolution (e.g. 10 or 60 s as shown in Fig. 2) as well as on course resolution of space and time (Gari da Silva Fonseca et al. 2018). In high resolution, second maxima at one are partially seen—the change from totally cloudy to sunny and back. The spatial average shows much smaller variations—especially in fine time resolution (dotted lines in Fig. 2).

Figure 3 shows the correlation of ramp rates in relation to the distance between the measurement points for Oahu, USA. This relation can be relatively well modelled with exponential or hyperbolic function, as we will show in the following.

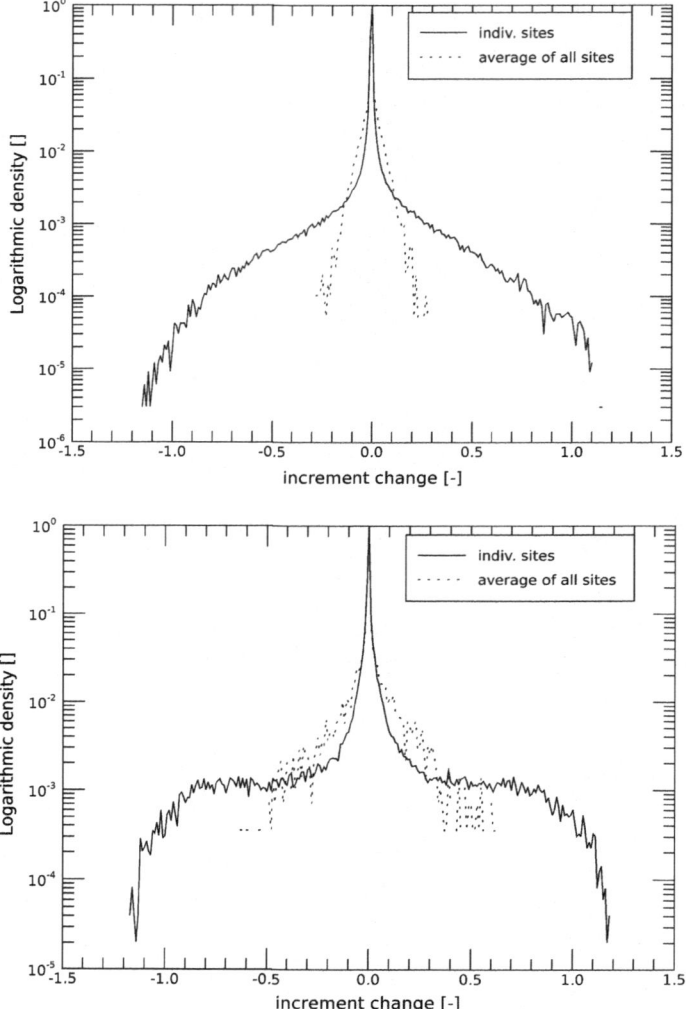

Fig. 2 Distribution (logarithmic) of 10 s (top) and 60 s (bottom) relative ramp rates in Oahu, Hawaii, for all 17 individual sites and the average

4 Variability Mitigation—The Smoothing Effect

The bigger the area and time resolution, the smoother the time series. Figure 4 shows as an example that this effect is based on the high-resolution network of Oahu, Hawaii.

The smoothing effect is based mainly on the cloud—their size, speed and transmission—as well as on frontal systems on a larger scale. Variability, therefore,

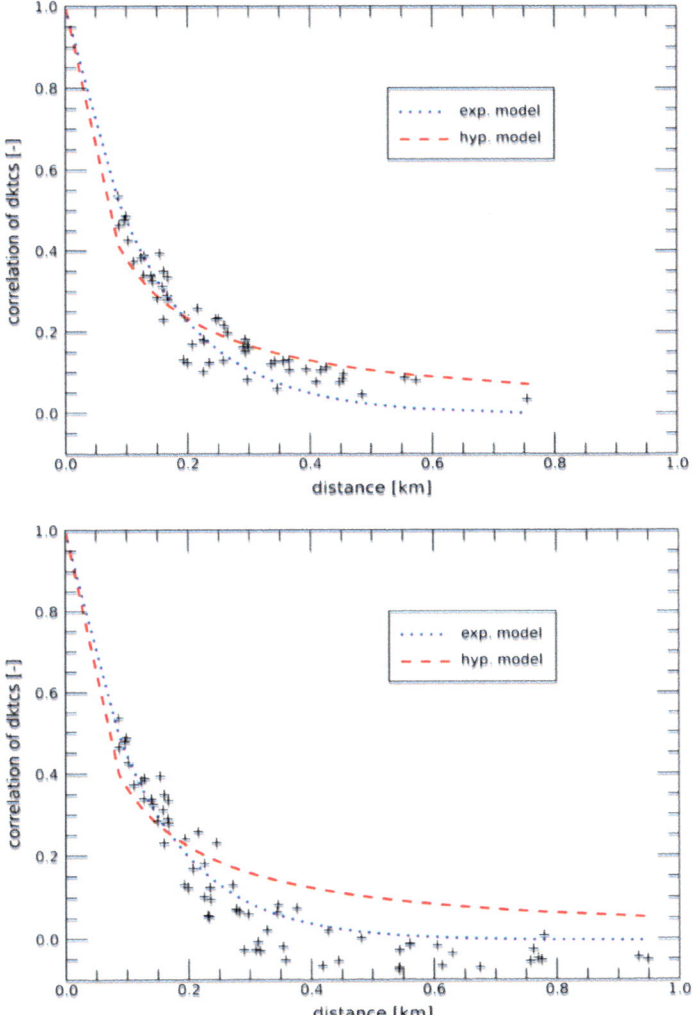

Fig. 3 Correlation of ramp rates versus distance in Oahu, Hawaii, for 30 s averages. Top: north–south; bottom: east–west direction

depends heavily on weather conditions. In clear sky conditions and overcast conditions, variability is low—and not critical. Strong and critical intermittency is happening mainly in scattered cloud situations (Fig. 5).

The distinction between the types of cloudiness is not very clear in the case of Oahu data—as fully overcast and cloudless situations are scarce. Nevertheless, the more frequent small changes in sunny and cloudy situations and less frequent high variations are visible.

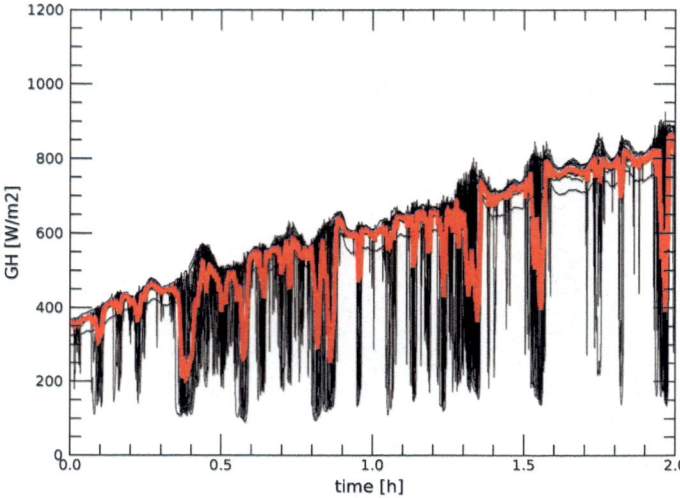

Fig. 4 Time series of 17 pyranometers (black lines) and average (red line) in 1 s time resolution (Oahu, Hawaii)

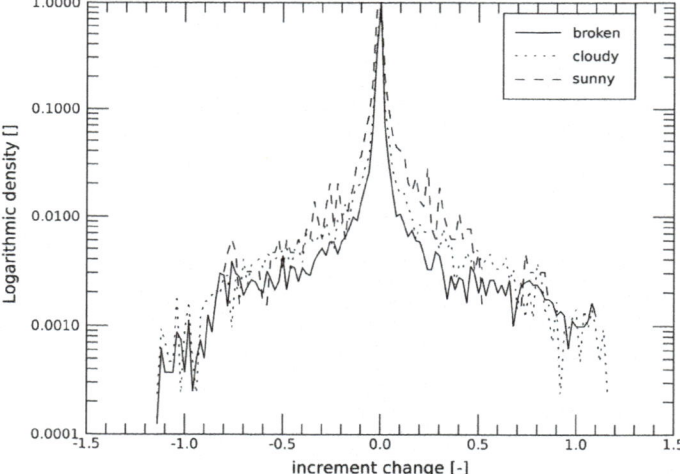

Fig. 5 Distribution (logarithmic) of 30 s relative ramp rates in Oahu, Hawaii, for all broken, cloudy and sunny conditions

5   Applied Tools and Models for Power Grid Management

As seen before, variability is mitigated in space and time. One of the simplest and nonetheless adequate models to describe the smoothing effect of ramps was introduced by (Hoff and Perez 2010, 2012).

The (normalized) output variation of the fleet is dependent on the output variation of each location and the correlation between them.

$$\sigma_{\Delta t}^{\text{Fleet}} = \frac{1}{C^{\text{Fleet}}} \sqrt{\sum_{i=1}^{N} \sum_{j=1}^{N} \sigma_{\Delta t}^{i} \sigma_{\Delta t}^{j} \rho_{\Delta t}^{i,j}} \tag{3}$$

In the special case when the change in output between locations is uncorrelated, fleet capacity is equally distributed and the variance at each location is the same (Hoff and Perez 2010), showed that fleet output variability equals the output variability at a single location divided by the square root of the number of locations:

$$\sigma_{\Delta t}^{\text{Fleet}} = \frac{\sigma_{\Delta t}^{1}}{\sqrt{N}} \tag{4}$$

Empirical data revealed that a model for the correlation between sites, based on Eq. (5), fitted well, as long as using location-specific parameters.

Optionally, another fitting equation presented by Perez et al. (2016) Eq. (6) is used widely:

$$\rho = \frac{1}{1 + \frac{d}{(\Delta t)(\text{CS}_1)}} \tag{5}$$

$$\rho = e^{\frac{d \ln(0.2)}{1.5(\Delta t)(\text{CS}_2)}} \tag{6}$$

where CS_1 and CS_2 refer to the cloud speed (km/h).

Two main drivers govern, therefore, the fleet variability: single-point variability (σ) and cloud speed (CS). Hoff and Perez showed also that the distance of de-correlation (set here deliberately at 0.25) is depending linearly from the time resolution. The gradient of this linear function depends on cloud speed (Fig. 5).

On large scales, the smoothing equations work fine as (Perez and Fthenakis 2013) shown in their papers.

An alternative widely used model is the one of (Lave et al. 2012), which is based on a wavelet model including the same correlation functions as Hoff and Perez. The variability maps for the USA available on the PV performance modelling website[1] are based on this model (Lave et al. 2017).

---

[1]https://pvpmc.sandia.gov/applications/wavelet-variability-model/.

The variability of a single plant can be estimated based on either existing plants or a pyranometer measurement in the region or modelled based on (Remund et al. 2015). Cloud speed can be estimated by a set of nearby measurement stations, by cloud motion vector analysis (Hammer et al. 2000) or can be estimated by the wind speed at altitudes between 1500 and 3000 m (which corresponds to 850–700 hPa pressure levels of weather models), a parameter easily available from numerical weather prediction models.

Hoff and Perez's work was based mainly on satellite data, which they partly upscaled with cloud motion vectors to 1-min data. Finer temporal resolutions below 60 s were not investigated. During the last years, some networks with extremely high resolutions (10–100 m and 10 Hz or 1 s time resolutions) were installed and analysed (Lohmann et al. 2016). This leads to new observations and models and showed some shortcomings of the simple models based on cloud speed.

These investigations showed that Hoff and Perez's model differs from reality in some cases. Here an overview of concerns and add-ons observed:

- The cloud speed is not the same for all spatial scales. (David et al. 2014) noticed that for systems in the scale of 10's or 100 of km, the clouds tend to form frontal systems, which move at another (lower) speed.
- Cross-wind and along wind need to be separated (Hinkelman 2013).
- For very high time resolutions below 1 min, the functions of Hoff and Perez do not match the observations for the cloudy situation (Lohmann et al. 2016).
- Some researchers noticed negative correlation values (e.g. Widén 2015); however, this was not confirmed by Lohmann. In our analysis, slightly negative values are seen, but not as pronounced as in Widen's (Fig. 5).

Overall, Lohmann's fractal model (Lohmann et al. 2017) seems to be the best model for modelling variability in high resolution. The drawback is the complexity of the model.

Figure 5 shows the de-correlation distance versus the time resolution for a data set in Oahu, Hawaii, separated in east–west and north–south directions (exponential function Eq. 6). Hawaii is located in the trade wind zone with prevailing easterly winds. The difference for the 15 days analysed was, however, not that pronounced for the different directions.

The function is not linear—also here the cloud speed seems to get lower looking at lower resolutions (some clustering of clouds could be the reason). Only averages between 10 s and above 2 min have been analysed; as for higher frequencies, the minimal distance between the pyranometers is too big (80 m) and for the lower frequencies, the area of the measurement field is too low (Fig. 6).

6   Conclusions

The headline regarding variability for grid operators is "don't panic about intermittency". For point observations and 1 s time resolution, intermittency is high—but strongly smoothed in space and time. The smoothing effect depends on climate,

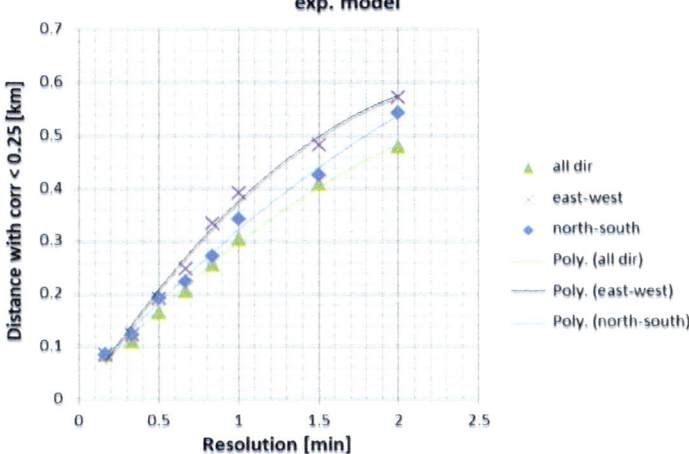

Fig. 6 De-correlation distance (<0.25) versus time resolution for data in Oahu, Hawaii, (Sengupta and Andreas 2011). Wind speed in east–west is generally higher, which can be seen in a steeper function of the distance on the resolution

cloud speed, cloud size and transparency. The higher the cloud speed, the bigger the smoothing effects.

The second good news is that variability can be modelled. For a start—not going into 1 s data and sub-1000 m scales—relatively simple models do a good job. For higher temporal and spatial resolutions, fractal models seem to be best.

Spatial and temporal scales of global radiation are always linked and need to be taken into account when modelling PV. It makes, e.g., no sense to model a transmission system in 1 s time resolution—as all ramps in this resolution are de-correlated after a distance of 100 m.

Based on Eq. 4, adequate distances can be determined based on time resolution and typical cloud speed (Table 2). Two PV installations with a distance larger than the adequate distance are independent largely.

Table 2 Adequate time and space scales (rounded) based on the distances with correlation <0.1–0.25 and cloud speeds between 10 and 50 km/h (Remund et al. 2015)

Time resolution	Adequate distances' average (km)	Adequate distances' ranges (km)
1 s	0.05	0.01–0.10
15 s	0.2	0.04–0.40
1 min	3	0.6–6.0
5 min	15	3–30
15 min	40	10–100
1 h	170	40–400
3 h	500	100–1000

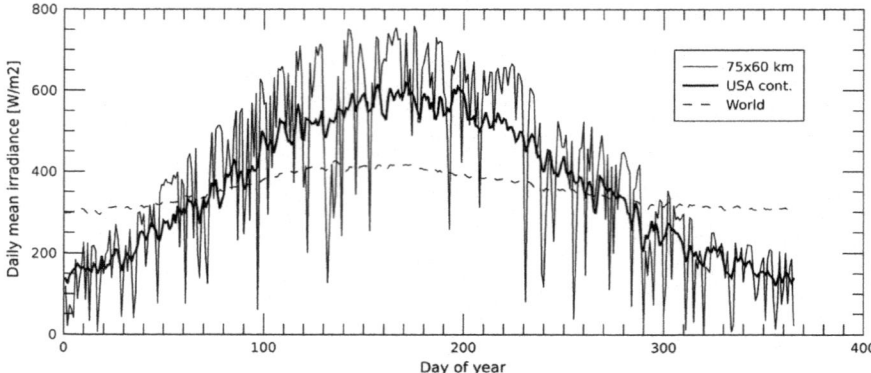

Fig. 7 Comparing the variability of daily global irradiance time series as a function of the considered footprint based on ERA-Interim data (Dee et al. 2011) of the year 2006 and Perez et al. (2016). The area of the single grid point 75 × 60 km lies nearby Detroit, USA; the world includes all areas above the sea level between 85°S and 85°N

With higher penetration, the absolute variability may grow—the relative, however, shrinks. Building PV all over the world with a good grid connection will lead to very low variability as Perez nicely showed (2016) (Fig. 7).

References

David M, Andriamasomanana FHR, Liandrat O (2014) Spatial and temporal variability of PV output in an insular grid: case of Reunion Island. Energy Procedia 57:1275–1282

Dee DP, Uppala SM, Simmons AJ, Berrisford P, Poli P, Kobayashi S, Vitart F (2011) The ERA-Interim reanalysis: configuration and performance of the data assimilation system. Q J Roy Meteor Soc 137(656):553–597. http://doi.org/10.1002/qj.828

Fritz R (2016) How an energy supply system with a high PV share handled a solar eclipse

Gari da Silva Fonseca J, Nishitsuji Y, Udagawa Y et al (2018) Regional photovoltaic power fluctuations within frequency regulation control time frames: a study with high-resolution data. Prog Photovoltaics Res Appl. https://doi.org/10.1002/pip.2999

Hammer A, Heinemann D, Lorenz E, Lückehe B (2000) Short-term forecasting of solar radiation based on image analysis of meteosat data

Hinkelman LM (2013) Differences between along-wind and cross-wind solar irradiance variability on small spatial scales. Sol Energy. https://doi.org/10.1016/j.solener.2012.11.011

Hoff TE, Perez R (2010) Quantifying PV power output variability. Sol Energy 84:1782–1793. https://doi.org/10.1016/j.solener.2010.07.003

Hoff TE, Perez R (2012) Modeling PV fleet output variability. Sol Energy 86:2177–2189

Lave M, Kleissl J, Arias-Castro E (2012) High-frequency irradiance fluctuations and geographic smoothing. Sol Energy 86:2190–2199. https://doi.org/10.1016/j.solener.2011.06.031

Lave M, Broderick RJ, Reno MJ (2017) Solar variability zones: satellite-derived zones that represent high-frequency ground variability. Sol Energy. https://doi.org/10.1016/j.solener.2017.05.005

Lohmann GM, Monahan AH, Heinemann D (2016) Local short-term variability in solar irradiance. Atmos Chem Phys. https://doi.org/10.5194/acp-16-6365-2016

Lohmann GM, Hammer A, Monahan AH et al (2017) Simulating clear-sky index increment correlations under mixed sky conditions using a fractal cloud model. Sol Energy. https://doi.org/10.1016/j.solener.2017.04.048

Perez MJR, Fthenakis VM (2013) Long-distance interconnection as solar resource intermittency solution: optimizing the use of energy storage and the geographic dispersion + interconnection of solar generating facilities. In: Conference record of the IEEE photovoltaic specialists conference

Perez R, David M, Hoff TE et al (2016) Spatial and temporal variability of solar energy. Found Trends® Renew Energy 1:1–44. https://doi.org/10.1561/2700000006

Remund J, Calhau C, Perret L, Marcel D (2015) Characterization of the spatio-temporal variations and ramp rates of solar radiation and PV

Sengupta M, Andreas A (2011) Oahu solar measurement grid (1-year archive): 1-second solar irradiance; Oahu, Hawaii (Data)

Widén J (2015) A model of spatially integrated solar irradiance variability based on logarithmic station-pair correlations. Sol Energy. https://doi.org/10.1016/j.solener.2015.10.043

.

Chapter 13
Demand-Side Management for PV Grid Integration

Islam Safak Bayram

Abstract Over the last two decades, solar photovoltaic (PV) systems have evolved from a small-scale niche market application to a major electricity source. Although the share of most annual PV production is less than ten per cent even for the most aggressive adopters, PV systems have started to create operational and planning issues for utility operators. In this chapter, we discuss PV integration issues both at low- and medium-voltage levels. Distributed PV systems at low-voltage networks lead to power quality issues, while large-scale PV farms have impacts on transmission network and generation assets. Then, we present a comprehensive overview of demand-side management (DSM) techniques and examine how DSM can be used to aid PV integration. In the last section, we present two experimental studies: (1) direct-load control of an air conditioner unit and (2) load shifting of a water heater. It is shown that such flexible loads can aid PV integration and increase PV self-sufficiency levels.

1 Introduction

Power systems are designed to deliver electricity to end-users in a reliable, continuous, clean and cost-effective way. Grid-connected PV systems offer a unique set of benefits to both consumers and grid operators. Such benefits relate to cleaner, and sustainable energy production, reduced transmission, and distribution network losses, lowered generation cost, improved resilience and protection, and boosted energy independence. Variable renewable energy source, both solar and wind, penetrations have seen a considerable push over the last few years. In countries like Ireland, Denmark and Germany, annual renewable integration levels, the percentage of electricity production from renewable in a year, have passed 20% at the national level. On the other hand, instantaneous penetration levels, the fraction of

I. S. Bayram (✉)
Qatar Environment and Energy Research Institute,
Hamad Bin Khalifa University, P.O. Box 5825, Doha, Qatar
e-mail: ibayram@hbku.edu.qa

© Springer Nature Switzerland AG 2019
J. Polo et al. (eds.), *Solar Resources Mapping*, Green Energy and Technology,
https://doi.org/10.1007/978-3-319-97484-2_13

electricity production from intermittent sources at a given instance, could be much higher; for instance, in Portugal, instantaneous penetration levels have reached 100% in 2018 (Portugal 2018). Such high penetration rates could threaten power grid stability as current control methods cannot handle high variability and uncertainty in the power generation. Similarly, PV rooftop systems also create bidirectional power flow which can degrade power quality. To match supply with demand in the presence of PV systems, a more responsive demand-side management (DSM) is required. In DSM, customer demand is monitored in real-time and *flexible* customer loads are curtailed in magnitude or shifted in time, hence managed in lieu of energy imbalance. To that end, in this chapter, we present the role of demand-side management for enhanced PV integration into power grids.

The idea of engaging demand-side activities in power system operations is not recent. DSM programs were first discussed in the 1970s as a remedy to global energy crises. However, it took almost a decade for DSM to be officially introduced to the electric utility business. DSM was first defined by Clark Gellings of Public Service Electric and Gas Company in Newark, NJ, as follows: *Demand-side management is the planning, implementation and monitoring of those utility activities designed to influence customer use of electricity in ways that will produce desired changes in the utility's load shape, e.g. changes in the time pattern and magnitude of a utility's load.* Over the last years, DSM programs have become popular utility application, and various measures have been deployed across different countries. For instance, in the PJM region,[1] nearly 11 GW of demand-side resources is active in the market for delivery in 2019. In general, DSM programs can be grouped into two parts. The first group of efforts aims to achieve *energy* savings by improving the energy efficiency and system design, using better materials, or developing novel processes for electrical loads. Such measures typically require capital investment but show an immediate effect on consumption patterns.

The second group of DSM programs targets to obtain *demand* savings typically during peak hours for a relatively short duration, i.e. 15–30 min., by influencing consumption patterns either via pricing or incentives. Pricing-based programs include time of use (TOU) rates, critical peak pricing (CPP), and real-time pricing (RTP). TOU pricing divides the day into multiple periods and specifies a price for each duration and keeps it constant for a given season, i.e. summer or winter rates. CPP and RTP, on the other hand, are dynamic pricing regimes in which electricity prices are determined based on market conditions and system load, and prices are updated every hour or even every five minutes. Incentive-based programs aim to control specific customer loads. At the residential sector, typical applications include direct-load control of air-conditioners during hot summer seasons. Such applications are widely popular in states like Florida, California and Texas. At commercial and industrial sectors, customers are offered incentive credits in their bills and asked for load shedding during peak hours. Applications in this group

---

[1]PJM is a regional transmission operator in the USA that has operations in Indiana, Illinois, Kentucky, Maryland, Michigan and nine more states.

target non-essential machines, cooling and air-conditioning units, motors, furnaces, pumps and compressors. The savings pertinent to demand reduction are related to avoided or deferred infrastructural capital investments which would have been required in the absence of DSM.

Furthermore, demand-side management could be a very effective tool to aid seamless integration of PV systems both at a low-voltage level, close to customer side typically at rooftops and at medium-voltage level at utility scale. DSM is a relatively affordable way to increase power system *flexibility* which is very essential in renewable energy integration. DSM programs can adjust customer loads in accordance with the variations in renewable output. For instance, sudden changes in large-scale PV farms may impact system frequency (e.g. 50 Hz) and DSM can act as an ancillary services mechanism to restore system balance. Similarly, uncontrolled PV generation will lead to voltage rises beyond an acceptable tolerance. In the rest of the chapter, we present the issues related to PV integration at a low- and medium-voltage level and show how DSM can be effective in overcoming them.

2 PV Integration into Low-Voltage Grid

Power systems around the globe are becoming more decentralized as the generation mix integrates distributed PV systems which are typically adopted by various financial support schemes such as subsidies, feed-in tariffs, green certificates and tax exemptions. One of the most popular applications of PV systems is the integration of PV modules at the low-voltage level such as rooftops, communities or microgrids, and distributed grid level. Distributed PV rooftop systems are particularly preferred in regions with limited land availability and allow adopters to enjoy financial benefits offered from various subsidies and incentive programs. For instance, in Germany, nearly 85% of the installed solar capacity (40 GW) is produced by PV systems with less than 1 MW installed capacity. Besides, 90% of such instalments are composed of systems with a capacity less than 30 kW.

There are a number of technical and economic challenges that are related to the integration of PV systems into power grids. Challenges in the distribution network level relate to planning and operation of distribution network concurrently. Traditionally, network planning was based on regional or local peak demand forecasting over a planning horizon, and the goal was to make sure that no physical constraints were violated during the system operation. Hence, in most distribution networks there was little or no monitoring and control, demand was assumed to be unresponsive, and there were no distributed generators. This approach, *fit-and-forget*, needs to be updated as situations like high solar power injections that happen a few times a year require active management approach. Otherwise, there would be a need for grid reinforcements which increase overall system cost.

The main body of the issues is related to power quality which is defined as the set of operating boundaries that allow specific electrical equipment to function in its intended manner without performance degradation. Some of the most common power qualities are voltage variations and unbalance, harmonics, grid islanding protection, and flicker and stress on transformers. Such issues arise due to bidirectional power flow, depending mainly on distribution grid architecture and load profiles. Next, we discuss power quality issues related to PV integration at distribution networks.

2.1 Procedure and Tasks of the Visit

The impact of PV on distribution networks can be summarized as below.

2.1.1 Voltage Issues

Voltage issues arise due to reverse power flow when PV generation and customer demand fluctuate and voltage levels deviate from the intended region. Related issues can be listed as (1) *voltage fluctuation*; (2) *voltage rise*; (3) *voltage unbalance* (Karimi et al. 2016). Voltage fluctuations are stemmed from intermittency of PV sources due to environmental conditions such as shading due to clouds, aerosols and dust deposition. To that end, voltage fluctuations can lead to voltage rise, unbalance and flickers in the network.

Voltage rise is one of the major operational challenges for distributed PV deployment. In a typical distribution network, voltage levels at customer sites are allowed to flow within ±5%. As shown in Fig. 1, in a circuit with no PV, voltage

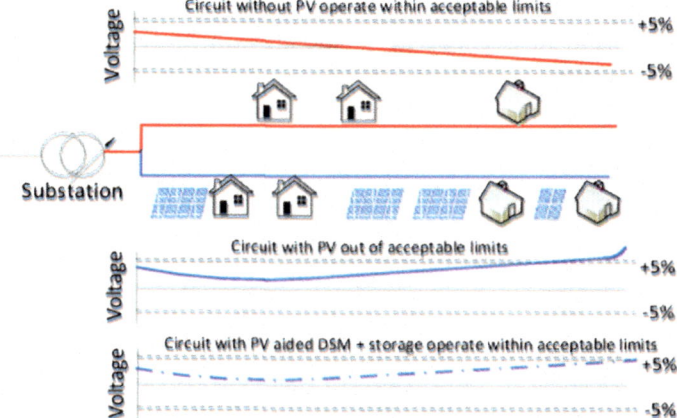

Fig. 1 Voltage rise in a distribution feeder

levels decrease as the distance from the substation increases. On the other hand, if PV systems inject power to a grid, the voltage levels will start to rise and exceed operation conditions. There are various ways, including demand-side management, to address voltage rise issues, and details will be discussed later in this chapter. Voltage unbalance occurs in a 3-phase network when there is a difference in phase angle or magnitude of voltages due to fluctuations of PV generation. Voltage unbalance estimated by voltage unbalance factor (VUF) which is defined as $\text{VUF} = \frac{V^-}{V^+} \times 100\%$, where V^- and V^+ are negative and positive voltage sequence, respectively. Typically, utility companies limit unbalances more than 2, or 3% and special attention is paid when PV generation reaches its peak typically between 11 am and 2 pm.

2.1.2 Harmonics

Harmonic distortion of voltage and current waveform is a growing issue due to the high penetration of PV generation. Distortion occurs due to the conversion of DC power that is generated by PV panels into AC power system is done via inverters which are the main source of harmonics. Injected harmonics can cause increased power losses due to heating, shorten equipment lifetime, and lead to power outages (Dartawan and Hui 2012). Therefore, harmonic distortion could be a bottleneck for PV integration and typical countermeasures include installing harmonic filters and repurposing capacitors. Furthermore, Institute of Electrical and Electronics Engineers has published standards for harmonics in IEEE 519-1992, and overall PV integration guidelines in IEEE 1547-2003 and an overview of requirements are given in Table 1.

Table 1 Feeder limitations for PV integration

Category	Criteria	Limit
Voltage	Overvoltage	≥ 1.05 V
	Voltage deviation	≥ 3
		$\geq$ half bandwidth at voltage regulators
	Unbalance	≥ 3
Protection	Forward flow fault contr	≥ 10
	Sympathetic breaker trip	≥ 50 A
	Anti-islanding	≥ 50
	Breaker/fuse coordination	≥ 100 A increase
Loading		≥ 100
Harmonics	Individual	≥ 3
	Total harmonic distortion	≥ 5

2.1.3 Islanding Detection and Operation

Grid-connected PV systems may experience more frequent abnormal operating conditions such as voltage shutdown, short-circuit or equipment failure. In order to avoid large-scale blackouts and system failures, a portion of the network is disconnected from the main grid, but distributed PV generators continue to provide power and maintain the scheduled voltage and frequency within the operating limits (Teoh and Tan 2011). Two types of islanding exist: planned (intentional) and unplanned (unintentional) modes. As the name suggests, intentional islanding is planned and typically used for maintenance purposes. During the maintenance, the PV system provides power. Unplanned islanding mode, on the other hand, can be severe as loss of grid synchronization may lead to instabilities. Therefore, islanding detection techniques are critical. Detection methods can be classified into two groups: remote techniques and local methods, and details can be found in references (Karimi et al. 2016) and (Dartawan and Hui 2012).

2.1.4 Other Issues

In addition to the issues mentioned above, several technical factors are affected by PV integration. One of them is the *thermal rating* which refers to the maximum current carrying capacity of network elements such as lines and transformers. PV generation under minimum load and maximum generation conditions may violate loading levels. In addition, reverse power flow may degrade the performance of tap chargers and the operation of voltage control schemes (CIGRE 2014).

2.2 PV Hosting Capacity

As discussed in the previous section, many limiting factors affect the penetration rates of PV system at the distribution network. Therefore, the term PV hosting capacity is used to define the maximum amount of PV that can be accommodated on a given distribution system without violating aforementioned operating limits and with no feeder modifications or investments (Ding et al. 2016). Similar to *fit-and-forget* approach, early assumptions on PV hosting capacity are 15% of the peak load which means that up to this point, no power quality issues are expected.

On the other hand, PV hosting capacity is affected by an array of factors which can be categorized into two groups. First set of factors are related to PV characteristics such as PV size, location, inverter control, and intermittency and variability due to factors like weather and aerosols. For instance, in general, PV hosting capacity can be increased if the locations closer to substations are restricted. The second set of factors are related to customer loads such as load variability, coincident load with PV or self-consumption levels, and non-coincidental load with PV. If the customer load is in line with PV generation, then PV impacts will be less. On

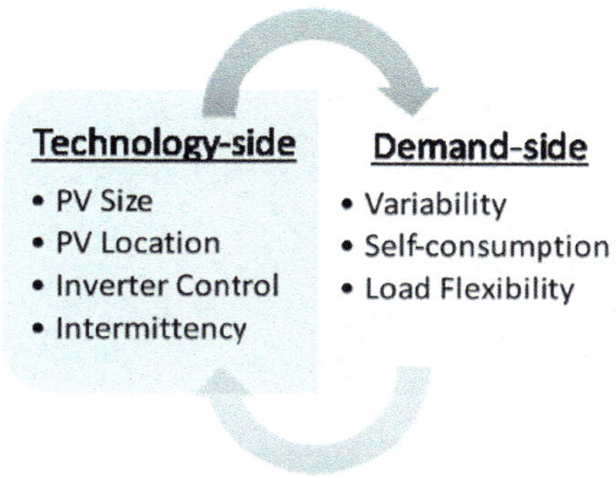

Fig. 2 Factors affecting PV hosting capacity

Technology-side
- PV Size
- PV Location
- Inverter Control
- Intermittency

Demand-side
- Variability
- Self-consumption
- Load Flexibility

the other hand, if there is a mismatch, the reverse power flow will contribute to the issues mentioned above. For instance, peak consumption in most European and North American cities occurs at night when there is no PV generation, while in countries with hot arid climate summer peaks take place during afternoons due to air-conditioning load inline with PV production. An overview is presented in Fig. 2. To that end, most PV hosting studies (Ding et al. 2016) employ stochastic simulation methods to generate and assess various scenarios. It is noteworthy that the first group of factors relate to technology, while the second group relates to demand-side activities. There are various ways to improve PV hosting capacity of a given feeder (Li et al. 2015). In order to aid the system to deliver power to higher voltages, *reactive power control* and *on-load tap-changing transformer* methods are applied. Moreover, increasing local consumption of produced PV power further increases hosting capacity. To align consumption with production, *decentralized storage units*, a costly option, and *demand-side management* are widely used.

3 Utility-Scale PV Integration

Large-scale PV farms, also called utility-scale systems, are typically connected to a medium-voltage level and supply electricity by following a power purchasing agreement made with the corresponding utility company. Even though there are diverging definitions regarding the minimum size of a power plant to be called utility scale, the most commonly accepted value is 1 MW. The advantage of utility-scale power plants over distributed ones is the economies of scale. For instance, a study conducted in the USA (Tsuchida et al. 2015) shows that the cost of generating electricity from a 300 MW PV farm is nearly one-half the unit cost of power produced from an equivalent 300 MW of 5 kW residential solar rooftops.

Unlike the previous case where PV rooftop systems affect distribution system components, utility-scale systems interact with the transmission network, generation system, and hence energy markets. In current power systems, transmission networks are deployed to deliver electricity generated from optimally sited power

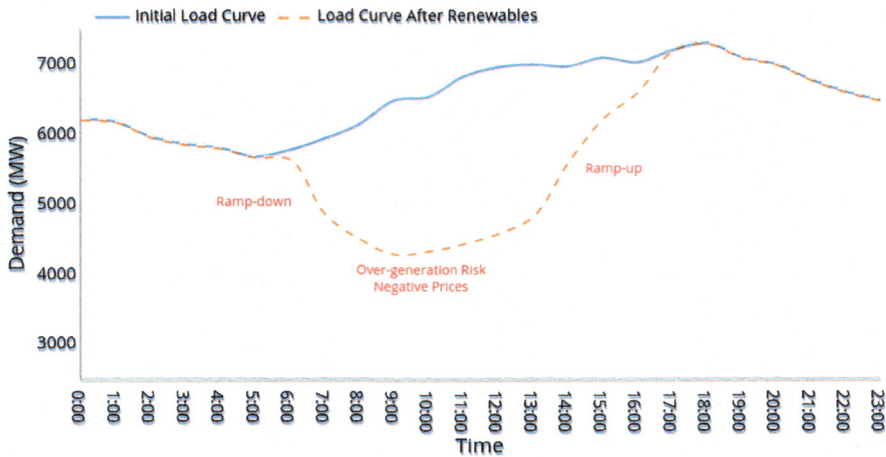

Fig. 3 A typical duck curve

plants. On the other hand, the optimal locations of the PV systems may not be perfectly in line with the transmission system, and hence an expansion to the transmission network may be required. This situation further poses a dilemma because if the cost of transmission expansion outweighs the benefits of the PV generation, then PV farms may need to be located at a suboptimal location.

Furthermore, utility-scale PV integration poses new challenges for grid operators in generation dispatch. One of the famous issues is "duck curve" which is named after its resemblance to a duck. Basically, a duck curve is the difference between customer demand and the generated solar power from the PV farm. The issue with a duck curve is the following. When power generation from PV farms increase as the sun shines, system operators have to ramp-down their conventional generators.

Similarly, when the production gradually reduces in the afternoon, this time, system operators have to ramp-up other generators to match supply with demand. This is illustrated in Fig. 3. The first issue is that if existing generators do not have enough flexibility, which is measured in terms of MW per min output change, the system operators face with over-generation risk. For instance, in regions with high shares of nuclear power plants,[2] significant influxes of PV generation will case over-generation, hence prices to be negative causing monetary losses (The MIT Energy Initiative 2016). It is noteworthy that if peak PV generation and peak customer demand do not align, then ramping requirements will increase.

Germany has experienced this phenomenon a noticeable amount of times over the last years. During the last few years, electricity prices have reached as low as negative 320 EUR/MWh and remain below zero for hours at a time. In Fig. 4, we present average electricity prices during April in Germany. The dramatic swings

---

[2]Nuclear power plants have low ramping capabilities compared to hydro or natural gas plants.

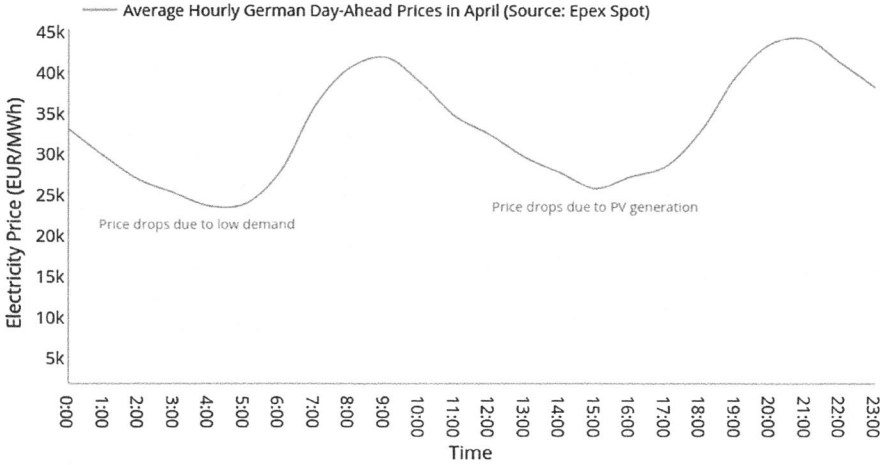

Fig. 4 Average hourly German day-ahead prices in April. *Source* Epex Spot

indicate the level of challenges that the grid managers face in smoothing power output of renewable plants. Similar challenges occur in the USA; in 2011, 18% of the solar generating hours lead to negative prices in the Electric Reliability Council of Texas region, while the duration of negative prices was nearly 6% of generating hours in California. In addition to financial losses, steep change of ramping requirements implies higher generation cost and faster ageing of generators.

4 The Role of Demand-Side Management

As discussed in the previous sections, PV integration is divided into two groups as low-voltage and medium-voltage integration options. Similarly, demand response applications that control *flexible* appliances are used to address the integration issues mentioned above for both integration options. Power systems are uniquely critical infrastructures as they provide an enabling function across several critical infrastructures such as transportation, industry, manufacturing and communications. For wide-area power system control, one of the main operating challenges of the power grids is to keep the supply in balance with demand, so that system frequency (e.g. 60 Hz in the USA) is kept within limits. As the production from intermittent sources increases, there is a need for spinning reserves to compensate variability and keep the system in sync. As shown in Fig. 5, there are different balancing issues with different timescales. As circled in this figure, automated demand response programs such as incentive-based programs can be highly effective. These programs often involve a hardware and communication system that automates load switching and control according to signals sent from utility operator. Therefore,

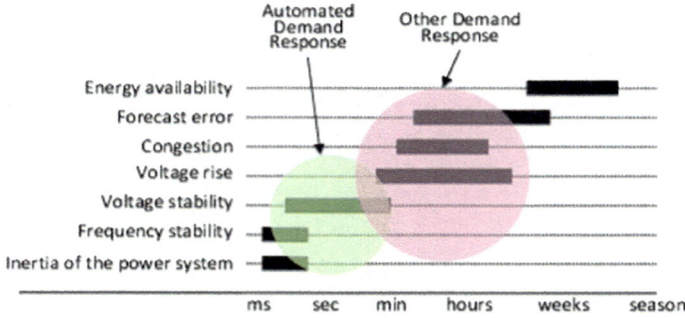

Fig. 5 Power grid balancing issues

incentive-based programs can be part of short-term ancillary services. Curtailing commercial and industrial loads such as cooling in data centres, facilities with electrochemical processing, compressing or pumping units are the most common applications. For medium timescale, pricing-based programs such as RTP and CPP can be an efficient way to handle PV intermittency. In such methods, there is no direct interaction between the grid operator and customers. However, pricing changes influence customer consumption patterns. Hence, such methods are effective in a longer timescale. Furthermore, DSM can be a part of a strategy to address ramping requirements for duck curve issues. For instance, most of the residential appliances can be shifted during the day when PV generation is high. Similarly, with smart thermostats output of thermal loads such as water heaters and AC units can be reduced and coordination of a significant number of residential units can lower ramping requirements.

For distributed PV systems, the main idea is to align PV production with customer demand so that most of the produced electricity is consumed locally. This way, reverse power flow is limited and issues discussed in section. To illustrate this we present a measurement study conducted in three-bedroom villa in the State of Qatar to measure appliance usage, a Smappee energy monitor was installed at the electrical distribution board of the house (Bayram et al. 2017). The energy monitor uses current clamps to measure consumption patterns in a non-intrusive manner. Customer demand is measured at every five minutes, and recorded values are uploaded to a cloud server using wireless home area network. Solar measurements, on the other hand, were recorded at Hamad Bin Khalifa University's solar test facility (Alrawi et al. 2018). In Fig. 6, average load and solar generation profile in August are presented. Rated PV size is assumed to be 15 kWp. It is important to note that self-consumption can be calculated as follows (Fig. 7):

$$\text{Self Consumption} = \frac{\text{SC}}{\text{SC} + \text{RP}}, \tag{1}$$

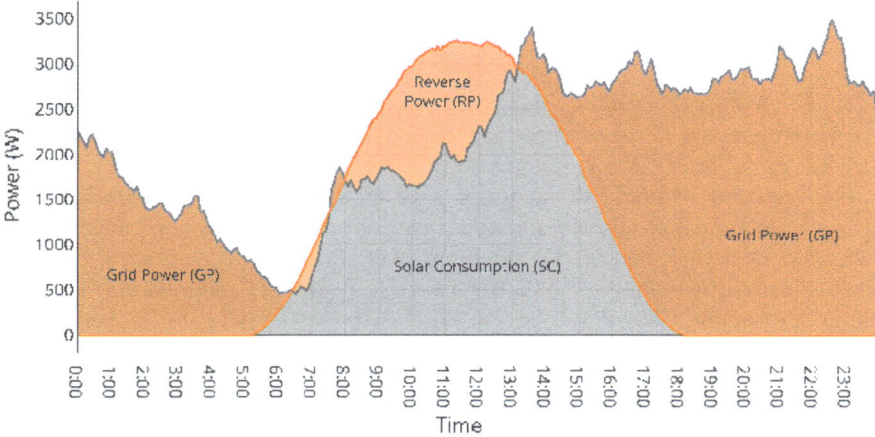

Fig. 6 Demand-PV generation profile in a three-bedroom villa in Qatar (August 2017)

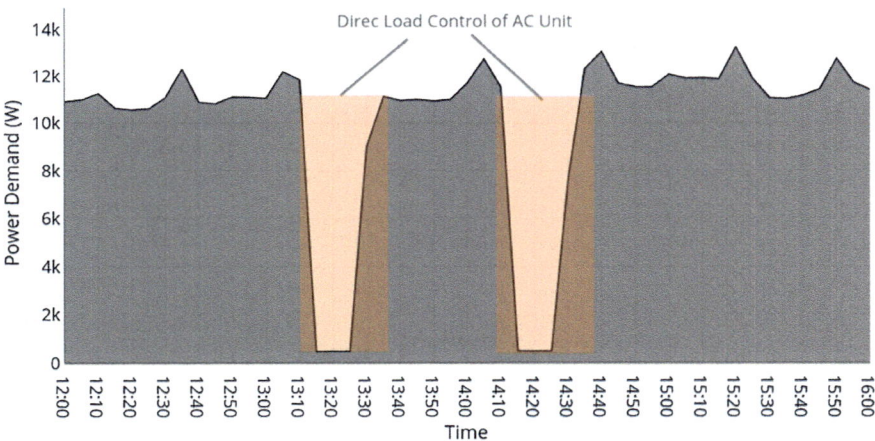

Fig. 7 Direct-load control of air-conditioner units in Qatar

where SC is the energy consumption from solar production during the day, and RP is the amount of energy sent back to the grid. In this specific example, daily electricity demand is 51.67 kWh, the total generated solar energy is 24.02 kWh, and 4.82 kWh energy is sent back to the grid. Moreover, using the formula above self-consumption can be calculated as 79.9%. The high self-consumption ratio is due to the high daily demand for air-conditioning during the day. For the same residential unit, self-consumption ratios become less than 30% in winter months. In a similar manner, *self-sufficiency* can be calculated by the formula given below,

$$\text{Self Sufficiency} = \frac{SC}{SC + GP}. \qquad (2)$$

It is noteworthy that *self-sufficiency* reflects the percentage of daily demand met by solar production. In the example given above, self-sufficiency can be calculated as 38%.

As mentioned before, PV integrated networks may need a flexible load. In regions residing in hot climates such as Qatar, air-conditioning (AC) load dominates residential consumption. Hence, direct-load control of AC units will provide enough flexibility for the distribution network to overcome abnormalities such as voltage deviations. To that end, a DLC experiment was conducted in a two-bedroom villa in Qatar during the month of August. AC unit was turned off for 30 min two times after 1 pm when AC consumption was at maximum capacity of 11 kW. This short duration is chosen so that customer comfort related to indoor temperature is not negatively affected. Moreover, instead of completely turning off the AC system, the fan could be kept on so that cool air would circulate in the building. In this case, demand reduction would be 9 kW, as the fan uses 2 kW of electrical power.

Another common demand response application is the control of electric water heaters. The residential building mentioned above also contains a water heater with a tank of 50 gallons and a rated power of 4 kW. On the left of Fig. 8, we show a typical daily consumption of the water heater. Around 7 a.m., occupants take a shower, so the water heater is on for more than half an hour. When it is not in use, water heater periodically turns on for 6–7 min. As shown in the right of this figure, depending on the PV generation, water heater cycles can be modified to increase self-consumption levels. In this example, the water heater is used to warm up the

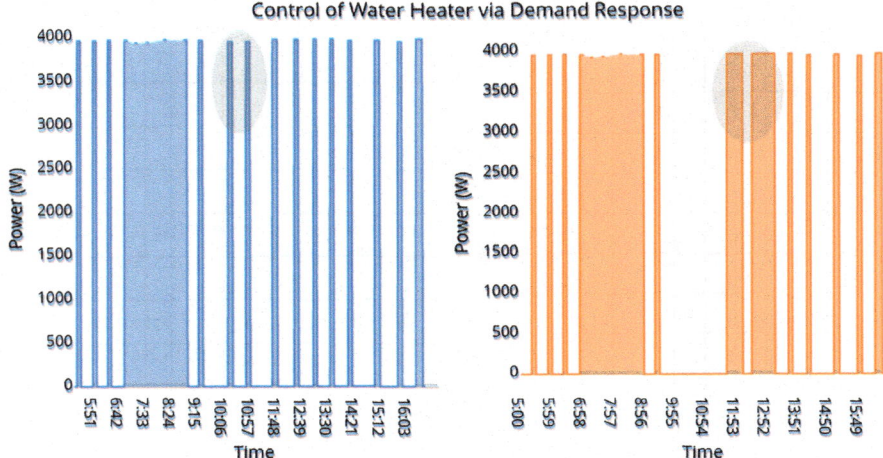

Fig. 8 Load-shifting of water heater cycles

water more during the noon to exploit extra PV generated by the solar panels (circled and shown in the right figure). Moreover, if there is no active occupant, water heating process can be suspended to help grid operations (circled and shown in the left figure). Such applications are even more common in regions with cold weather where the demand for water heating is high (Li et al. 2015).

5 Conclusions

In this chapter, we have presented a detailed overview of the role of demand-side management activities on PV integration. At a low-voltage level, we showed that PV integration is limited by several operating limits such as voltage levels, harmonics distortion and thermal loading. To overcome such issues, we showed that customer load can be shifted to the times of high PV production. Moreover, we discussed the issues related to PV farm integration such as negative prices and increase ramping requirements of baseline generators. In the last section, we presented two experimental studies: direct-load control of AC units and load shifting of electric water heaters.

References

Alrawi O, Bayram IS, Koc M (2018) High-resolution electricity load profiles of selected houses in Qatar. In: IEEE 12th international conference on compatibility, power electronics and power engineering. Doha (Qatar)

Bayram IS, Koc M, Alrawi O, Al-Naimi H (2017) Direct load control of air conditioners in Qatar: an empirical study. IEEE 6th international conference on renewable energy research and applications (ICRERA). San Diego, CA (United States), pp 1007–1012

CIGRE (2014) Capacity of distribution feeders for hosting distributed energy resources

Dartawan K, Hui L (2012) Harmonics issues that limit solar photovoltaic generation on distribution circuits. In: SOLAR 2012, world renewable energy forum (WREF 2012). Colorado Convention Center in Denver, 13–17 May

Ding F, Mather B, Gotseff P (2016) Technologies to increase PV hosting capacity in distribution feeders. In: IEEE power and energy society general meeting

Karimi M, Mokhlis H, Naidu K et al (2016) Photovoltaic penetration issues and impacts in distribution network—a review. Renew Sustain Energy Rev

Li X, Borsche T, Andersson G (2015) PV integration in low-voltage feeders with demand response. In: 2015 IEEE Eindhoven Powertech, PowerTech 2015

Portugal (2018) Portugal looks to renewables as March output tops mainland power demand. https://www.reuters.com/article/portugal-energy-renewables/portugal-looks-to-renewablesas-%0Amarch-output-tops-mainland-power-demand-idUSL5N1RG35T

Teoh WYT, Tan CW (2011) An overview of islanding detection methods in photovoltaic systems. World Acad Sci Eng Technol

The MIT Energy Initiative (2016) Utility of the future

Tsuchida B, Sergici S, Mudge B et al (2015) Comparative generation costs of utility-scale and residential-scale PV in Xcel energy colorado's service Area

Chapter 14
Concentrating Solar Power and Desalination Plants

Patricia Palenzuela and Diego C. Alarcón-Padilla

Abstract Many developing countries face serious water and electricity shortage, which make water and energy supply a matter of national security that requires cost-effective and reliable processes. Thermal desalination processes have been an answer for the shortage of drinkable water especially in the Middle East and North Africa. The problem is that these processes are intensive energy consumers, so if fossil fuels are used as a primary energy source, it will lead to a negative impact on the environment due to the CO_2 emissions. Improving the efficiency of this technology will be a good solution for the growth of these countries, which, in turn, have high insolation levels. In this scenario, freshwater and electricity cogeneration by integrating desalination plants into concentrating solar power plants (CSP+D) is proposed as one of the most sustainable options to solve water and energy. It seems clear that the best integration concepts should be chosen in order to maximize the freshwater and power production. This chapter shows a state of the art of desalination processes powered by concentrating solar technologies, considering the most promising desalination technologies: multi-effect distillation and reverse osmosis. The impacts of the location site and the choice of the cooling system on the overall efficiency of the CSP+D plant and their electricity and water costs are discussed and compared between each other. Though no CSP+D plant has been built yet, the current ongoing research can help to decide which desalination process is more suitable to be coupled to a CSP plant, depending on the location in which the project will be implemented.

1 Introduction

Water shortage is becoming a more serious threat to many countries. Even regions that had no severe water scarceness are now struggling to have a freshwater supply. Industrial desalination plants are among the best technological solutions to provide

P. Palenzuela (✉) · D. C. Alarcón-Padilla
Plataforma Solar de Almería (PSA-CIEMAT), Ctra. de Senés, s/n,
04200 Tabernas, Almería, Spain
e-mail: patricia.palenzuela@psa.es

© Springer Nature Switzerland AG 2019   327
J. Polo et al. (eds.), *Solar Resources Mapping*, Green Energy and Technology,
https://doi.org/10.1007/978-3-319-97484-2_14

freshwater from seawater. Regions such as the Middle East and North Africa have been solving the water scarcity by thermal desalination plants driven by fossil fuels so far, which is neither sustainable nor economically feasible in a long-term perspective, as fuels are increasingly becoming expensive and scarce (IEA-ETSAP and IRENA 2012). Although the case of south of Europe is not as drastic as the Middle East, it increasingly suffers from a serious lack of freshwater supplies and it depends all the more on seawater desalination to solve this problem. Approximately 50% of the population in these regions lives within 200 km of the coast and many locations are good candidates for the development and installation of solar energy plants due to the high values of solar irradiation available. Therefore, it is clear that there is a nexus between energy and water that makes the coupling between desalination processes and solar energy technologies a real alternative to fossil fuel-powered desalination.

Among all desalination technologies, reverse osmosis (RO), multi-effect distillation (MED), and multi-stage flash evaporation (MSF) are the most widespread (Water Desalination Report 2017). RO is the process with the largest worldwide installed capacity, followed by MSF. However, MED technology is more preferable than MSF, since it has higher overall efficiency, higher heat transfer coefficients, and less water recycling (Sommarva 2008; Gastli et al. 2010). When considering the solar energy application in desalination processes, two different concepts can be distinguished: (a) direct solar desalination, in which the desalination unit and the solar collector are integrated within a unique device and (b) indirect solar desalination, in which a conventional desalination system is coupled to a solar collector field that provides the energy (power or thermal energy) required by the desalination process. The most usual application for large capacities is indirect solar desalination systems. In the case of thermal energy, it is possible to establish a classification of the different solar thermal collectors depending on the sun-tracking motion and the operating temperature: (a) stationary solar collectors (flat-plate collectors, compound parabolic concentrators, evacuated tube collectors) that have no sun-tracking motion and operate at a temperature level between 40 and 200 °C, (b) single-axis tracking (parabolic-trough collectors and linear Fresnel collectors) with a temperature level between 125 and 500 °C and (c) two-axis tracking (parabolic dishes and central receiver systems) that operate between 100 and 2000 °C. The last two solar systems operate with high concentration factors and are widely used in concentrating solar power (CSP) plants to produce electricity.

2 State of the Art of Desalination Processes Powered by Concentrating Solar Technologies

The combination of CSP plants and desalination processes is of great interest due to the inherent synergy existing between both systems. In the case of thermal desalination plants, the use of common facilities and further utilization of process heat available in CSP plants makes this integration very attractive. Generally speaking, the combined freshwater and electricity production, so-called CSP+D, provides the following advantages: reduces the cost of combined power and desalination production against independent plants; improves the cost-effectiveness of the cogeneration plants by making better use of a common infrastructure and the economy of scale of the steam turbine; and allows additional savings of greenhouse gas emissions coming from the production of freshwater. Also, it can reduce conflicts due to water and energy scarcity and cut down on the economic risks related to the cost increase of non-renewable energy sources (Fichtner 2011; Debele-Negewo 2012). Besides these advantages, taking into account the difficulties for the market introduction of CSP plants and desalination processes as stand-alone power plants, the combination of both could represent an opportunity to facilitate their accessibility.

One of the main issues in CSP+D is the choice of the most adequate desalination technology to be coupled with the CSP plant. In the case of thermal desalination technologies, as MED, they normally use vapor from the power cycle as the heat source for the desalination process. There are two main configurations in MED: low-temperature operation mode (LT-MED) and thermal vapor compression operation mode (MED-TVC). In the first one (see Fig. 1), the exhaust steam from the turbine is used to drive the desalination plant and, in this way, the energy that would otherwise be dissipated in the cooling of the power cycle is used to produce freshwater. In the second case, a steam ejector is coupled to the MED plant (see Fig. 2), which increases the thermal efficiency of the desalination unit recovering part of the low-pressure steam (also called entrained vapor) generated in the process by the use of a high-pressure steam (also called motive steam) extracted, in this case, from the power cycle. In both operation modes, the power cycle suffers from a decrease in the efficiency in this layout, due to the higher temperature required in the MED plant (around 70 °C) in the first case, and due to the use of high-pressure steam from the cycle, in the second case. An alternative scenario is to feed the electricity produced by the CSP plant into an RO plant (see Fig. 3) with the further advantage that, in this case, the power and the desalination plants can be away from

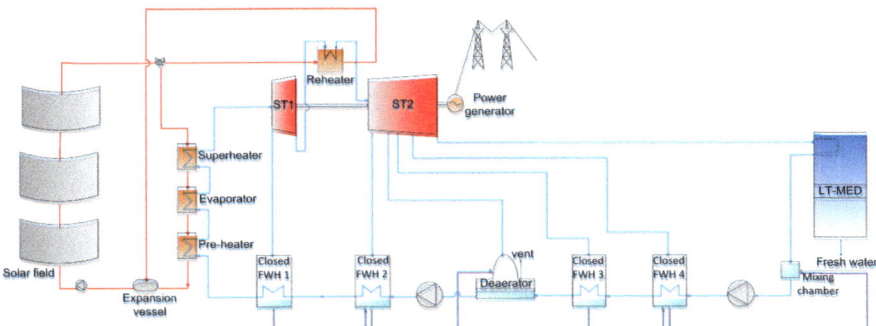

Fig. 1 Flow diagram of an LT-MED integrated into a CSP plant

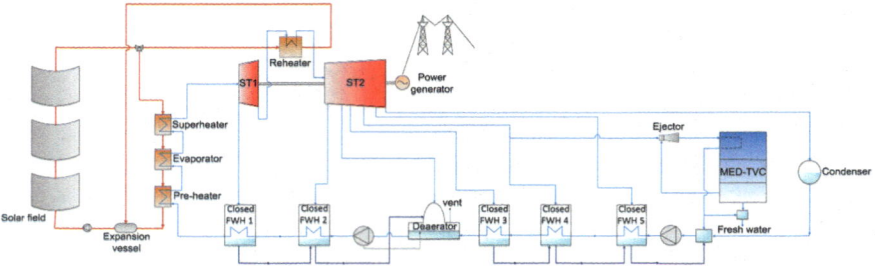

Fig. 2 Flow diagram of a MED-TVC integrated into a CSP plant

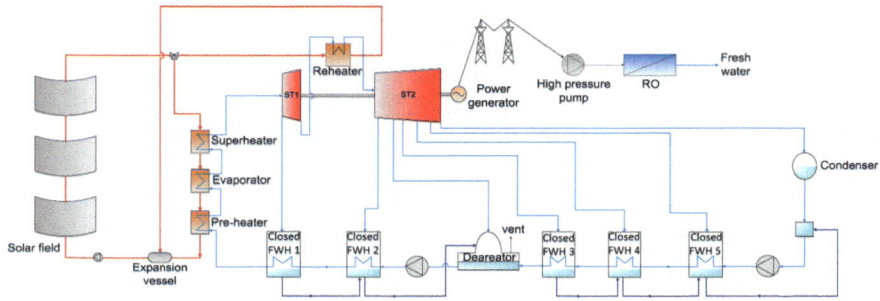

Fig. 3 Flow diagram of an RO unit connected to a CSP plant

each other. In this layout, a penalty in the power production can also occur owing to the significant electricity consumption of the RO plant. The selection of the best desalination option depends on many factors such as the required power to water ratio, cost of fuel energy charged to the desalting process, electricity sales, capital costs, and local requirements that have to be taken into account when choosing the most suitable combination (Mussati et al. 2003). Another sensitive issue is the land cost and availability in the case of the CSP plant being located near the sea due to the desalination demands. Solar direct normal irradiance (DNI) is normally lower in these areas, so the most optimal locations of CSP plants would be far from the coast. The great water consumption in the CSP plant due to the refrigeration system (in case of wet cooling) and to the mirror cleaning is another aspect to be considered. On the other hand, solar energy, and indeed renewable energies in general, are resources that typically vary depending on the weather. This is a problem for desalination plants because they need to be in operation continually, even when working at partial load.

Though no commercial CSP+D plants have been built yet, the current ongoing research can help to decide which desalination process is more suitable to be coupled to a CSP plant. Feasibility studies for the deployment of CSP+D plants started in 2002, when (Trieb et al. 2002; F et al. 2004) presented the economic perspectives for the combination of solar power and desalination plants for

Southern Europe. In 2008 and later in 2009, the same author (Trieb and Müller-Steinhagen 2008; Trieb et al. 2009) gave a long-term scenario for solving the demand of freshwater in MENA (the Middle East and North Africa) by seawater desalination powered with solar energy. In these works, the authors concluded that CSP+D systems are a safe and sustainable solution, capable of addressing the growth deficits present in these regions. Other works investigated the CSP+D potential in specific locations. Hamdan et al. (2008) proposed to solve the depletion of freshwater in Gaza Strip, in this case by a cogeneration (power and water) solar-powered plant, consisting in a parabolic-trough solar power plant combined with an MSF plant. Gastli et al. (2010) evaluated the possibilities of coupling between CSP plants and desalination plants in Duqum (Oman) considering MED and RO as desalination alternatives. They highlighted the advantages of the MED coupling arrangements in comparison with the RO option. However, the previous studies did not evaluate the economic and technical performance of the complete system in detail. Only a deep study in specific locations that accommodates most of the previously mentioned factors could provide the best combination for CSP and desalination plants. The solar desalination R&D Unit at the Plataforma Solar de Almería has been working in the CSP+D research line since 2008 and they have performed lots of exhaustive techno-economic studies in different locations under several boundary conditions. These and other complementary studies published in the scientific literature have provided interesting conclusions that help to select the best CSP+D combination, which is detailed hereinafter.

Generally speaking, in locations with low seawater salinity (such as those ones in the Mediterranean area), the configuration CSP+RO has shown to be more favorable than CSP+MED (Palenzuela et al. 2011; Olwig et al. 2012). Although the electricity consumption of MED plants is quite lower than RO, the higher steam pressure at the outlet of the turbine (in case of CSP+MED) penalizes the power production with respect to the case with RO, and this penalty is large enough to overcome the saving in gross power production due to the LT-MED having less electrical consumption than RO. The required modifications at the power cycle in the case of CSP+MED can decrease the power cycle efficiency from 38 to 33% (Schmitz et al. 2009). The difference is even higher with small-scale MED plants due to their too low GOR and too much electricity consumption to be competitive with RO on an energetic basis (Ghobeity et al. 2011). The economic results are also in favor of CSP+RO, although the difference in the electricity costs is almost negligible (0.4%) (Palenzuela et al. 2011). The difference in water costs is higher (3% lower water costs than CSP+MED (Palenzuela et al. 2011)) which is mainly due to the lower investment costs for RO (Schmitz et al. 2009; Gastli et al. 2010). On the other hand, locations in the Mediterranean basin present lower ambient temperatures, which allow low outlet turbine temperatures. In these locations, the most frequent cooling method for CSP plants is evaporative and the difference in the overall efficiency of a power cycle using this cooling system with respect to the use of dry cooling increases in favor of the former (Blanco-Marigorta et al. 2011). Also, the difference in the combination of an RO with a CSP plant using evaporative cooling with respect to the combination of a MED unit with a CSP plant becomes higher (Olwig et al. 2012; Palenzuela et al. 2013, 2015b). The CSP+RO system using evaporative cooling in the power cycle presents 7% higher overall efficiency (Palenzuela et al. 2013, 2015b) than the rest of the cooling methods. In terms of electricity costs, the CSP+RO using this wet cooling method in the power block is also the best (around 6% lower costs (Palenzuela et al. 2013, 2015b)). However, the water consumption in wet cooling systems is an issue. Around 4000 m^3/day of freshwater is needed in the condenser for a CSP+RO plant with 50 MW$_e$ and around 50000 m^3/day (Palenzuela et al. 2013, 2015b), which increases the size of the RO unit about 10%, affecting the investment costs (higher LWC). The other wet cooling method, once-through, seems to be also suitable for Mediterranean areas due to the low exhaust steam temperatures required (the steam condensation temperature depends on the seawater temperature). The problem is that this cooling method requires large amounts of seawater to be pumped from the sea, adding to the environmental problems that limit the seawater temperature at the outlet of the condenser. On the other hand, the pumping of seawater means an increase in the levelized electricity cost (LEC) between 4 and 9%, depending on the boundary conditions (Palenzuela et al. 2013, 2015b). In the case that a CSP+RO system using once-through or dry cooling (power consumption from the funs) as cooling methods are compared with the CSP+MED option for a location in the Mediterranean basin, the electricity costs of the former become higher than the latter provided the specific electricity consumptions of the RO unit is above 4.5 kWh/m^3 (Palenzuela et al. 2015d).

Regions such as the Middle East/Arabian Gulf have seasonal high seawater temperatures and saline concentrations, which make the use of steam to drive the MED process more attractive against the RO option (Moser et al. 2010; Blanco et al. 2013; Palenzuela et al. 2013; Fylaktos et al. 2015). The higher salinity leads to a higher electrical consumption of the RO plant overcoming the lower thermal efficiency of the turbine in the MED case. In these locations, the water availability is lower and dry cooling is preferred to condense the exhausted vapor in the power cycle. Moreover, in these locations, the ambient conditions imply to increase the temperature level at the turbine outlet, so the penalty in the power production that results from using steam to drive a MED process is substantially reduced with respect to the case of CSP+RO using dry cooling for the power cycle (Blanco et al. 2013). From the thermodynamic studies, it has been found that the coupling of a MED unit with a CSP plant is 9% better than the RO option with dry cooling in terms of overall efficiency. Also, the thermal system results preferable from an economic point of view, being the electricity costs 13% lower (Palenzuela et al. 2015c). This coupled system is also more favorable than the RO option using once-through, mainly as a result of the extra power that the CSP must generate for this cooling system (Palenzuela et al. 2015d). In the case of using evaporative cooling for the CSP+RO system, only for specific electricity consumptions higher than 4 kWh/m^3 in the RO plant, the CSP+MED option results more favorable (Palenzuela et al. 2015d). In the case of the water costs, the RO option with dry cooling performs better than the MED one as a result of the lower investment costs of RO. However, the difference between the two options is not that high (less than

5%, (Palenzuela et al. 2015a)) and might not be a strong enough reason for choosing this system. Also, in these regions, RO is heavy penalized from the O&M point of view due to the raw water quality and possible red algae blooms that force to the use of very sophisticated pretreatments in order to avoid the damage of the membranes (Gastli et al. 2010). Besides, the possible substitution of the cooling system in the CSP+MED could lead to a 10% lower investment for the power block (Gastli et al. 2010).

Nevertheless, the full condenser replacement in the CSP+MED configuration can be a great risk for the implementation of a CSP+D demonstration plant, since the power production is therefore entirely dependent on the desalination system. Other configurations of MED integrated into the power cycle of the CSP plant could avoid these aspects allowing the use of the power cycle condenser in case of failure or maintenance in the desalination plant and providing further advantages such as allowing the exhaust steam to fully expand and decreasing the fluctuations of the available heat for the MED plant that could destabilize the freshwater production. The alternative MED configurations can be a MED desalination plant driven either by a thermocompressor or by a mechanical vapor compressor, namely MED-TVC and MED-MVC. The combination of a MED-TVC unit with a concentrating solar plant can be either by using the hot oil produced by the solar field and a boiler to provide the steam for the thermocompressor or by integrating the MED-TVC unit into a CSP plant. In the case of the combination of a MED-MVC with CSP, the vapor generated in a boiler is used to drive an organic rankine cycle (ORC) and the power output to feed the MED-MVC unit. Such a combination is not very usual due to the higher specific electricity consumption of the desalination plant, high water price and product costs (Sharaf et al. 2011). On the other hand, in the case of the coupling of a MED-TVC unit with a CSP plant, the thermal efficiency of the coupled plants is higher than that of both plants separately. The coupled system leads to a reduction of roughly 3% in the aperture area with respect to a system composed of two independent plants (which means cost saving) (Palenzuela et al. 2011). As already mentioned, the main problem of the CSP+MED-TVC system is the penalty suffered in the power production with respect to the CSP+RO option due to the extraction of steam from the turbine to be used as high-pressure steam for the steam ejector. Therefore, the integration process between a MED-TVC unit and a CSP plant is not a straightforward issue but it requires an optimization process as a function on the motive steam pressure and the position of the steam ejector in order to maximize the overall efficiency and minimize the costs of the full system. Regarding the motive steam pressure, the lower the steam pressure extracted from the turbine to feed the ejector as motive steam, the higher the overall efficiency (Ortega-Delgado et al. 2016a), which allows to decrease the differences regarding the case of RO (Palenzuela et al. 2015d). With respect to the position of the steam ejector, there is an optimal location for each motive steam pressure that maximizes the thermal efficiency of the MED-TVC and minimizes the specific heat transfer area (i.e., the capital costs of the desalination plant). These positions are closer to the last effect the higher the motive steam pressure and to intermediate effects for low motive steam pressures (Ortega-Delgado et al. 2016b).

As the combination of an RO process with a CSP plant is better with respect to any CSP+MED system the lower the exit turbine temperature, the CSP+MED-TVC option will be more penalized in those locations with lower average ambient temperature. In (Palenzuela et al. 2015b), it was found that the differences in the overall efficiency and electricity costs were around 15 and 10%, respectively. In warmer locations, the difference in overall efficiency and electricity costs found was not that high (11 and 7%, respectively) (Palenzuela et al. 2015b).

A new concept for the integration of multi-effect distillation with thermal vapor compression can be considered. In this case, part of the exhaust steam and steam extracted from the turbine are used as entrained vapor and motive steam, respectively, in the ejector that produces medium-pressure vapor for driving the low-temperature MED plant (LT-MED), a design known as LT-MED-TVC (see Fig. 4). This scheme has the same advantages than the ones mentioned for MED-TVC and also is more efficient. For high exhaust temperatures and specific electricity consumptions above 4.5 kWh/m^3 for the RO plant, the option with LT-MED-TVC could result more favorable than the one with RO (Palenzuela et al. 2015d). As a matter of fact, in case of using dry cooling in the power cycle, the overall efficiency of the thermal distillation scheme resulted 3% higher and the electricity costs 5% lower than the case with RO (Palenzuela et al. 2015b). In locations with low exhaust steam temperature and low salinity, this novel concept has similar results as the CSP+RO system in terms of overall efficiency and electricity costs, if dry cooling is considered in the power cycle. Therefore, it could be contemplated as a further option considering the saving in the cooling requirements of the CSP+LT-MED-TVC scheme (part of the exhaust steam is used to feed the steam ejector instead of being condensed through the power cycle).

On the other hand, the combination of membrane and thermal desalination processes by an effective integration could also reduce the water and power costs when they are coupled to a CSP plant (Ludwig 2004). These processes are characterized by the flexibility in the operation, low specific energy consumption, high plant availability, and a better power to water ratio (De Gunzbourg and Larger 1999). Also, the hybridization permits to use the same intake and outfall

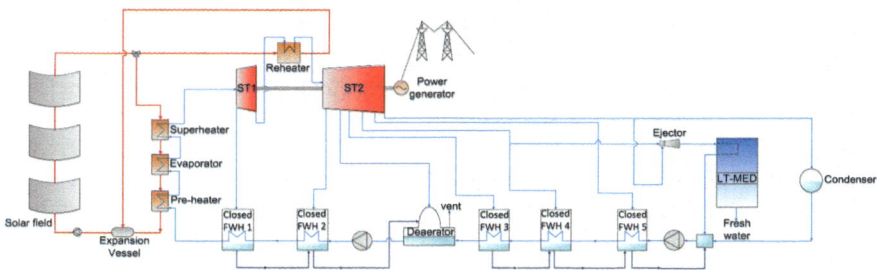

Fig. 4 Flow diagram of an LT-MED-TVC integrated into a CSP plant

installations with the subsequent benefit in pumping energy, compared with the stand-alone cases. Besides, it allows the feed to the membrane process to be warmed up and to increase the permeate production, leading up to a 25% of water costs reduction in RO/MED hybrid systems coupled to a CSP plant (Iaquaniello et al. 2014). In this kind of hybrid desalination systems, it is recommended to use a low-pressure steam extraction for the MED unit in case any subsidy for power and water generation is applied. On the contrary, in case that only an electricity feed-in tariff is accounted, it is not recommended to extract any steam for the distillation unit (Ghobeity et al. 2011).

Although the CSP technology included in most of the mentioned studies is parabolic trough due to its higher maturity and proven operation in real plants, other CSP technologies can be considered for the thermoelectric production. The one with the highest potential in terms of the electricity and water production costs is the central receiver (Kalogirou 2013), especially in regions with the highest solar irradiation levels (Kouta et al. 2016). On the other hand, according to the temperature levels, a convenient combination could be a Fresnel CSP plant with a MED-TVC unit, which even is projected to obtain similar results in terms of water costs compared with the use of fossil fuels for the desalination plant, especially in locations with high annual direct normal irradiation (Hamed et al. 2016).

3   Worldwide Experiences in CSP+MED Plants

3.1   PROTEAS Field Facility

The pilot/experimental plant consists of a heliostat-central receiver system for solar harvesting, thermal energy storage in molten salts (60–40% b.w. of $NaNO_3$–KNO_3) followed by a Rankine cycle for electricity production and a multi-effect distillation unit for desalination. Some pictures of the facility are depicted in Fig. 5 and a simplified schematic of the concept is shown in Fig. 6 (Papanicolas et al. 2016). The solar field consists of 50 heliostats, each with a reflective area of 5 m^2 and

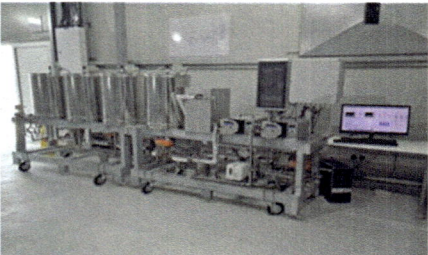

Fig. 5 Overall view of the PROTEAS field facility (left), Multi-effect distillation unit (right)

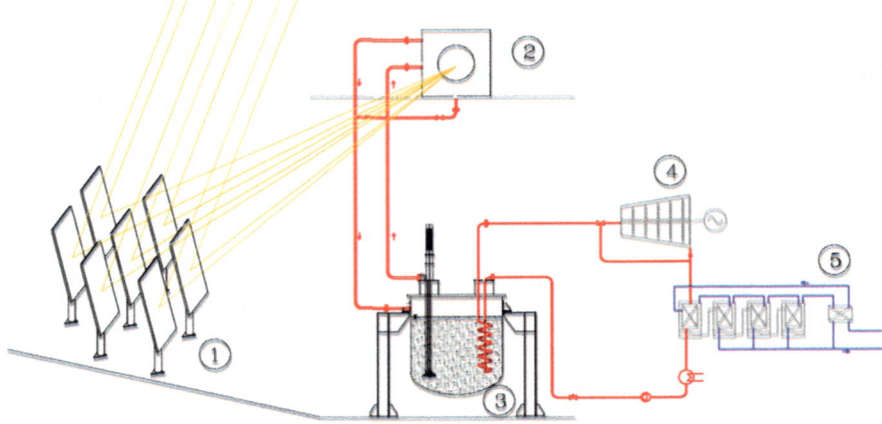

Fig. 6 Simplified schematic of the plant layout, with the following components schematically indicated: (1) the heliostat field; (2) the central receiver; (3) the molten salt storage tank; (4) the steam turbine; and (5) the MED desalination unit (Papanicolas et al. 2016)

constructed out of a single mirror facet. Each mirror has a reflectivity of 93%. The central receiver is a novel design that integrates the receiver and storage functions, which allows reducing complexity, operational and capital costs. It is placed on a 14 m tower. The CSP plant has a nominal thermal power of 45 and 150 kW$_{th}$ maximum. There is one storage tank that has a height of 2.8 m and a volume of 8 m^3. It has been designed to operate at temperatures up to 600 °C in a non-pressurized environment, resulting in a total thermal storage capacity up to 0.6 MWh. Also, five electrical heaters, with a total capacity of 45 kW, are installed as a backup to maintain the salt in a molten state all the time. A superheated coil is in contact with the thermal energy storage lid, which raises the steam temperature by 5–10 °C above the saturation temperature, generating the necessary superheated state for the turbine operation. The steam turbine is fed with 10 kW$_{th}$ of the superheated steam producing 1 kW$_e$ at the design point. The exhaust steam leaving the turbine is used as thermal input to the desalination plant. This desalination plant is a custom designed four-effect distillation unit (MED) constructed to operate either in series or in parallel. The unit is designed to produce 1 m^3 of distillate product per day. This facility has been recently installed and no test campaigns have been performed and published yet.

3.2 MATS Research Project

The MATS project is focused on the innovative CSP technology developed by ENEA as an improvement of its solar thermodynamic technology based on molten salts as heat transfer fluid. This technology, referred as TREBIOS, allows combined heat and power production from solar source integrated with renewable fuels, such as biomass, biogas, and industrial residues by means of standardized units that provide high performances and limited cost.

The objective of the proposal is the full-scale demonstration of TREBIOS technology through the industrial development, the realization, and the experimental operation of a multipurpose facility to be installed in Egypt. The thermal energy produced by this plant will be used as an energy source in a desalination unit included in the installation, as well as for district heating and cooling. The use of suitable heat storage systems enhances mismatch of power production from the instantaneous solar radiation availability. These features enable electrical energy production "on demand" and the optimized utilization of captured solar energy by additional loads like desalination. The integration with a backup gaseous fuel, from either biomass or natural gas, makes the system flexible and enables continuous power production. The plant has been recently implemented and has started the experimental campaign within the demonstration phase of the project. The main specifications of the demonstration plant in MATS project are shown in Table 1.

3.3 Sundrop Farms

This is the first real application for the cogeneration of electricity and desalinated seawater from concentrated solar thermal power all over the world. In 2010, it began the operation of the Sundrop Farms' first commercial greenhouse facility in South Australia (see Fig. 7). The greenhouse is situated at the top of the Spencer Gulf, near the city of Port Augusta. Given the lack of freshwater and the harsh climates, traditional horticulture is not feasible in this area. It uses a state-of-the-art solar tower to produce energy to power the plant growing systems and to heat and cool the greenhouses as required. It is 115 m high and has 23,000 mirrors pointed at it. The CSP system has been designed and delivered by Danish renewable energy specialist, Aalborg CSP, and it is the first large-scale CSP-based technology in the world to provide multiple energy streams—heating, freshwater, and electricity—for horticultural activities. The 51,500 m^2 solar field comprises the solar collector system. Commissioned in October 2016, the heat production rate is 20,000 MWh/year and produces 250,000 m^3/year of desalinated water while producing 1700 MW$_e$/year of electricity.

The water comes from the Spencer Gulf with a salinity of around 47,000 ppm via an inlet channel, about 5 km away. It is heated to make steam which powers the

Table 1 Main specifications of the demonstration plant in MATS project		
	Electric power	1.0 MW$_e$
	Outlet thermal power	4.0 MW$_{th}$
	Inlet thermal power	5.7 MW$_{th}$
	Desalination unit	250 m^3/day

Fig. 7 Overall view of the Sundrop Farms Facility

greenhouses and provides controlled temperatures of 22–26 °C during the day and 16 °C at night. The water is kept in storage tanks before being delivered via a pipe system throughout the greenhouses. It is then recycled and headed to the solar tower, where it is heated and used for desalination by a MED plant with a capacity of 1000 m^3/day and a specific thermal consumption of 110 kW$_{th}$/m^3. The desalinated seawater does not return waste brine to the Spencer Gulf. The brine is collected in ponds from which salt could be harvested. The expanded facility discharges its brine into the cooling water outflow channel at the existing coal-fired Port Augusta power stations. Sundrop Farms continues to investigate commercially viable solutions for the recovery of minerals from brine at a large scale.

References

Blanco-Marigorta AM, Victoria Sanchez-Henríquez M, Peña-Quintana JA (2011) Exergetic comparison of two different cooling technologies for the power cycle of a thermal power plant. Energy. https://doi.org/10.1016/j.energy.2010.09.033

Blanco J, Palenzuela P, Alarcón-Padilla D et al (2013) Preliminary thermoeconomic analysis of combined parabolic trough solar power and desalination plant in Port Safaga (Egypt). Desalin Water Treat. https://doi.org/10.1080/19443994.2012.703388

De Gunzbourg J, Larger D (1999) Cogeneration applied to very high efficiency thermal seawater desalination plants. Desalination. http://dx.doi.org/10.1016/S0011-9164(99)00139-3

Debele-Negewo B (2012) Renewable energy desalination an emerging solution to close the water gap in the middle east and the north of Africa, Washington (USA)

F T, S K, V Q et al (2004) SOKRATES-Projekt Solarthermische Kraft-werkstechnologie für den Schutz des Erdklimas, DLR, ISE, PSE

Fichtner (2011) MENA regional water outlook part II desalination using renewable energy

Fylaktos N, Mitra I, Tzamtzis G, Papanicolas CN (2015) Economic analysis of an electricity and desalinated water cogeneration plant in Cyprus. Desalin Water Treat. https://doi.org/10.1080/19443994.2014.940219

Gastli A, Charabi Y, Zekri S (2010) GIS-based assessment of combined CSP electric power and seawater desalination plant for Duqum—Oman. Renew Sustain Energy Rev 14:821–827. https://doi.org/10.1016/j.rser.2009.08.020

Ghobeity A, Noone CJ, Papanicolas CN, Mitsos A (2011) Optimal time-invariant operation of a power and water cogeneration solar-thermal plant. Sol Energy. https://doi.org/10.1016/j.solener.2011.06.023

Hamdan LK, Zarei M, Chianelli RR, Gardner E (2008) Sustainable water and energy in Gaza Strip. Renew. Energy

Hamed OA, Kosaka H, Bamardouf KH et al (2016) Concentrating solar power for seawater thermal desalination. Desalination. https://doi.org/10.1016/j.desal.2016.06.008

Iaquaniello G, Salladini A, Mari A et al (2014) Concentrating solar power (CSP) system integrated with MED-RO hybrid desalination. Desalination. https://doi.org/10.1016/j.desal.2013.12.030

IEA-ETSAP, IRENA (2012) Water desalination using renewable energies

Kalogirou SA (2013) Solar thermoelectric power generation in Cyprus: selection of the best system. Renew Energy. https://doi.org/10.1016/j.renene.2012.01.014

Kouta A, Al-Sulaiman F, Atif M, Marshad S Bin (2016) Entropy, exergy, and cost analyses of solar driven cogeneration systems using supercritical CO2Brayton cycles and MEE-TVC desalination system. Energy Convers Manag. https://doi.org/10.1016/j.enconman.2016.02.021

Ludwig H (2004) Hybrid systems in seawater desalination - Practical design aspects, present status and development perspectives. Desalination. https://doi.org/10.1016/s0011-9164(04)00151-1

Moser M, Trieb F, Kern J (2010) Combined water and electricity production on industrial scale in the MENA countries with concentrating solar power. Tel Aviv, Israel

Mussati S, Aguirre P, Scenna N (2003) Dual-purpose desalination plants. Part II. Optimal configuration. Desalination. http://doi.org/10.1016/S0011-9164(02)01126-8

Olwig R, Hirsch T, Sattler C et al (2012) Techno-economic analysis of combined concentrating solar power and desalination plant configurations in Israel and Jordan. Desalin Water Treat. https://doi.org/10.1080/19443994.2012.664674

Ortega-Delgado B, Palenzuela P, Alarcón-Padilla DC, García-Rodríguez L (2016a) Quasi-steady state simulations of thermal vapor compression multi-effect distillation plants coupled to parabolic trough solar thermal power plants. Desalin Water Treat. https://doi.org/10.1080/19443994.2016.1173377

Ortega-Delgado B, Palenzuela P, Alarcón-Padilla DC (2016b) Parametric study of a multi-effect distillation plant with thermal vapor compression for its integration into a Rankine cycle power block. Desalination. https://doi.org/10.1016/j.desal.2016.04.020

Palenzuela P, Alarcón-Padilla D-C, Zaragoza G (2015a) Concentrating solar power and desalination plants : engineering and economics of coupling multi-effect distillation and solar plants

Palenzuela P, Alarcón-Padilla DC, Zaragoza G (2015b) Large-scale solar desalination by combination with CSP: techno-economic analysis of different options for the Mediterranean Sea and the Arabian Gulf. Desalination. https://doi.org/10.1016/j.desal.2014.12.037

Palenzuela P, Alarcón-Padilla DC, Zaragoza G, Blanco J (2015b) Comparison between CSP +MED and CSP+RO in mediterranean area and MENA region: techno-economic analysis. Energy Procedia 69:1938–1947. https://doi.org/10.1016/j.egypro.2015.03.192

Palenzuela P, Zaragoza G, Alarcón-Padilla DC (2015d) Characterisation of the coupling of multi-effect distillation plants to concentrating solar power plants. Energy. https://doi.org/10.1016/j.energy.2015.01.109

Palenzuela P, Zaragoza G, Alarcón-Padilla DC, Blanco J (2013) Evaluation of cooling technologies of concentrated solar power plants and their combination with desalination in the mediterranean area. In: Applied thermal engineering

Palenzuela P, Zaragoza G, Alarcón D, Blanco J (2011) Simulation and evaluation of the coupling of desalination units to parabolic-trough solar power plants in the Mediterranean region. Desalination. https://doi.org/10.1016/j.desal.2011.08.014

Papanicolas CN, Bonanos AM, Georgiou MC et al (2016) CSP cogeneration of electricity and desalinated water at the Pentakomo field facility. In: AIP conference proceedings

Schmitz K, Riffelmann K, Thaufelder T (2009) Techno-economic evaluation of the cogeneration of solar electricity and desalinated water. In: Proceedings of the solarPACES conference. SolarPaces, Berlin (Germany)

Sharaf MA, Nafey AS, García-Rodríguez L (2011) Thermo-economic analysis of solar thermal power cycles assisted MED-VC (multi effect distillation-vapor compression) desalination processes. Energy. https://doi.org/10.1016/j.energy.2011.02.015

Sommarva C (2008) Utilisation of power plant waste heat steams to enhance efficiency in thermal desalination. Desalination. https://doi.org/10.1016/j.desal.2007.01.122

Trieb F, Müller-Steinhagen H (2008) Concentrating solar power for seawater desalination in the middle east and north Africa. Desalination 220:165–183. https://doi.org/10.1016/j.desal.2007.01.030

Trieb F, Müller-Steinhagen H, Kern J et al (2009) Technologies for large scale seawater desalination using concentrated solar radiation. Desalination. https://doi.org/10.1016/j.desal.2007.04.098

Trieb F, Nitsch J, Kronshage S et al (2002) Combined solar power and desalination plants for the Mediterranean region—sustainable energy supply using large-scale solar thermal power plants. Desalination. https://doi.org/10.1016/s0011-9164(02)01091-3

Water Desalination Report market profile, IDA desalination yearbook 2016–2017

Chapter 15
Solar Water Detoxification

Alejandro Cabrera, Sara Miralles and Lucas Santos-Juanes

Abstract Persistent organic pollutants cannot be treated by biological processes as these compounds are non-biodegradable and mostly toxic. In this context, advanced oxidation processes (AOP) have evolved as an emerging alternative. Time is the independent variable used to represent the process of evolution in chemical engineering. Nevertheless, time is not an effective variable to represent the evolution of a solar driven process since solar irradiation is also changing during the operation of the solar photoreactor. For this reason, accumulated UV solar energy has become one of the typically used variables to represent process evolution. Besides, the modeling of photocatalytic reactors requires an analysis of the radiation field in the photoreactor. UV irradiation has to be monitored during the solar treatment by means of UV radiometer. The absorbance spectral range of the catalyst determines the measurement range; being 320–400 nm the most used one for solar photo-Fenton. Yearly and/or monthly average values are also needed for plant scaling up, consequently, solar UV monitoring is essential.

A. Cabrera (✉)
Escuela Universitaria de Ingeniería Mecánica, Universidad de Tarapacá,
Avd. General Velásquez 1775, Arica, Chile
e-mail: acabrera@limza.cl

S. Miralles
Solar Energy Research Centre (CIESOL), Ctra de Sacramento s/n,
Almería, Spain
e-mail: sara.miralles@ual.es

S. Miralles
Chemical Engineering Department, University of Almería,
Ctra de Sacramento s/n, Almería, Spain

L. Santos-Juanes
Departamento de Ingeniería Textil, Papelera de la Universitat
Politécnica de Valencia, Valencia, Spain
e-mail: lusanju1@txp.upv.es

© Springer Nature Switzerland AG 2019
J. Polo et al. (eds.), *Solar Resources Mapping*, Green Energy and Technology,
https://doi.org/10.1007/978-3-319-97484-2_15

1 Introduction: Advanced Oxidation Processes

World water consumption has grown over the last century at a rate twice that of the population. Water scarcity affects all continents and more than 40% of the population of our planet. By 2025, 1.8 billion people will live in countries or regions with a drastic lack of water, and two-thirds of the world's population could be in conditions of scarcity. This problem is being aggravated by climate change, especially in the arid regions of the world, where more than 2,000 million people currently live, and half of the total poor population. Climate change has also intensified storms and floods that destroy crops, pollute freshwater and disable the infrastructure used to store and transport it (Organization World Health and United United Nations Children's Fund 2017).

On the other hand, agriculture is the main consumer of fresh water in the world, about 70% of the fresh water that is extracted from the lakes, watercourses and aquifers of the entire planet is destined for this purpose. The figure is close to 95% in many developing countries, where about three-quarters of the world's irrigated land is found. Food consumes water, between 1000 and 2000 L of water are needed to produce a kilo of wheat, and between 13,000 and 15,000 L to produce the same amount of grain-fed beef. Without water, there is no agricultural production. In addition, it should be also highlighted the unstoppable urbanization and the increasing consumption at the domestic and industrial level of people living in the most developed areas of the planet. Furthermore, due to increased consumption, the generation of waste due to the use of water also increases and water-related diseases such as malaria, cholera, typhoid fever and schistosomiasis affect and kill millions of people every year. The excessive use and contamination of water supplies are also inflicting serious damage to the natural environment and present increasing risks to many biological species (Genthe et al. 2013). When contaminated waters reach the environment, it supposes the release of these compounds to the environment, including the reserves of groundwater, rivers, lakes, and seas. For centuries, man has used water as a place to pour different compounds resulting from human and industrial activity (Rivera-Utrilla et al. 2013).

Urban wastewater can be decontaminated by conventional biological treatments by means of activated sludge activity, a mixture of microorganisms. This process has been successfully applied for more than one hundred years, it is highly efficient and extremely inexpensive, in the range of several cents of euro per cubic meter of treated water. In spite of this, this process still presents some important limitations. Persistent organic pollutants cannot be treated by biological processes as these compounds are non-biodegradable and mostly toxic (Langenbach 2013; Cabrera Reina et al. 2015a). Toxic and non-biodegradable compounds are typically present in industrial wastewater and wastewater treatment plant (WWTP) effluents. In the case of industrial wastewater, the pollutants are present in the range of milligrams per litre while for WWTP effluents these are present in the range of micrograms or nanograms per litre.

In this context, advanced oxidation processes (AOP) have evolved as an emerging alternative. These processes are based on the generation of hydroxyl radical, highly reactive species that can oxidize almost any organic pollutant present in water. Within these AOPs, those that are able to take advantage of solar radiation are of special interest since the energy cost would be zero when the radiation is supplied by the sun and, therefore, the use of artificial radiation from lamps is not necessary. These processes are heterogeneous and homogeneous (photo-Fenton) photocatalysis (Malato et al. 2009).

During the solar photo-Fenton cycle, Fe(II) reacts with hydrogen peroxide yielding Fe(III) and hydroxyl radical and then Fe(III) can be photo-reduced by UV radiation so that Fe(II) is again generated. In this way, hydrogen peroxide, iron and UV radiation drive the photo-Fenton process. As far as hydrogen peroxide is present in the solution, the interaction between catalyst load and UV radiation will determine the amount of radicals produced and, consequently, the decontamination process (Cabrera Reina et al. 2017).

The heterogeneous solar photocatalytic detoxification process consists of making use of the near-ultraviolet (UV) band of the solar spectrum (wavelength shorter than 400 nm), to photo-excite a semiconductor catalyst in contact with water and in the presence of oxygen. Oxidizing species (hydroxyl radicals, $HO^{\bullet}$, produced due to the photogenerated holes), which attack oxidizable contaminants, are generated producing a progressive break-up of molecules yielding CO_2, H_2O and diluted inorganic acids (Malato et al. 2009).

It is important to note that the study of the efficiency of AOPs solar processes (photocatalysis and photo-Fenton) must be carried out based on the time elapsed and the radiation incident on the reactor. As explained below, it is an approach that integrates the radiant energy density of the solar spectrum useful for solar photocatalysis (Malato et al. 2003) and that is why solar mapping is also important in this field. The incident solar radiation on the photoreactor is included in the kinetic calculations of the degradation experiments by means of a mathematical approach that allows to compare and combine experiments carried out on different days with different meteorological conditions, as explained below. In the next sections, the importance of UV radiation in AOPs will be reviewed, evaluating the types of solar photoreactors as well as photo-reactor scaling-up procedure based on solar UV radiation measurements and mathematical models considering photo-catalyst/UV radiation interaction.

2 Solar PhotoReactors

The design of solar photoreactors applied to water/wastewater treatment is complex, mainly, due to the presence of photocatalyst, which may be suspended, immobilized or dissolved. Obviously, to use a photoreactor with good contact between reagents and catalyst is a key aspect to improve process performance but, additionally, an optimum light distribution inside the photoreactor will greatly influence the decontamination treatment (Blanco 2005). In heterogeneous processes, the presence of the catalyst causes photon absorption and scattering, which occurs along the trajectories of light beams, while for homogeneous processes, as solar photo-Fenton, only absorption takes place. In this way, during solar water treatments, other conventional operating parameters, such as temperature or pressure, share their importance in the process performance with light distribution in the photoreactor. Photocatalytic reactors cannot use a standard glass cover because of two undesired effects: (i) absorption of solar radiation between 300 and 400 nm and (ii) further decrease of the UV-transmissivity due to the damaging caused by solar radiation. The iron content of the glass is the main component causing both effects (Blanco et al. 1999) so that low-iron borosilicate glass, which presents good transmittance in the solar range to about 285 nm (Pyrex or Duran glass), is usually the choice for the construction of treatment plants. With regard to the reflecting/concentrating materials, when used, aluminium is the best option due to its low cost and high reflectivity in the solar UV spectrum on earth surface (Malato et al. 2009).

The first solar photocatalysis photoreactors appeared at the end of the eighties and were based on parabolic trough collectors (PTCs), probably, because this technology was already developed for thermal energy production applications (Bechtel Corporation 1991). Theoretically, concentrated solar energy augments the detoxification process because more high-energy photons are projected directly into the stream of water, which flows through a glass tube placed in the focal line of the reflector instead of the metal absorber tube used for other applications (Fernández-García et al. 2010). Different pilot plants with different solar collector concentration factors using aperture areas that range from 0.174 to 465 m^2 (Matthews 1986; Gupta and Anderson 1991; Pacheco et al. 1993) have been tested for both heterogeneous (TiO$_2$) (Jiménez et al. 2000; Parra et al. 2002) and homogeneous (solar photo-Fenton) photocatalysis (Rodriguez et al. 2002; Rodríguez et al. 2005). Although the PTC has been proved effective for wastewater treatment (Spasiano et al. 2015), its use has been neglected due to high cost. Non-concentrating solar reactors were then investigated due to their simplicity and lower manufacturing cost (Dillert et al. 1999). Several types of solar pilot plants were studied (free-falling film collector, pressurized flat plate collector, solar ponds, etc.) where compound parabolic collector (CPC) photoreactor was found to be the most adequate option. CPC is made of two halves of a parabola with closely located focal points and their axes inclined to each other. Incident rays within the angle between the two axes are reflected with single or multiple reflections towards the region between the two focal points and get concentrated in that region, where the glass tubes are located (Blanco et al. 1999). Reflector designs for CPC have the ability to collect all direct and diffused UV radiation (Goswami et al. 1997), resulting in more efficient UV based wastewater treatment. The CPC photoreactors have been satisfactorily used to treat toxic and non-biodegradable wastewater, to remove contaminants of emerging concern and for disinfection repeatedly (Malato et al. 2009). Lately, raceway pond reactors (RPR) have appeared as a new approach to solar photo-Fenton when it is applied as a tertiary treatment to remove contaminants of emerging concern (Carra et al. 2014).

3 PhotoReactor Scaling-up Based on Solar UV Radiation Measurements

3.1 General Considerations

Scale up consists in developing the quantitative rules that describe the operation of a chemical reactor at different scales (Donati and Paludetto 1997). Normally, a process is proved and optimized at the laboratory scale; then, the upscaling consists of several steps before the industrial plant is built. Mathematical modeling is, theoretically, the main tool for scaling up chemical processes but this is not an easy way. It requires the knowledge of the complete chemical process mechanism and a comprehensive representation of the effects of the different variables to describe the reaction rate independently of the shape and configuration of the laboratory reactor. For these reasons, the construction of a pilot plant is an adequate step to confirm all the processes and measure the data by simulating the industrial scale process (Piccinno et al. 2016).

The scaling up of solar photocatalytic reactors is a critical issue because some extra considerations (apart from the conventional complications) must be considered. As expressed above, the physical geometry of the reactor is of critical importance in ensuring that photons are collected effectively to achieve efficient exposure of the catalyst to solar irradiation, so axial and radial scale-up are essential parameters for maximizing the surface areas exposed per unit of reactor volume and making distribution of sunlight inside the reactor uniform. Another important factor related to a natural solar photoreactor design is its optical-path length (OPL), since it must be ensured that it is uniform for both the flow and distribution of the photon flux at all times, everywhere inside the reactor also when the dispersion of light (heterogeneous catalyst) is present (Spasiano et al. 2015). In this scale-up procedure, the availability of solar light data is essential.

3.2 Irradiance Versus Time

Time is the independent variable used to represent the process evolution in chemical engineering and is usually expressed as 'residence time' or 'reaction time'. Nevertheless, time is not an effective variable to represent the evolution of a solar driven process since solar irradiation is also changing during the operation of the solar photoreactor. For this reason, accumulated UV solar energy has become one of the typically used variables to represent process evolution. In several cases, this value of energy has to be corrected because the reactor can consist of illuminated and non-illuminated elements or can concentrate solar radiation. In each case, solar measurements are unavoidable.

Irradiance is the radiant flux (power) received by a surface per unit area. The SI unit of irradiance is the watt per square meter (W/m^2) and is the value commonly provided by pyranometers and radiometers. When this value is registered online, the global amount of energy received by the photoreactor can be easily calculated. It is important to mention that different commercial UV radiometers can employ different absorption ranges so, at the same irradiation conditions, different models of UV radiometers can give different values (Navntoft et al. 2009). The selection of the most adequate UV radiometer depends on the absorbance spectral range of the catalyst employed during the process. For instance, when the iron is used as a photocatalyst (photo-fenton), the UV measurements are usually obtained in the 320–400 nm range.

A convenient way to represent solar photocatalytic processes is the use of the accumulated UV energy per unit volume, normally expressed as Q_{UV} (J/L) (Malato et al. 2002) which represents the amount of energy needed to achieve the treatment goal. The same variable has also been represented as E_{tot}/V (J/L) (Blanco et al. 1999).

The Q_{UV} value can be calculated by integrating irradiance value (I_{UV}, W/m^2) during the reaction time (t, s) for the illuminated surface (A, m^2) and dividing by the treated volume (V_{tot}, L) (Eq. 1) (Malato et al. 2009):

$$Q_{UV} = \frac{A}{V_{tot}} \int_0^t I_{UV} \mathrm{d}t \tag{1}$$

Nevertheless, in chemical engineering, the reaction time is the main independent variable used and the use of other variables can be less intuitive. For this reason, a time variable employing a standardized illumination time was developed (Eq. 2). Using a similar procedure that was used for obtaining Q_{UV} this time variable can be calculated by selecting reference irradiance and normalizing the values. For this purpose, a standard irradiation value under clear skies for sunny countries was selected; $I_0 = 30$ W/m^2 and the resulting variable is expressed as t_{30W} (min) (Malato et al. 2001).

$$t_{30W} = \frac{1}{I_0} \int_0^t I_{UV} \mathrm{d}t \tag{2}$$

Again, normalized time must be corrected in case the photo-reactor consists of illuminated and non-illuminated elements.

Other proposals were also used but today are not employed because they are less handy. These are the cases when the unit Einstein was employed (Ein/L) (Curcó et al. 1996; Malato et al. 1998), (Ein/L s) (Minero et al. 1996) or a combination of the residence time by the corresponding instantaneous global flux, $t_r \cdot \Phi$ (min W/m^2) (Guillard et al. 1999; Herrmann et al. 2002).

3.3 Scaling up from Pilot Plants

As commented before, the use of pilot plants is an adequate step to confirm all the processes tested in laboratory scale and measure data simulating the industrial scale process. Once a pilot plant is tested, the scaling up process can be easily done by using the amount of accumulated radiation (expressed as Q_{UV}) necessary to achieve a certain goal in the polluted water. This objective can be different depending on the process: total removal of pollutants, percentage of Total Organic Carbon (TOC) or Chemical Oxygen Demand (COD) descent etc. When this value is assessed by experimentation, the size of the solar field can be calculated only if the yearly average local irradiance and the sun hours of the year are available.

The area of illuminated photo-reactor (m^2) is calculated by Eq. 3, where V_{tot} is the total volume of water to be treated, t_{op} is the number of hours of operation, I_{UV} is the average local global solar UV irradiation (sunrise to sunset) and Q_{UV} is the solar energy necessary to achieve the selected goal (Malato et al. 2009). This procedure can be done for a yearly average or for achieving the objective in the month with the worst environmental conditions.

$$A = \frac{Q_{UV} V_{tot}}{t_{op} I_{UV}} \tag{3}$$

Finally, it is important to highlight that this procedure is independent of the type of pilot plant that will be scaled up but is important to make a profit of all incident radiation. If the experiments in the pilot plant are carried out with an excess of irradiance, which cannot be absorbed by the system, the scaling up process will oversize the industrial plant.

3.4   Scaling up from Mathematical Models

The traditional, purely empirical methodology for scaling-up starts from laboratory experiments and gradually increases the size of the proposed reactor up to the desired commercial size device. This approach is simple but requires significant investments. Scaling-up procedures employing mathematical models based on the fundamentals of chemical engineering can reduce expensive and time-consuming steps (Marugán et al. 2009).

The modelling of photocatalytic reactors requires an analysis of the radiation field in the photoreactor. This analysis, linked to the modeling of the fluid dynamics and the reaction kinetics, results in integro-differential equations, which almost invariably require demanding numerical solutions. Obviously, the construction of the designed photoreactor and the comparison of its experimental conversions with the simulated values would allow the validation of the whole scaling-up procedure (Li Puma 2003).

To include the rate of photon absorption in the kinetics models, the Local Volumetric Rate of Photon Absorption (LVRPA) was defined. This parameter includes the spectral incident radiation and the absorptivity of the sample; thus, the spectral absorbed incident radiation can be calculated. This methodology was described in depth for homogeneous and heterogeneous systems employing different types of photoreactors at laboratory scale (Cassano et al. 1995) and including scattering effects (Alfano et al. 1995). This procedure requires a complete knowledge of the system and a complex mathematical formulation. Instead of the difficulties of the method, several studies have been also published using different types of lamps including solar simulators (Toepfer et al. 2006; Satuf et al. 2007; Motegh et al. 2014; Cabrera Reina et al. 2015b).

In homogeneous processes, the rate of photochemical reactions can be expressed by the following kinetic equation (Kusic et al. 2011):

$$r_m = \phi \cdot I_0 \cdot [1 - \exp(2.303 \cdot L \cdot \varepsilon_i \cdot c_i)] \tag{4}$$

where Φ and ε_i represent quantum yield and extinction coefficient of specie i, respectively. I_0 and L stand for the UV irradiation and reactor configuration properties, i.e. the incident photon flux by reactor volume unit and the effective optical path, for used reactor configuration.

Another simpler approach is the direct use of the solar UV measurement in the kinetic expression of the photochemical reaction (Cabrera Reina et al. 2012):

$$r = k[\text{Fe(III)}][I_0] \tag{5}$$

where k stands for the kinetic constant, [Fe(III)] the iron concentration and I_0 the UV irradiation. Again, the solar resource monitoring is essential to determine process performance.

One-step beyond consisted of using direct solar radiation and developing the mathematical model including average radiation values. The resolution of a semi-empirical model allowed predicting the area of the irradiated plant (CPC based) needed to treat industrial wastewater based on different environmental (temperature and average UV radiation) conditions Fig. 1 (Cabrera Reina et al. 2014). For plants in which the liquid depth can vary (raceway) a good model of the

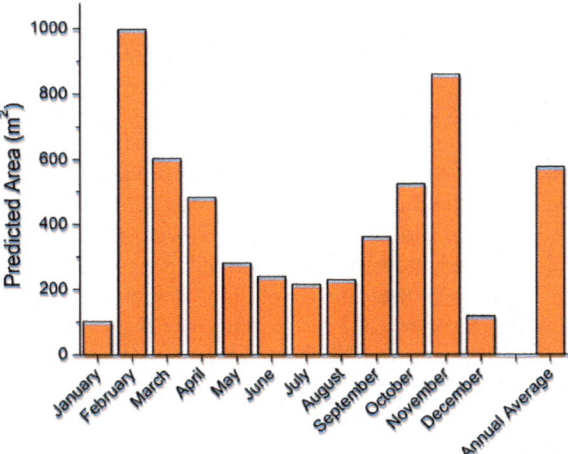

Fig. 1 Predicted CPC surface area for the environmental conditions of Almería, Spain. Adapted from (Cabrera Reina et al. 2014)

process can allow selecting the optimal height of volume and consequently estimating the treatment capacity for different irradiation conditions (Rivas et al. 2015).

4 Conclusions

Solar water treatments have emerged as a promising alternative to conventional biological processes. In heterogeneous processes, the catalyst causes photon absorption and scattering, which occurs along the trajectories of light beams, while for homogeneous processes, as solar photo-Fenton, the only absorption takes place. In this way, the design of solar photoreactors applied to water/wastewater treatment is complex and has been deeply studied. The CPC photoreactor has been found to be the most adequate system, although lately, new options have appeared depending on the type of water (municipal or industrial matrix).

Time is the independent variable used to represent the process of evolution in chemical engineering; nevertheless, it is not effective to represent the evolution of a solar driven process. A convenient way to represent solar photocatalytic processes is the use of the accumulated UV energy per unit volume, normally expressed as Q_{UV} (J/L), which represents the amount of energy needed to achieve the treatment goal. To calculate this variable, UV irradiation has to be monitored during the solar treatment by means of UV radiometer. The absorbance spectral range of the catalyst determines the measurement range; being 320–400 nm the most used one.

The area of illuminated photo-reactor needed can be calculated based on the Q_{UV} (J/L) value, which can be obtained by means of pilot plant scale experiments or using mathematical models that include special equations describing the interaction between the catalyst and UV irradiation. The average local UV irradiation and the number of sun hours available are also needed, consequently, the solar mapping is essential to photoreactor scale-up procedures.

Acknowledgements Ph.D. Sara Miralles-Cuevas acknowledges MINECO for his Juan de la Cierva-formación grant (No. FJCI-2016-28965). Ph.D. Alejandro Cabrera wishes to thanks FONDECYT/Iniciación/11160680 and SERC Chile (FONDAP/15110019).

References

Alfano OM, Negro AC, Cabrera MI, Cassano AE (1995) Scattering effects produced by inert particles in photochemical reactors. 1. Model and experimental verification. Ind Eng Chem Res. https://doi.org/10.1021/ie00041a009
Bechtel Corporation (1991) Conceptual design of a photocatalytic wastewater treatment plant, Report SAND 91-7005
Blanco J (2005) El reactor solar fotocatalítico: estado del arte. In: Solar safe water project book
Blanco J, Malato S, Fernández P et al (1999) Compound parabolic concentrator technology development to commercial solar detoxification applications. Sol Energy. https://doi.org/10.1016/s0038-092x(00)00078-5

Cabrera Reina A, Santos-Juanes Jordá L, García Sánchez JL et al (2012) Modelling photo-fenton process for organic matter mineralization, hydrogen peroxide consumption and dissolved oxygen evolution. Appl Catal B Environ. https://doi.org/10.1016/j.apcatb.2012.02.021

Cabrera Reina A, Casas López JL, Maldonado Rubio MI et al (2014) Effects of environmental variables on the photo-fenton plant design. Chem Eng J 237:469–477. https://doi.org/10.1016/J.CEJ.2013.10.046

Cabrera Reina A, Santos-Juanes Jordá L, Casas López J et al (2015a) Biological oxygen demand as a tool to predict membrane bioreactor best operating conditions for a photo-Fenton pretreated toxic wastewater. J Chem Technol Biotechnol. https://doi.org/10.1002/jctb.4295

Cabrera Reina A, Santos-Juanes L, García Sánchez JL et al (2015b) Modelling the photo-fenton oxidation of the pharmaceutical paracetamol in water including the effect of photon absorption (VRPA). Appl Catal B Environ. https://doi.org/10.1016/j.apcatb.2014.11.023

Cabrera Reina A, Miralles-Cuevas S, Casas López JL, Sánchez Pérez JA (2017) Pyrimethanil degradation by photo-fenton process: influence of iron and irradiance level on treatment cost. Sci Total Environ. https://doi.org/10.1016/j.scitotenv.2017.06.217

Carra I, Santos-Juanes L, Acién Fernández FG et al (2014) New approach to solar photo-Fenton operation. Raceway ponds as tertiary treatment technology. J Hazard Mater. https://doi.org/10.1016/j.jhazmat.2014.07.010

Cassano AE, Martin CA, Brandi RJ, Alfano OM (1995) Photoreactor analysis and design: fundamentals and applications. Ind Eng Chem Res. https://doi.org/10.1021/ie00046a001

Curcó D, Malato S, Blanco J, Giménez J (1996) Photocatalysis and radiation absorption in a solar plant. Sol Energy Mater Sol Cells. https://doi.org/10.1016/0927-0248(96)00059-1

Dillert R, Cassano AE, Goslich R, Bahnemann D (1999) Large scale studies in solar catalytic wastewater treatment. Catal Today. https://doi.org/10.1016/s0920-5861(99)00188-1

Donati G, Paludetto R (1997) Scale up of chemical reactors. Catal Today 34:483–533. https://doi.org/10.1016/S0920-5861(96)00069-7

Fernández-García A, Zarza E, Valenzuela L, Pérez M (2010) Parabolic-trough solar collectors and their applications. Renew Sustain Energy Rev. https://doi.org/10.1016/j.rser.2010.03.012

Genthe B, Le Roux WJ, Schachtschneider K et al (2013) Health risk implications from simultaneous exposure to multiple environmental contaminants. Ecotoxicol Environ Saf. https://doi.org/10.1016/j.ecoenv.2013.03.032

Goswami DY, Sharma SK, Mathur GD, Jotshi CK (1997) Analysis of solar detoxification systems. Sol Energy Eng 119

Guillard C, Disdier J, Herrmann JM, et al (1999) Comparison of various titania samples of industrial origin in the solar photocatalytic detoxification of water containing 4-chlorophenol. Catal Today. https://doi.org/10.1016/s0920-5861(99)00184-4

Gupta BP, Anderson JV (1991) Solar detoxification of hazardous waste—an overview of the U.S. Department of Energy program. Sol Energy Mater 24:40–61. https://doi.org/10.1016/0165-1633(91)90047-O

Herrmann JM, Guillard C, Disdier J et al (2002) New industrial titania photocatalysts for the solar detoxification of water containing various pollutants. Appl Catal B Environ. https://doi.org/10.1016/s0926-3373(01)00265-x

Jiménez AE, Estrada CA, Cota AD, Román A (2000) Photocatalytic degradation of DBSNa using solar energy. Sol Energy Mater Sol Cells 60:85–95. https://doi.org/10.1016/S0927-0248(99)00066-5

Kusic H, Peternel I, Ukic S et al (2011) Modeling of iron activated persulfate oxidation treating reactive azo dye in water matrix. Chem Eng J. https://doi.org/10.1016/j.cej.2011.05.076

Langenbach T (2013) Persistence and bioaccumulation of Persistent Organic Pollutants (POPs). In: Applied bioremediation—active and passive approaches. InTech

Li Puma G (2003) Modeling of thin-film slurry photocatalytic reactors affected by radiation scattering. Environ Sci Technol. https://doi.org/10.1021/es0300362

Malato S, Blanco J, Richter C et al (1998) Enhancement of the rate of solar photocatalytic mineralization of organic pollutants by inorganic oxidizing species. Appl Catal B Environ. https://doi.org/10.1016/s0926-3373(98)00019-8

Malato S, Caceres J, Agüera A et al (2001) Degradation of imidacloprid in water by photo-fenton and TiO_2 photocatalysis at a solar pilot plant: a comparative study. Environ Sci Technol. https://doi.org/10.1021/es000289k

Malato S, Blanco J, Vidal A, Richter C (2002) Photocatalysis with solar energy at a pilot-plant scale: an overview. Appl. Catal. B Environ

Malato S, Blanco J, Vidal A, et al (2003) Applied studies in solar photocatalytic detoxification: an overview. Sol Energy. https://doi.org/10.1016/j.solener.2003.07.017

Malato S, Fernández-Ibáñez P, Maldonado MI et al (2009) Decontamination and disinfection of water by solar photocatalysis: recent overview and trends. Catal Today

Marugán J, van Grieken R, Cassano AE, Alfano OM (2009) Scaling-up of slurry reactors for the photocatalytic oxidation of cyanide with TiO_2 and silica-supported TiO_2 suspensions. Catal Today. https://doi.org/10.1016/j.cattod.2008.12.026

Matthews RW (1986) Photo-oxidation of organic material in aqueous suspensions of titanium dioxide. Water Res. https://doi.org/10.1016/0043-1354(86)90020-5

Minero C, Pelizzetu E, Malato S, Blanco J (1996) Large solar plant photocatalytic water decontamination: effect of operational parameters. Sol Energy. https://doi.org/10.1016/0038-092x(96)00029-1

Motegh M, Van Ommen JR, Appel PW, Kreutzer MT (2014) Scale-up study of a multiphase photocatalytic reactor—Degradation of cyanide in water over TiO_2. Environ Sci Technol. https://doi.org/10.1021/es403378e

Navntoft C, Dawidowski L, Blesa MA et al (2009) UV-A (315–400 nm) irradiance from measurements at 380 nm for solar water treatment and disinfection: comparison between model and measurements in Buenos Aires. Sol Energy, Argentina and Almería, Spain. https://doi.org/10.1016/j.solener.2008.10.010

Organization World Health (WHO), United United Nations Children's Fund (UNICEF) (2017) Progress on drinking water, sanitation and hygiene: 2017 update and SDG baselines

Pacheco JE, Prairie MR, Yellowhorse L (1993) Photocatalytic destruction of chlorinated solvents in water with solar energy. J Sol Energy Eng 115:123. https://doi.org/10.1115/1.2930038

Parra S, Malato S, Pulgarin C (2002) New integrated photocatalytic-biological flow system using supported TiO_2 and fixed bacteria for the mineralization of isoproturon. Appl Catal B Environ. https://doi.org/10.1016/s0926-3373(01)00293-4

Piccinno F, Hischier R, Seeger S, Som C (2016) From laboratory to industrial scale: a scale-up framework for chemical processes in life cycle assessment studies. J Clean Prod. https://doi.org/10.1016/j.jclepro.2016.06.164

Rivas G, Carra I, García Sánchez JL et al (2015) Modelling of the operation of raceway pond reactors for micropollutant removal by solar photo-fenton as a function of photon absorption. Appl Catal B Environ. https://doi.org/10.1016/j.apcatb.2014.09.015

Rivera-Utrilla J, Sánchez-Polo M, Ferro-García MÁ et al (2013) Pharmaceuticals as emerging contaminants and their removal from water a review. Chemosphere

Rodriguez M, Timokhin V, Michl F et al (2002) The influence of different irradiation sources on the treatment of nitrobenzene. Catal Today. https://doi.org/10.1016/s0920-5861(02)00227-4

Rodríguez M, Malato S, Pulgarin C et al (2005) Optimizing the solar photo-Fenton process in the treatment of contaminated water. Determination of intrinsic kinetic constants for scale-up. Sol Energy

Satuf ML, Brandi RJ, Cassano AE, Alfano OM (2007) Scaling-up of slurry reactors for the photocatalytic degradation of 4-chlorophenol. Catal Today. https://doi.org/10.1016/j.cattod.2007.06.056

Spasiano D, Marotta R, Malato S et al (2015) Solar photocatalysis: materials, reactors, some commercial, and pre-industrialized applications. A comprehensive approach. Appl Catal B Environ

Toepfer B, Gora A, Li Puma G (2006) Photocatalytic oxidation of multicomponent solutions of herbicides: reaction kinetics analysis with explicit photon absorption effects. Appl Catal B Environ 68:171–180. https://doi.org/10.1016/j.apcatb.2006.06.020

Chapter 16
Solar Nowcasting

Antonio Sanfilippo

Abstract With the increasing ubiquity of solar energy systems, solar nowcasting is needed to redress short-term power system imbalances emerging from solar energy integration and normalize electricity markets in near real time. In this chapter, we provide an overview of the applications, solar resource data, evaluation procedures, modeling methods, and emerging technologies in solar nowcasting.

1 Introduction

Solar nowcasting, the ability to forecast solar irradiance up to six hours ahead,[1] is crucial in managing power network operations and regulating electricity markets when solar energy is integrated into power grids. Solar irradiance is subject to abrupt changes due to meteorological change (e.g., clouds, haze, dust storms). With increasing solar energy penetration, this variability can cause critical imbalances in the short-term generation, transmission, distribution, and pricing of electricity that can disrupt power networks and cause uncertainty in energy markets (Diagne et al. 2013; Anderson and Leach 2004; Moreno-Munoz et al. 2008; Sanfilippo et al. 2016a; Voyant et al. 2017). Solar nowcasting can help redress these imbalances by providing insights into solar variability that can be used to optimize power network operations and normalize electricity markets in near real time (Paulescu et al. 2013).

Data used in solar nowcasting may change depending on the application. For example, Global Horizontal Irradiation (GHI) predictions address the performance characteristics of solar photovoltaics (PV) applications, while Direct Normal Irradiance (DNI) predictions are more relevant to concentrated solar power (CSP) applications (Pelland et al. 2013). Variations of these basic solar

---

[1] http://www.wmo.int/pages/prog/amp/pwsp/Nowcasting.htm.

A. Sanfilippo (✉)
Qatar Environment and Energy Research Institute, Doha, Qatar
e-mail: asanfilippo@hbku.edu.qa

© Springer Nature Switzerland AG 2019
J. Polo et al. (eds.), *Solar Resources Mapping*, Green Energy and Technology,
https://doi.org/10.1007/978-3-319-97484-2_16

measurements are often used to streamline the predictive impact of solar geometry vs. cloudiness and aerosols, and factor in air mass diversity (see Chapter 5 this book). The source of data may differ according to the nowcasting approach adopted, to include measurements from ground solar stations, or modeled data from satellite and whole sky camera imagery. Exogenous meteorological parameters such as air temperature, relative humidity, and atmospheric turbulence are also occasionally used.

Machine learning and stochastics techniques have been shown to be better suited for solar nowcasting than physics-based approaches, such as numerical weather prediction (NWP) models, which lack the temporal resolution needed for intra-hour forecasts (Diagne et al. 2013; Inman et al. 2013). Various machine learning algorithms have been used in solar nowcasting (Diagne et al. 2013; Inman et al. 2013; Pelland et al. 2013; Mellit and Kalogirou 2008; Mellit et al. 2006). Several techniques have been developed to optimize the performance of these algorithms. These techniques include: wavelet transforms (Lyu et al. 2014), ensemble and multi-modeling modeling strategies (Chaouachi et al. 2010; Mohammed et al. 2015; Sanfilippo et al. 2016a), detrending (Jain 1984; Baig et al. 1991; Kaplanis 2006; Ji and Chee 2011; Akarslan and Hocaoglu 2016; Sanfilippo et al. 2016b, 2018), and multivariate forecasting (Sfetsos and Coonick 2000).

A number of evaluation techniques are used to evaluate nowcasting results, including root-mean-square error, mean bias error, mean absolute percentage error, skill score, and their normalized counterparts. Results depend strongly on the number of time steps predicted, the duration of each time steps, and geographical location.

Solar nowcasting performance can be greatly improved through multi-modeling and detrending techniques. Potential ways of achieving further gains in nowcasting performance have recently been introduced by a host of emerging solar forecasting technologies (Yang et al. 2018).

2 Applications

Solar nowcasting offers several technical and economic benefits in the management of solar energy. For example, it provides predictions about the solar power generation fleet's behavior in near real time to help grid managers ensure power grid stability and optimize operating costs by committing appropriate amount of energy resources and reserves, mixing and matching storage and electricity cogeneration strategies, to meet specific demand profiles.

With the rising use of energy storage and demand response to optimize the retention and discharge of excess renewable energy, the integration of solar nowcasting becomes crucial in ensuring power system safety and reliability (Anderson and Leach 2004; Moreno-Munoz et al. 2008; Diagne et al. 2013; Paulescu et al. 2013; Sanfilippo et al. 2016a; Voyant et al. 2017). For example, storage-based bridging power solutions can be effectively applied to peak shaving and frequency regulation to assure the continuity and reliability of service during short periods (minutes to a few hours) of transition from renewable to nonrenewable energy (Voyant et al. 2017). By predicting when such energy source shifts may occur, solar nowcasting offers a valuable management tool for storage-based bridging power operations.

Solar nowcasting can also support economic goals (Hirsch et al. 2014). For example, knowledge about short-term electricity production from solar energy can help fine-tune electricity sales by providing a reliable match between announced and delivered electricity for solar plants that lack storage capacity.

Overall, solar nowcasting is increasingly becoming a focus of research on the areas of operation, management, and integration of solar power plants into the grid, and government agencies such as the European Commission and the US Department of Energy are making significant research investment in this area (DNICast; Solar Forecasting 2).

3   Data

Training material for the development of solar nowcasting algorithms is typically derived from solar monitoring stations that use a pyrheliometer for measuring Direct Normal Irradiance (DNI) and pyranometers for Global Horizontal Irradiance (GHI) and Diffuse Horizontal Irradiance (DHI) measurements (see Chaps. 2 and 3 in this book). Data from the monitoring station are sampled in near real time (e.g., every second) and aggregated as time interval averages in Watt per square meter (W/m^2), varying in duration according to the granularity of the forecasting horizon targeted. Data quality is enforced following the Baseline Surface Radiation Network (BSRN) recommendations (Long and Dutton 2002; McArthur 1998; Chap. 4 in this book) to ensure that the irradiation measurements collected are viable for the intended modeling applications. Some examples of the tests applied to irradiance data to verify their quality are given below:

- GHI lies outside the physically possible (maximum and minimum) values.
- GHI reflects "extremely rare" but physically possible values.
- The ratio of calculated GHI (from measured diffuse and direct irradiances) to measured GHI lies outside some limits, and the diffuse to global ratio exceeds a possible limit.

The solar measurements collected can be normalized in a number of ways. For example, the clearness index (K_T) is often used ad as a GHI-based measure, in order of separate forecasting complexity into the prediction of solar geometry and the prediction of cloudiness and aerosol. K_T is calculated as the ratio of GHI to the incoming solar radiation on a horizontal surface at the top of the earth's atmosphere (Black et al. 1954; Duffie and Beckman 1991). It characterizes the attenuating impact of the atmosphere on solar irradiance by specifying the proportion of extraterrestrial solar radiation that reaches the surface of the earth.

K_T in turn can be normalized to alleviate K_T's dependency on the zenith angle, which changes the traversed air mass during the course of a day. For example, it can be transformed into the zenith angle-independent K_T^* measure, as proposed by Perez et al. (1990), where K_T is normalized with respect to a standard clear-sky global irradiance profile for a relative air mass of one, as shown in Eq. (1), where K_T and AM are the clearness index and air mass at the (same) corresponding minute. The air mass is calculated according to the formula in (2), where SZA is the solar zenith angle (measured from the vertical overhead) in degrees at the corresponding minute (Kasten and Young 1989).

$$K_T^* = \frac{K_T}{0.1 + 1.031 * \exp[-1.4/(0.9 + 9.4/\text{AM})]} \tag{1}$$

$$\text{AM} = \frac{1}{\cos(\text{SZA}) + 0.50572 * (96.07995 - \text{SZA})^{-1.6364}} \tag{2}$$

Other measurements used for training solar nowcasting models include modeled data from images collected through sky-imaging systems such as Total Sky Imagers (TSI) (Chow et al. 2011; Marquez and Coimbra 2013; Chu et al. 2013) and Whole Sky Imagers (WSI) (Long et al. 2006), and satellites (Hammer et al. 1999; Lorenz et al. 2004).

TSI and WSI images provide a photograph of the entire sky from below which is then analyzed to support various meteorological applications, including the measurement of cloud cover with reference to solar forecasting. An example of algorithm used for deriving numerical cloud cover time series data from TSI is described by Marquez and Coimbra (2013). The TSI image is first projected onto a flat rectangular grid to remove any distortion from the TSI convex mirror. Pairs of the consecutive projected images are then processed with the particle image velocimetry (PIV) algorithm[2] to compute the flow direction of the clouds. Next, a classification algorithm (Li et al. 2011) is used to identify clouds. Finally, a set of pixels representing the sun's position is placed over the processed TSI images and cloud indices are calculated as the percentage of pixels classified as cloud elements in the combined image. Other approaches for deriving cloud indices from TSI and WSI images include the use of machine learning techniques such artificial neural networks and binary decision trees are described in Buch and Sun (1995) and Cazorla et al. (2008).

Models that derive surface solar irradiance measurements from satellite imagery are described in Chap. 6 of this book.

Additional data used in solar nowcasting include exogenous meteorological variables such as air temperature, relative humidity, and atmospheric turbulence (Gordon 2009; Mellit et al. 2010; Mandal et al. 2012; Li et al. 2011).

---

[2]MATLAB PIV Toolbox. http://www.oceanwave.jp/softwares/mpiv.

4 Evaluation

Several measures have been used to evaluate solar forecasting algorithms. The root-mean-square error (RMSE) and mean bias error (MBE), with their normalized counterparts (rRMSE) and (rMBE) shown in (3) and (4), are perhaps most widely used; see Diagne et al. (2013); Inman et al (2013) and references therein. Other measures, such as the mean average percentage error (MAPE) shown in (6), have also been used (Mandal et al. 2012). The (relative) skill score (SS/rSS) is sometimes used to assess the improvement over a reference/baseline model (Beyer et al. 2009).

$$\text{rRMSE} = \frac{\sqrt{\frac{\sum_{t=1}^{n}\left(x_{\text{predicted value}_t} - x_{\text{observed value}_t}\right)^2}{n}}}{\text{mean}\left(\text{observed values}_{1,\dots,n}\right)} \tag{3}$$

$$\text{rMBE} = \frac{\frac{\sum_{t=1}^{n}\left(x_{\text{predicted value}_t} - x_{\text{observed value}_t}\right)}{n}}{\text{mean}\left(\text{observed values}_{1,\dots,n}\right)} \tag{4}$$

$$\text{MAPE} = \frac{1}{N}\sum_{t=1}^{N}\left(\frac{x_{\text{predicted value}_t} - x_{\text{observed value}_t}}{x_{\text{observed value}_t}}\right) * 100 \tag{5}$$

$$\text{rSS}(\text{model}_i) = \left(1 - \frac{\text{rRMSE}(\text{model}_i)}{\text{rRMSE}(\text{reference model})}\right) * 100 \tag{6}$$

In addition to evaluation metrics, a persistence model is often used as a baseline for comparison. The persistence model is based on the assumption that the value for each step-ahead in the forecasting horizon chosen is always the same as the present value, as shown in (7), where $x_{\text{now}+k}$ is the predicted solar irradiation at each k step (s)-ahead and x_{now} is the observed solar irradiation at the current time.

$$x_{\text{now}+k} = x_{\text{now}}, \quad \text{for } k > 0 \tag{7}$$

5 Methodology

5.1 Time Series Analysis and Forecasting

Stochastic and machine learning approaches used for solar nowcasting typically rely on regression methods to learn coefficients that provide the basis for prediction by measuring the relationship between an observation at time t and the observations at previous times. Depending on the methodology adopted, the ensuing nowcasting approach may differ in the way regression is used.

Within an autoregressive modeling approach, $AR(N)$, the current term x_t of a time series $x_t, \ldots, x_N$, for $N > t >$, can be estimated as the linear weighted sum of previous terms $x_{t-1}, \ldots, x_{t-N}$ in the series, as shown in (8) where c is a constant, ε_t is white noise, and $a_i, \ldots, a_N$ are the autoregression coefficients that function as the weights in the sum. AR coefficients can be estimated using Yule–Walker equations or regression techniques such as least squares estimation or maximum entropy (Brockwell and Davis 2006).

$$x_t = c + \sum_{i=1}^{N} a_i x_{t-i} + \varepsilon_t \tag{8}$$

Within a support vector regression (SVR) approach to time series, forecasting (Smola et al. 1997) is to find a function that for each vector $\vec{x}_i \in \mathbb{R}^n$ representing a time series within a dataset with N training time series sequences approximates its value $y_i (i \geq 0 \leq N)$ as closely as possible. The resulting function provides the basis for forecasting. When the input data are amenable to linear regression, SVR is expressed by the equation in (9), where:

- $\vec{w}$ is the weight vector, i.e., a linear combination of training patterns that supports the regression function.
- $\vec{x}_i$ is the input vector, e.g., a time series training data sample.
- y_i is the value for the input vector, e.g., the solar irradiation values to be predicted.
- b is the bias; i.e., $\frac{b}{\|\vec{w}\|}$ is the perpendicular distance from the origin of the vector space to the hyperplane that separates the data points in the vector space.

$$y_i = \vec{w} \cdot \vec{x}_i + b \tag{9}$$

The objective of regression is to estimate the weight vector $\vec{w}$ with the smallest possible length to avoid over-fitting. To ease the regression task, a given margin of deviation ε is allowed with no penalty, and a given margin ξ is specified where deviation is allowed with increasing penalty. The minimal length of the weight vector $\vec{w}$ is obtained by minimizing the loss function (10) subject to the constraint in (11) or (12), for $\xi_i, \xi_i^* \geq 0$. The solution is given by constructing a Lagrange function from the loss function and the associated constraints, as shown in (13) where α_i and α_i^* are Lagrange multipliers—see Smola and Schoölkopf (1998) for details. The training vectors giving nonzero Lagrange multipliers are called *support vectors* and are used to construct the regression function. If the input data are not amenable to linear regression, then the vector data are mapped into a higher-dimensional feature space using a kernel function Φ, such as the polynomial kernel:

$$\Phi(\vec{w}) \cdot \Phi(\vec{x}_i) = 1 + (\vec{w} \cdot \vec{x}_i)^3.$$

$$\frac{1}{2}\|\vec{w}\|^2 + C \sum_{i=1}^{n} \left(\xi_i + \xi_i^* \right) \tag{10}$$

$$y_i - (\vec{w} \cdot \vec{x} + b) \leq \varepsilon + \xi_i \tag{11}$$

$$y_i - (\vec{w} \cdot \vec{x} + b) \geq -\varepsilon - \xi_i^* \tag{12}$$

$$y_i = \sum_{i=1}^{n} \left(a_i - a_i^* \right) (\vec{w} \cdot \vec{x}_i) + b, \quad \text{for } i \geq 0 \leq n \tag{13}$$

Other stochastic and machine learning techniques have been used for solar nowcasting. These include: autoregressive (integrated) moving average (ARMA, ARIMA), autoregressive integrated moving average (ARIMA), coupled autoregressive and dynamical system (CARDS), several varieties of artificial neural network (ANN), k-nearest neighbor (k-NN), regression tree, boosting, bagging, random forests—see Diagne et al. (2013); Inman et al. (2013); Mellit et al. (2006); Mellit and Kalogirou 2008; Voyant et al. (2017), and references therein. Results vary considerably depending on location, horizon, predicted variable (e.g., GHI, DNI, DHI), and the choice of single vs. multi-modeling approach, as shown in Voyant et al. (2017) where top and bottom rRMSE ranging from of 5 and 24% were reported.

5.2 Multi-modeling

While single machine learning methods may outperform one another, others depending on data location and prediction horizon (Voyant et al. 2017: Table 3), there is strong evidence that ensemble approaches which combine diverse machine learning methods tend to rival single algorithms (Voyant et al. 2017: Table 4). For example, while SVR is reported to rival both ANN and k-NN (Ferrari et al. 2012), the combination of ANN and SVM is shown to outperform both ANN and SVM (Prokop et al. 2013).

Multi-modeling is motivated by the observation that when comparing different single model solutions, while a single modeling approach may perform better than others, even the worst performing model provides correct results in a significant number of instances, as shown in Fig. 1 for two autoregressive models (AR3 and AR11) and SVR- and persistence-based (PER) models (Sanfilippo et al. 2016a). The ability to leverage the results of different models should therefore provide an effective way to improve solar nowcasting.

The use of multi-modeling and ensemble modeling techniques has been making headway in solar nowcasting. One approach is to merge the results of diverse models. For example, Bayesian model averaging (BMA) is proposed in Lauret et al.

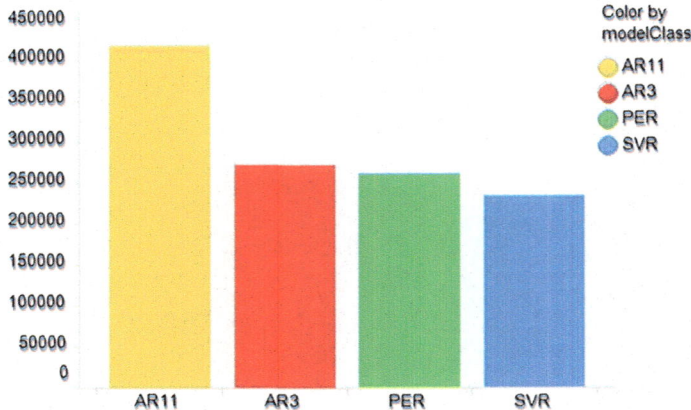

Fig. 1 Number of times each model yielded the lowest rRMSE for solar prediction in Doha (Qatar) of 15 min in one-minute intervals as compared to the other models. Adapted from Sanfilippo et al. (2016a)

(2012) to combine the results of two models (autoregressive moving average and neural networks) to form a model committee, by weighing the prediction of the two models according to their posterior model probabilities (PMPs), as shown in (14). The PMP of a model M_k is calculated as shown in (15) where $p(M_k)$ is the prior probability of M_k and $p(D|M_k)$ is the integrated or marginal model likelihood—see Gibbons et al. (2008) for details. As shown in Table 1, the BMA committee approach outperforms both models and the persistence baseline.

$$y_{t,\text{committee}} = \text{PMP}_{\text{arma}} \times y_{t,\text{arma}} + \text{PMP}_{nn} \times y_{t,nn} \qquad (14)$$

$$p(M_k|D) = \frac{p(D|M_k)p(M_k)}{\sum_{l=1}^{K} p(D|M_l)p(M_l)} \qquad (15)$$

Another approach to multi-modeling is to select one out of a series of available models according to the context of prediction. The study referred in Fig. 1 (Sanfilippo et al. 2016a) provides an example of such an approach by using supervised classification of forecasting evaluation results from four models to select the best predictions, according to their expected superiority in terms of lower error rate. The advantages of the proposed multi-modeling approach are demonstrated in an experimental evaluation where its application with the four models shows a relative skill score improvement of 44.92% over the baseline model and 19.06% over the best performing model (AR11), as shown in Table 2. Moreover, any combination of four and three models consistently outperforms a single model solution, as shown in Table 3.

Table 1 Evaluation results for single (ARMA, NN, persistence) and combined (BMA committee) models

Model	rRMSE	MBE
ARMA (2,1)	25.01	−4.08
NN (12,3)	23.1	3.06
Persistence	24.96	5.97
BMA committee	22.6	−0.29

Adapted from Lauret et al. (2012)

Table 2 rRMSE and rSS of nowcasting results by modeling approach

	SVR (%)	AR3 (%)	AR11 (%)	Multi-modeling (%)	Persistence (%)
rRMSE	6.70	4.65	3.62	2.93	5.32
rSS over PER	−25.94	12.59	31.95	44.92	−
rSS over AR11	−85.08	−28.45	−	19.06	−46.96

Adapted from Sanfilippo et al. (2016a)

Table 3 Comparison of solar nowcasting results across single models and four-/three-wise multi-model combinations

Single models	rRMSE (%)	Multi-model combinations	rRMSE (%)
SVR	12.21	SVR/PER/AR3/AR11	4.75
PER	9.04	PER/AR3/AR11	5.90
AR3	7.89	SVR/AR3/AR11	5.95
AR11	6.04	SVR/PER/AR11	5.95
		SVR/PER/AR3	5.62

Adapted from Sanfilippo et al. (2016a)

Other examples of multi-modeling applications to solar nowcasting include the studies presented in Chaouachi et al. (2010), Mohammed et al. (2015), and Chu et al. (2013).

5.3 Detrending

In general, the use of larger historical datasets promotes model accuracy through the development of more robust regression coefficients in nowcasting algorithms. However, fitting a solar nowcasting model to a training dataset that is more stationary can help increase nowcasting accuracy, since most stochastic approaches to time series analysis assume that the input time series data are stationary or at least weakly stationary (Inman et al. 2013).

Several detrending methods have been developed in the solar nowcasting literature. These can be grouped into two classes according to whether stationarity is induced through "deseasoning," i.e., algorithmic removal of seasonal patterns from the time series data, or by partitioning the time series data into more stationary data subsets (e.g., seasons instead of a whole year).

One of the earliest deseasoning methods was a model aimed at fitting solar irradiation time series data to a Gaussian function (Jain 1984). This approach was later modified to provide a better fit with the recorded data during the start and end periods of a day (Baig et al. 1991). A variant of this approach was also developed in Kaplanis (2006) by providing a calculation of the standard deviation of the Gaussian curve which would yield a closer match with recorded solar radiation data during the day length. An approach similar to Jain (1984) and Baig et al. (1991) based on Fourier series techniques was also proposed by Borland (1995, 2008), while the moving average filter is used in Severiano et al. (2017) as a way of smoothing the solar data time series. In all these cases, detrending is used as a way of transforming the solar time series data to make them more amenable to forecasting algorithms. A different approach is adopted in Ji and Chee (2011) where the augmented Dickey–Fuller test is used to measure the stationarity of a detrended series to establish when to switch from a linear to a nonlinear solar prediction model.

Detrending by way of partitioning a time series data is used in Sanfilippo et al. (2016b, 2018) where clustering by machine learning techniques such as K-means, expectation maximization, and learning vector quantization is proposed as an additional criterion to break down a dataset into more stationary data subsets. Both studies find that cluster-based detrending offers a way to reduce error in solar nowcasting. More specifically, the best performing clustering method in Sanfilippo et al. (2018) (learning vector quantization clustering) rivals both the partitioning of a whole year dataset into seasons and no detrending (i.e., training the model on a whole year dataset), as shown in Fig. 2. At the same time, Sanfilippo et al. (2018) observe that even the worst detrending method can provide the correct solution in a significant number of times, and endeavor to develop an *ensemble* approach to detrending where several ways of partitioning training data into more cohesive subsets are used in parallel to improve predictions. After detrending a solar irradiance training set covering a whole year according to season-based and cluster-based detrending criteria, a classification model is developed which predicts the most appropriate detrending option for each choice of time series input to forecasting. The ensuing ensemble detrending approach shows overwhelming improvements as compared to all other methods, with twofold to over threefold reductions in forecasting error rates (Fig. 2).

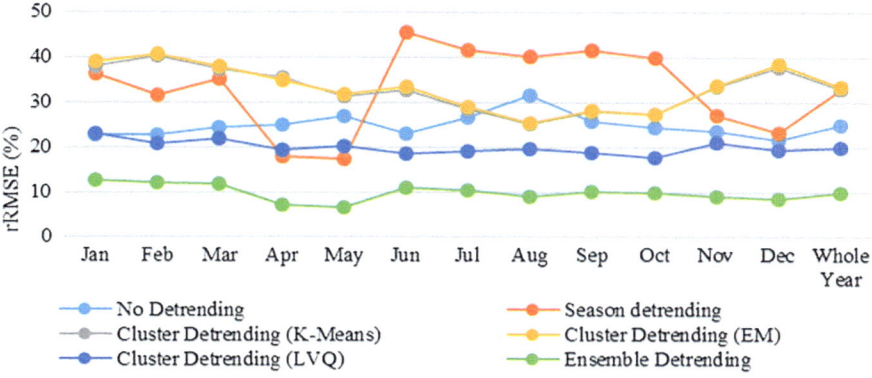

Fig. 2 rRMSE average and standard deviation across forecasting results for diverse detrending methods. Adapted from Sanfilippo et al. (2018)

6 Emerging Technologies

A recent study on the history and trends in solar forecasting (Yang et al. 2018) identifies six top-ranked emerging technologies in solar forecasting:

1. Advection with ground sensors (Inage 2017)
2. Standardized forecast evaluation (Vallance et al. 2017)
3. Hierarchical forecasting (Yang et al. 2017)
4. Forecasting with multi-modeling (Sanfilippo et al. 2016a)
5. Quality control for PV power output (Killinger et al. 2017)
6. Shadow cameras (Kuhn et al. 2017).

The solar prediction approach pioneered in Inage (2017) uses a form of stochastic partial differential equation (advection equation) with irradiance measurements from an irregular monitoring network interpolated onto a regular grid to model the geostatistical process as a lattice process so that a discrete time and space approximation of the partial differential equation can be used.

Vallance et al. (2017) propose two new evaluation metrics that enable the evaluation of solar forecasting with less data sensitivity: namely temporal distortion mix (TDM) and ramp score. TDM is based on a dynamic time warping capable of assessing the aptitude of forecast lags. The ramp score uses the swinging door algorithm to measure prediction accuracy for ramping events.

The hierarchical forecasting approach presented by Yang et al. (2017) provides a technique for reconciling forecasts at the PV plant and transmission levels.

The multi-modeling forecasting approach developed by Sanfilippo et al. (2016a) is described in the previous section.

Killinger et al. (2017) provide a method for validating and completing metadata needed to use irradiance forecasting algorithms with data from PV system sensors (azimuth, tilt, and degradation) and perform quality control on power output data.

Kuhn et al. (2017) present a method for deriving DNI measurements by comparing current images from shadow cameras with images taken during a sunny period.

Further developments of these emerging technologies and their integration with other solar forecasting techniques hold the promise of further improvements in the solar nowcasting area.

7 Conclusions

With the increasing ubiquity of solar energy systems, solar forecasting is becoming one of the most important applications of solar energy resources for its ability to redress power system imbalances resulting from the integration of solar energy, optimize power network operations, and normalize electricity markets in near real time. Current treatments of solar nowcasting rely on time series analysis algorithms based on machine learning and stochastics techniques. The source of data may differ according to the nowcasting approach adopted to include measurements from ground solar stations or modeled data from satellite and sky camera imagery. Exogenous meteorological parameters such as air temperature, relative humidity, and atmospheric turbulence are also occasionally used. Solar nowcasting performance can be greatly improved through multi-modeling and detrending techniques. Moving forward, further gains in nowcasting performance may be achieved by further development of emerging solar forecasting technologies and their integration with other techniques, such as the combination of multi-modeling and detrending.

References

Akarslan E, Hocaoglu FO (2016) A novel adaptive approach for hourly solar radiation forecasting. Renew Energy 87:628–633

Anderson D, Leach M (2004) Harvesting and redistributing renewable energy: on the role of gas and electricity grids to overcome intermittency through the generation and storage of hydrogen. Energy Pol 32:1603–1614

Baig A, Akhter P, Mufti A (1991) A novel approach to estimate the clear day global radiation. Renew Energy 1(1):119–123

Beyer HG, Martinez JP, Suri M (2009) MESOR report on benchmarking of radiation products. In: Deliverable 1.1. 3, Management and Exploitation of Solar Resource Knowledge (MESOR), European Commission 6th framework programme. Contract, (038665)

Black JN, Bonython CW, Prescott JA (1954) Solar radiation and the duration of sunshine. Q J Royal Meteorol Soc 80(344):231–235

Boland J (1995) Time series analysis of climatic variables. Sol Energy 55(5):377–388

Boland J (2008) Time series and statistical modelling of solar radiation. In: Recent advances in solar radiation modelling. Springer, pp 283–312

Brockwell PJ, Davis RA (2006) Introduction to time series and forecasting. Springer Science & Business Media

Buch KA Jr, Sun Chen-Hui (1995) Cloud classification using whole-sky imager data. N. p. Web, United States

Cazorla A, Olmo FJ, Alados-Arboleda L (2008) Development of a sky imager for cloud cover assessment. J Opt Soc Am A 25:29–39

Chaouachi A, Kamel RM, Nagasaka K (2010) Neural network ensemble-based solar power generation short-term forecasting. JACIII 14(1):69–75

Chow CW, Urquhart B, Lave M, Dominguez A, Kleissl J, Shields J, Washom B (2011) Intra-hour forecasting with a total sky imager at the UC San Diego solar energy testbed. Sol Energy 85 (11):2881–2893

Chu Y, Pedro HTC, Coimbra CFM, Hybrid intra-hour (2013) DNI fore-casts with sky image processing enhanced by stochastic learning. Solar Energy 98(Part C):592–603

Diagne M, David M, Lauret P, Boland J, Schmutz N (2013) Review of solar ir-radiance forecasting methods and a proposition for small-scale insular grids. Renew Sustain Energy Rev 27:65–76

DNICast, N.d. (2015) Direct normal irradiance nowcasting methods for optimized operation of concentrating solar technologies. In: Project funded by the European Commission under the 7th Framework Energy Research Programme. Retrieved from http://cordis.europa.eu/project/rcn/109593_en.html on Nov 12

Duffie JA, Beckman WA (1991) Solar engineering of thermal processes, 2nd ed. Wiley

Ferrari S, Lazzaroni M, Piuri V, Salman A, Cristaldi L, Rossi M, Poli T (2012) Illuminance prediction through extreme learning machines. In: 2012 IEEE Workshop Environment Energy Structural and Monitoring Systems EESMS, pp 97–103

Gibbons JM, Cox GM, Wood ATA, Craigon J, Ramsden SJ, Tarsitano D, Crout NMJ (2008) Applying Bayesian model averaging to mechanistic models: an example a comparison of methods. Environ Modell Softw23(8):973–985

Gordon R (2009) Predicting solar radiation at high resolutions: a comparison of time series forecasts. Solar Energy 83(3):342–349

Hammer A, Heinemann D, Lorenz E, Lackehe B (1999) Short-term forecasting of solar radiation: a statistical approach using satellite data. Sol Energy 67(13):139–150

Hirsch T, Martin Chivelet N, Gonzalez Martinez L, Biencinto Murga M, Wilbert S, Schroedter-Homscheidt M, Chenlo F, Feldhoff JF (2014) Technical report on the functional requirements for the nowcasting method. DNICast, Deliverable 2.1. grant #608623. Retrieved from http://www.dnicast-project.net/ on 12 Nov 2015

Inage S (2017) Development of an advection model for solar forecasting based on ground data first report: development and verification of a fundamental model. Sol Energy 153:414–434

Inman RH, Pedro HT, Coimbra CF (2013) Solar forecasting methods for re-newable energy integration. Prog Energy Combust Sci 39(6):535–576

Jain PC (1984) Comparison of techniques for the estimation of daily global irradiation and a new technique for the estimation of global irradiation. Solar Wind Technol 1:123–134

Ji W, Chee KC (2011) Prediction of hourly solar radiation using a novel hybrid model of ARMA and TDNN. Sol Energy 85(5):808–817

Kaplanis SN (2006) New methodologies to estimate the hourly global solar radiation; comparisons with existing models. Renew Energy 31(6):781–790

Kasten F, Young A (1989) Revised optical air mass tables and approximation formula. Appl Opt 28(22):4735–4738

Killinger S, Engerer N, Müller B (2017) QCPV: a quality control algorithm for dis-tributed photovoltaic array power output. Sol Energy 143:120–131

Kuhn P, Wilbert S, Prahl C, Schüler D, Haase T, Hirsch T, Wittmann M, Ramirez L, Zarzalejo L, Meyer A, Vuilleumier L, Blanc P, Pitz-Paal R (2017) Shadow camera system for the generation of solar irradiance maps. Sol Energy 157:157–170

Lauret P, Rodler A, Muselli M, David M, Diagne H, Voy-ant C (2012) A Bayesian model committee approach to forecasting global solar radiation. World Renewable Energy Forum, Denver, United States

Li Q, Lu W, Yang J (2011) A hybrid thresholding algorithm for cloud detection on ground-based color images. J Atmos Oceanic Technol 28:1286–1296

Long CN, Dutton EG (2002) BSRN global network recommended QC tests, V2.0 BSRN Technical Report

Long CN, Sabburg JM, Calbó J, Pagès D (2006) Retrieving cloud characteristics from ground-based daytime color all-sky images. J Atmos Ocean Technol 23:633–652

Lorenz E, Hammer A, Heinemann D (2004) Short term forecasting of solar radiation based on satellite data. EUROSUN2004 (ISES Europe Solar Congress)

Lyu L, Kantardzic M, Arabmakki E (2014) Solar irradiance forecasting by using wave-let based denoising. In: 2014 IEEE symposium on computational intelligence for engineering solutions (CIES), pp 110–116

Mandal P, Madhira STS, Haque AU, Meng J, Pineda RL (2012) Forecasting poweroutput of solar photovoltaic system using wavelet transform and artificial intelligence techniques. In: Proceeding of computer science, vol 12, pp 332–337

Marquez R, Coimbra CFM (2013) Intra-hour DNI forecasting based on cloud tracking image analysis. Sol Energy 91(2013):327–336

McArthur LJB (1998) Baseline surface radiation network. Operat Manual

Mellit A, Benghanem M, Kalogirou SA (2006) An adaptive wavelet network model for forecasting daily total solar-radiation. Appl Energy 83(7):705–722

Mellit A, Eleuch H, Benghanem M, Elaoun C, Pavan AM (2010) An adaptive model for predicting of global, direct and diffuse hourly solar irradiance. Energy Convers Manage 51(4):771–782

Mellit A, Kalogirou SA (2008) Artificial intelligence techniques for photovoltaic applications: a review. Progress in Energy and Combustion Science 34(5):574–632

Mohammed AA, Yaqub W, Aung Z (2015) Probabilistic forecasting of solar power: an ensemble learning approach. In: Neves-Silva R, Jain L, Howlett R (eds) Intelligent decision technologies. IDT 2017. Smart innovation, systems and technologies, vol 39. Springer, Cham

Moreno-Munoz A, De la Rosa JJG, Posadillo R, Bellido F (2008) Very short term fore-casting of solar radiation. In: 33rd IEEE Photovoltaic Specialists Conference PVSC 08, 2008, pp 1–5

Paulescu M, Paulescu E, Gravila P, Badescu V (2013) Weather modeling and forecasting of PV systems operation. Springer, London

Pelland S, Remund J, Kleissl J, Oozeki T, De Brabandere K (2013) Photovoltaic and solar forecasting: state of the art. IEA PVPS Task 14, Subtask 3.1. Report IEA-PVPS T14–01: October 2013

Perez R, Ineichen P, Seals R, Zelenka A (1990) Making full use of the clearness index for parameterizing hourly insolation conditions. Sol Energy 45:111–114

Prokop L, Misak S, Snasel V, Platos J, Kroemer P (2013) Supervised learning of photovoltaic power plant output prediction models. Neural Netw World 23:321–338

Sanfilippo A, Martin-Pomares L, Mohandes N, Perez-Astudillo D, Bachour D (2016a) An adaptive multi-modeling approach to solar nowcasting. Solar Ener-gy 125:77–85

Sanfilippo A, Pomares L, Perez-Astudillo D, Mohandes N, Bachour D (2016b) Optimal selection of training datasets for solar nowcasting models. In: Proceedings to the 32nd European photovoltaic solar energy conference and exhibition, pp 1482–1484

Sanfilippo A, Pomares L, Perez-Astudillo D, Mohandes N, Bachour D (2018) Ensemble detrending for solar nowcasting. In: Proceedings to the 35th European photovoltaic solar energy conference and exhibition

Severiano CA, Silva PCL, Sadaei HJ, Guimarães FG (2017) Very short-term solar forecasting using fuzzy time series. In: 2017 IEEE international conference on fuzzy systems (FUZZ-IEEE), Naples, pp 1–6

Sfetsos A, Coonick AH (2000) Univariate and multivariate forecasting of hourly solar radiation with artificial intelligence techniques. Solar Energy 68(2):169–178. https://doi.org/10.1016/S0038-092X(99)00064-X

Smola A, Schoölkopf B (1998) A tutorial on support vector regression. NeuroCOLT Tech. Rep. TR 1998–030, Royal Holloway College, London, U.K

Smola M, Smola AJ, Ratsch G, Scholkopf B, Kohlmorgen J, Vapnik V (1997) Predicting time series with support vector machines. In: Proceedings of ICANN '97, Springer LNCS 1327, pp 999–1004

Solar Forecasting 2. https://www.energy.gov/eere/solar/solar-forecasting-2

Vallance L, Charbonnier B, Paul N, Dubost S, Blanc P (2017) Towards a standardized procedure to assess solar forecast accuracy: a new ramp and time alignment metric. Sol Energy 150: 408–422

Voyant C, Notton G, Kalogirou S, Nivet M-L, Paoli C, Motte F, Fouilloy A (2017) Machine learning methods for solar radiation forecasting: a review. Renew Energy 105:569–582

Yang D, Quan H, Disfani VR, Liu L (2017) Reconciling solar forecasts: geographical hierarchy. Sol Energy 146:276–286

Yang D, Kleissl J, Gueymard CA, Pedro HTC, Coimbra CFM (2018) History and trends in solar irradiance and PV power forecasting: a preliminary assessment and review using text mining. Sol Energy 168:60–101

Printed by Printforce, the Netherlands